Contents

Chapter 4. Looking to the future: challenges for scientists studying climate change and health   61

Chapter 5. Impacts on health of climate extremes   79

Chapter 6. Climate change and infectious diseases   103

Preface

There is now widespread consensus that the Earth is warming at a rate unprecedented during post hunter-gatherer human existence. The last decade was the warmest since instrumental records began in the nineteenth century, and contained 9 of the 10 warmest years ever recorded. The causes of this change are increasingly well understood. The Third Assessment Report of the Intergovernmental Panel on Climate Change, published in 2001, goes further than its predecessors, stating that *"There is new and stronger evidence that most of the warming observed over the last 50 years is likely to be attributable to human activities"*, most importantly the release of greenhouse gases from fossil fuels.

Stresses on the climate system are already causing impacts on Earth's surface. These include not only rising surface temperatures, but also increasingly frequent floods and droughts, and changes in natural ecosystems, such as earlier flowering of plants, and poleward shifts in the distribution of several species. All of these changes are inextricably linked to the health of human societies. Climatic conditions affect human well-being both directly, through the physical effects of climatic extremes, and indirectly, through influences on the levels of pollution in the air, on the agricultural, marine and freshwater systems that provide food and water, and on the vectors and pathogens that cause infectious diseases.

As it is now widely accepted that humans are influencing global climate, decision makers are now focusing on the type and timing of actions to limit the rate of change. Attention is shifting to the balance between the possible impacts of climate change, and the economic costs, technological advances and societal adaptations that are necessary for mitigation.

International agreements, supported by hard science, are proving effective in combating wide-ranging environmental threats such as ozone depletion and long-range transboundary air pollution. Can similar agreements be implemented to address the more complex risks posed by global climate change? Scientific analysis in general, and the health sector in particular, need to inform and help advance ongoing policy discussions. Firstly, the scientific community must produce rigorous and balanced evidence not only of the breadth and magnitude of climate change effects, but also of how they are distributed across populations, and over time. Just and equitable decisions on appropriate responses to climate change can only be reached by giving consideration to all those affected by policy actions (or inactions), including future generations. Secondly, as some degree of continued climate change is now inevitable, it is necessary to identify vulnerable populations, and formulate policies and measures to help them adapt to changing conditions.

This book, prepared jointly by the World Health Organization, the World Meteorological Organization and the United Nations Environment Programme, works towards these ends. It provides a comprehensive update of a previous review, published in 1996. More importantly, it expands the scope of the review to include quantitative estimates of the total health impacts of climate change. It lays out the steps necessary to further scientific investigation and to develop strategies and policies to help societies adapt to climate change.

Dr Gro Harlem Brundtland
Director-General Emeritus
World Health Organization

Professor G.O.P. Obasi
Secretary General
World Meteorological
Organization

Dr Klaus Töpfer
Executive Director
United Nations
Environment Programme

Acknowledgements

The World Health Organization, World Meteorological Organization and United Nations Environment Programme wish to express their appreciation to all those whose efforts made this book possible.

The following provided review comments, ideas, or made a major contribution to analyses that underpin the work presented in this book: Natasha Andronova, Michael Schlesinger, (University of Illinois at Urbana-Champaign, USA); Martha Anker, Monika Blössner, Robert Bos, Charles Delacollette, Alan Lopez, Alessandro Loretti, Chris Murray, Mercedes de Onis, Anthony Rodgers, Colin Roy, Mike Ryan, Stephen Vander Horn (World Health Organization, Geneva, Switzerland); Bruce Armstrong (University of Sydney, Sydney, Australia); Ian Burton (Environment Canada, Toronto, Canada); Sandy Cairncross, Paul Coleman, Clive Davies, Bo Draser, Andy Haines (London School of Hygiene and Tropical Medicine, London, England), Elsa Casamiro (SIAM project, Portugal); Will Checkley (Johns Hopkins University, Baltimore, USA); Majid Ezzati (Resources for the Future, Washington DC, USA); Roger Few, Matthew Livermore, Tim Mitchell, David Viner (University of East Anglia, Norwich, England); Chuck Hakkarinen (Electrical Power Research Institute, Palo Alto, USA); Simon Hay (University of Oxford, Oxford, England); Frank de Hruijl (University Hospital Utrecht, Utrecht, the Netherlands); Saleemul Huq (International Institute for Environment and Development, London, England); John Last (University of Ottawa, Ottawa, Canada); Neil Leary (AIACC, Washington DC, USA); David Le Sueur, Frank Tanser (Medical Research Centre, Mtubatuba, South Africa); Pim Martens (International Centre for Integrative Studies, Maastricht, the Netherlands); Gordon McBean (University of Western Ontario, London, Canada); Robert Nicholls, Theresa Wilson (University of Middlesex, Enfield, England); Eric Noji (U.S. Centers for Disease Control, Atlanta, USA); Dieter Riedel (Health Canada, Ottawa, Canada); Roger Street (Environment Canada, Toronto, Canada); Richard Tol (University of Hamburg, Hamburg, Germany); Mark Wilson (University of Michigan, Ann Arbor, USA); Rosalie Woodruff (Australian National University, Canberra, Australia).

Particular thanks, to Jo Woodhead, for patient editing of the text, to Liza Furnival for preparing the index, and to Eileen Tawffik for administrative support.

CHAPTER 1

Global climate change and health: an old story writ large

A.J. McMichael[1]

Introduction

The long-term good health of populations depends on the continued stability and functioning of the biosphere's ecological and physical systems, often referred to as life-support systems. We ignore this long-established historical truth at our peril: yet it is all too easy to overlook this dependency, particularly at a time when the human species is becoming increasingly urbanized and distanced from these natural systems. The world's climate system is an integral part of this complex of life-supporting processes, one of many large natural systems that are now coming under pressure from the increasing weight of human numbers and economic activities.

By inadvertently increasing the concentration of energy-trapping gases in the lower atmosphere, human actions have begun to amplify Earth's natural greenhouse effect. The primary challenge facing the world community is to achieve sufficient reduction in greenhouse gas emissions so as to avoid dangerous interference in the climate system. National governments, via the UN Framework Convention on Climate Change (UNFCC), are committed in principle to seeking this outcome. In practice, it is proving difficult to find a politically acceptable course of action—often because of apprehensions about possible short-term economic consequences.

This volume seeks to describe the context and process of global climate change, its actual or likely impacts on health, and how human societies should respond, via both adaptation strategies to lessen impacts and collective action to reduce greenhouse gas emissions. As shown later, much of the resultant risk to human populations and the ecosystems upon which they depend comes from the projected extremely rapid rate of change in climatic conditions. Indeed, the prospect of such change has stimulated a great deal of new scientific research over the past decade, much of which is elucidating the complex ecological disturbances that can impact on human well-being and health— as in the following example.

The US Global Change Research Program (Alaska Regional Assessment Group) recently documented how the various effects of climate change on aquatic ecosystems can interact and ripple through trophic levels in unpredictable ways. For example, warming in the Arctic region has reduced the amount of sea ice, impairing survival rates for walrus and seal pups that spend part of their life cycle on the ice. With fewer seal pups, sea otters have become the alternative food source for whales. Sea otters feed on sea urchins, and with fewer sea otters sea

---

[1] National Centre for Epidemiology and Population Health, The Australian National University, Canberra, Australia.

urchin populations are expanding and consuming more of the kelp that provides breeding grounds for fish. Fewer fish exacerbate the declines in walrus and seal populations. Overall, there is less food available for the Yupik Eskimos of the Arctic who rely on all of these species.

Global climate change is thus a significant addition to the spectrum of environmental health hazards faced by humankind. The global scale makes for unfamiliarity—although most of its health impacts comprise increases (or decreases) in familiar effects of climatic variation on human biology and health. Traditional environmental health concerns long have been focused on toxicological or microbiological risks to health from local environmental exposures. However, in the early years of the twenty-first century, as the burgeoning human impact on the environment continues to alter the planet's geological, biological and ecological systems, a range of larger-scale environmental hazards to human health has emerged. In addition to global climate change, these include: the health risks posed by stratospheric ozone depletion; loss of biodiversity; stresses on terrestrial and ocean food-producing systems; changes in hydrological systems and the supplies of freshwater; and the global dissemination of persistent organic pollutants.

Climate change and stratospheric ozone depletion are the best known of these various global environmental changes. Human societies, however, have had long experience of the vicissitudes of climate: climatic cycles have left great imprints and scars on the history of humankind. Civilisations such as those of ancient Egypt, Mesopotamia, the Mayans, the Vikings in Greenland and European populations during the four centuries of Little Ice Age, all have both benefited and suffered from nature's great climatic cycles. Historical analyses also reveal widespread disasters, social disruption and disease outbreaks in response to the more acute, inter-annual, quasi-periodic ENSO (El Niño Southern Oscillation) cycle (1). The depletion of soil fertility and freshwater supplies, and the mismanagement of water catchment basins via excessive deforestation, also have contributed to the decline of various regional populations over the millennia (2).

Today, climate scientists predict that humankind's increasing emission of greenhouse gases will induce a long-term change in the world's climate. These gases comprise, principally, carbon dioxide (mostly from fossil fuel combustion and forest burning), plus various other heat-trapping gases such as methane (from irrigated agriculture, animal husbandry and oil extraction), nitrous oxide and various human-made halocarbons. Indeed, most climate scientists now suspect that the accumulation of these gases in the lower atmosphere has contributed to the strong recent uptrend in world average temperature. In its Third Assessment Report, published in 2001, the Intergovernmental Panel on Climate Change (IPCC) stated: "There is new and stronger evidence that most of the warming observed over the last 50 years is attributable to human activities" (3).

During the twentieth century, world average surface temperature increased by approximately 0.6 °C (Figure 1.1). There were, of course, natural influences on world climate during this time. These include an increase in volcanic activity between 1960 and 1991 (when Mount Pinatubo erupted) which induced a net negative natural radiative forcing for the last two (up to possibly four) decades; and a slight overall increase in solar activity in the first half of the century which may have accounted for around one-sixth of that century's observed temperature increase. The twentieth century's global warming has taken Earth's average surface temperature above the centuries-long historical limit of the amplitude of natural variations.

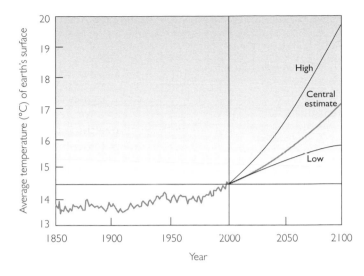

FIGURE 1.1 Global temperature record, since instrumental recording began in 1860, and projection for coming century, according to Intergovernmental Panel on Climate Change (*3*). The wide range around the projection reflects uncertainties about aspects of the climate system and future human economic activity and technology choices. World temperature has increased by around 0.4 °C since the 1970s and now exceeds the upper limit of natural (historical) variability. Climatologists consider around five-sixths of that recent increase to be due to human influence.

The unprecedented prospect of human-induced (rapid) changes to the global climate has prompted a large international scientific effort to assess the evidence. The IPCC, established within the UN framework in 1988, was charged with advising national governments on the causes and processes of climate change; likely impacts and their associated costs; and ways to lessen the impacts. The IPCC's Third Assessment Report (2001) projects an increase in average world surface temperature ranging from 1.4 to 5.8 °C over the course of the twenty-first century (see Figure 1.1). That estimation, with its wide range, is drawn from a large number of different global climate models and a range of plausible scenarios of greenhouse gas and sulphate aerosol precursor emissions. Those scenarios entail different future storylines of demographic, economic, political and technological change. A temperature increase anywhere within this range would be much more rapid than any naturally occurring increase that has been experienced by humans since the advent of agriculture, around 10 000 years ago.

Recognising the complexity of systems upon which life depends: an ecological perspective

As a human-generated and worldwide process, global climate change is a qualitatively distinct and very significant addition to the spectrum of environmental health hazards encountered by humankind. Historically, environmental health concerns have focused on toxicological or microbiological risks to health from local exposures. However, the scale of environmental health problems is increasing and various larger-scale environmental hazards to human population health have begun to appear.

Appreciation of this scale and type of influence on human health entails an *ecological* perspective. This perspective recognises that the foundations of long-term good health in populations reside in the continued stability and functioning of the biosphere's life-supporting ecological and physical systems. It also brings an appreciation of the complexity of the systems upon which we depend and moves beyond a simplistic, mechanistic, model of environmental health risks to human health.

FIGURE 1.2 Diagrammatic representation of the typical sequence of three categories of environmental impacts and hazards to health resulting from human social and economic development. As a society becomes wealthier, more literate and better able to exert legislative control, the profile of local and community-wide environmental hazards (categories A and B) is reduced. Meanwhile, however, as the ecological footprint of the population increases, so larger scale and less locally evident environmental changes (category C) accrue.

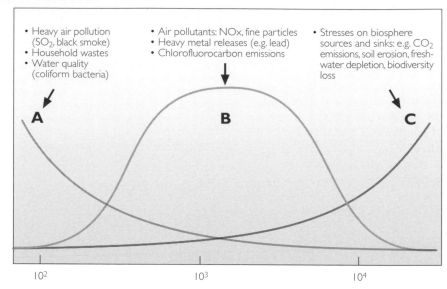

• Heavy air pollution
  (SO$_2$, black smoke)
• Household wastes
• Water quality
  (coliform bacteria)

• Air pollutants: NOx, fine particles
• Heavy metal releases (e.g. lead)
• Chlorofluorocarbon emissions

• Stresses on biosphere
  sources and sinks: e.g. CO$_2$
  emissions, soil erosion, freshwater depletion, biodiversity
  loss

Amount of emission, depletion, or other environmental damage

A   B   C

10^2 10^3 10^4

Increasing GNP/person (orders of magnitude, over time, in equivalent US$)

In simplified diagrammatic fashion, Figure 1.2 illustrates the approximate chronological succession of environmental hazards, as societies undergo economic growth and consequent increases in the scale of human activity and environmental impact (4). Historically, on a local scale, category A hazards have predominated. In the early years of the industrial revolution in Europe much of the environmental hazard was at household and neighbourhood level. In the middle decades of the twentieth century developed countries began to reduce the levels of category B hazards, often via environmental legislation—such as the clean air acts of European and North American countries.

Today, category C hazards are increasing, reflecting the great pressures that human societies collectively exert on the biogeophysical systems of this planet. Carbon dioxide emission is an important example of a category C hazard. Emission rates increased markedly (around twelve-fold) during the twentieth century, as worldwide industrialization proceeded and land-use patterns changed at an accelerating rate.

The scale of environmental health problems has expanded from household (e.g. indoor air pollution) to neighbourhood (domestic refuse) to community (urban air pollution) to regional (acid rain) to global level (climate change). This requires consideration of the "ecological footprint" and how to curtail its size within the limits of global ecological sustainability. Folke and colleagues have estimated that the cities around the Baltic Sea require an area of land and sea surface several hundred times larger than the sum of the areas of the cities themselves (5). This large ecological footprint, typical of modern industrialized societies, comprises the supplies of food, water and raw materials and the environmental "sinks" into which urban-industrial metabolic waste is emptied. The moral dilemma is clear: a world of six billion cannot live at that privileged level of environmental impact. There simply is not enough world available! A recent study has estimated that human demands on the biosphere have exceeded the world's "biocapacity" since the 1970s, and is currently about 25% beyond the sustainable capacity of Earth (6).

Further, not only can the actions of one population affect the health of distant populations—as with the environmental dissemination of chlorinated hydrocarbons (persistent organic pollutants: POPs)—but actions today may jeopardise the well-being and health of future generations. There is already in motion a process of sea level rise that will continue for many centuries as the extra heat trapped at Earth's surface by the human-amplified greenhouse effect progressively enters the deep ocean water. Similarly, it is likely that the continuing rapid extinction of populations and species of plants and animals will leave a biotically impoverished, less ecologically resilient and less productive world for future generations.

Despite global climate change currently being the most widely discussed of various recent global environmental changes, there is mounting evidence that humans, in aggregate, are overloading the planet's great biogeochemical systems. This has been well summarised by Vitousek and colleagues:

> "Human alteration of Earth is substantial and growing. Between one-third and one-half of the land surface has been transformed by human action; the carbon dioxide concentration in the atmosphere has increased by nearly 30% since the beginning of the Industrial Revolution; more atmospheric nitrogen is fixed by humanity than by all natural terrestrial sources combined; more than half of all accessible surface fresh water is put to use by humanity; and about one-quarter of the bird species on Earth have been driven to extinction. By these and other standards, it is clear that we live on a human-dominated planet." (7)

The long history of climatic fluctuations since the end of the last global glaciation around 15 000 years ago, along with the evidence of recent temperature rises and the IPCC's projected rapid warming in the current century, are summarized in Figure 1.3. Several of the rises and falls of great civilisations are shown. Note that the climatic variations before around 1850 essentially were due to natural forcing processes—cosmological alignments, volcanic activity, solar activity and so on. Since 1850 there has been an increasing influence via human emission of greenhouse gases in excess of the biosphere's capacity to absorb them without an increase in atmospheric concentration. That more recent period also is shown, in more detail, in Figure 1.1 above.

Climate change: overview of recent scientific assessments

The latest report from the Intergovernmental Panel on Climate Change (IPCC) makes several compellingly clear points (8). First, human-induced warming has apparently begun: the particular pattern of temperature increase over the past quarter-century has fingerprints that indicate a substantial contribution from the build-up of greenhouse gases due to human activities. Second, a coherent pattern of changes in simple physical and biological systems has become apparent across all continents—the retreat of glaciers, melting of sea ice, thawing of permafrost, earlier egg-laying by birds, polewards extension of insect and plant species, earlier flowering of plants and so on. Third, the anticipated average surface-temperature rise this century, within the range of 1.4 to 5.8 °C, would be a faster

FIGURE 1.3 Variations in Earth's average surface temperature, over the past 20 000 years. Prior to 1860, analogue measures of temperature are necessary (tree rings, oxygen isotope ratios in ice cores and lake sediments, etc.). Note the substantial natural fluctuations throughout the period. Note also that (with the logarithmic nature of the time axis) the anticipated rate of increase in world temperature this century is 20–30 times faster than occurred as the planet emerged from the last glaciation, from around 15 000 years ago.

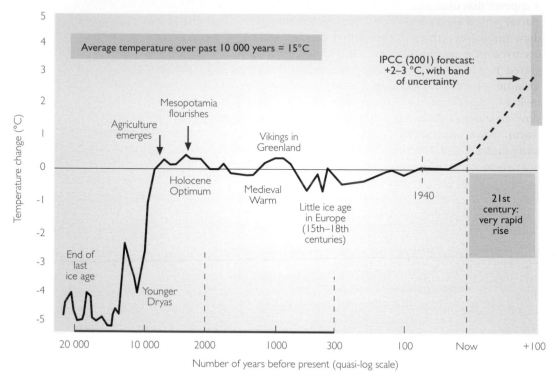

increase than predicted in the IPCC's previous major report, in 1996 (9). It is the rate of change in temperature that will pose a particular stress upon many ecosystems and species. The IPCC also reported that even if humankind manages to curb excess greenhouse gas emissions within the next half-century, the world's oceans will continue to rise for up to 1000 years, reflecting the great inertial processes as heat is transferred from surface to deep water (10). By that time the sea level rise would have approximated 1–2 metres.

The estimated rise in average world temperature over the coming century conceals various important details. Anticipated surface temperature increases would be greater at higher latitudes, greater on land than at sea, and would affect the daily minimum night-time temperatures more than daily maximum temperatures. Alaska, northern Canada and northern Siberia, for example, could warm by approximately 5 °C during the twenty-first century. Indeed, the temperature increases that have occurred already above the Arctic Circle have disrupted polar bear feeding and breeding, the annual migrations of caribou and the network of telephone poles in Alaska (previously anchored in ice-like permafrost). Global climate change also would cause rainfall patterns to change with increases over the oceans but a reduction over much of the land surface—especially in various low to medium latitude mid-continental regions (central Spain, the US midwest, the Sahel, Amazonia) and in already arid areas in north-west India, the Middle East, northern Africa and parts of central America. Rainfall events would tend to intensify with more frequent extreme events increasing the likelihood of flooding and droughts. Regional weather systems, including the great south-west Asian monsoon, could undergo latitudinal shift.

According to glaciologists there is a slight possibility that large sections of the Antarctic ice mass would melt, thus raising sea level by several metres. However, it appears that disintegration did not occur during the warm peak of the last interglacial period around 120 000 years ago, when temperatures were 1–2 °C higher than now. Nevertheless, substantial melting of Antarctic ice appears to have occurred in a previous interglacial, and several large ice-shelves have disintegrated in the past two decades (3). Another possibility is that the northern Atlantic Gulf Stream might weaken and eventually even shut down if increased melt-water from Greenland disturbs the dynamics of that section of the great, slow and tortuous "conveyor belt" circulation that distributes Pacific-equatorial warm water around the world's oceans (3). North-west Europe, relative to same-latitude Newfoundland, currently enjoys 5–7 °C of free heating from this heat-source. If weakening of the Gulf Stream does occur over the coming century or two, Europe may actually become a little colder even as the rest of the world warms.

As mentioned earlier in this chapter, global climate change is only one of a larger set of destabilising large-scale environmental changes that are now underway, each of them reflecting the increasing human domination of the ecosphere (3, 11). These include major global changes such as stratospheric ozone depletion, biodiversity loss, worldwide land degradation, freshwater depletion, and others such as the disruption of the elemental cycles of nitrogen and sulphur, and the global dissemination of persistent organic pollutants. All have great consequences for the sustainability of ecological systems: food production; human economic activities and human population health (12). Figure 1.4 illustrates (in simplified fashion) how part of this complex of interacting, large-scale environmental changes impinges on human health. Many of the pathways would of course be modulated by cultural and technological characteristics of human societies. That is, local populations vary in their vulnerability to these potential impacts.

There is growing realization that the sustainability of population health must be a central consideration in the public discourse on how human societies can make the transition to sustainable development (13, 14). Hence, public, policymakers and other scientists have an increasing interest in hearing from population health researchers, moving towards a view of population health as an ecological entity: an index of the success of longer-term management of social and natural environments (15). Indeed this recognition will assist in altering social and economic practices and priorities, to avert or minimize the occurrence of global environmental changes and their adverse impacts.

Change in world climate would influence the functioning of many ecosystems and the biological health of plants and creatures. Likewise, there would be health impacts on human populations, some of which would be beneficial. For example, milder winters would reduce the seasonal winter-time peak in deaths that occurs in temperate countries, while in currently hot regions a further increase in temperatures might reduce the viability of disease-transmitting mosquito populations. Overall, scientists consider that most of the health impacts of climate change would be adverse (16, 17). This assessment will be greatly enhanced by the accrual of actual evidence of early health impacts which epidemiologists anticipate will emerge over the coming decade.

FIGURE 1.4 Interrelationships between major types of global environmental change, including climate change. Note that all impinge on human health and—though not shown here explicitly—there are various interactive effects between jointly acting environmental stresses. The diverse pathways by which climate change affects health are the subjects of much of the remainder of this volume.

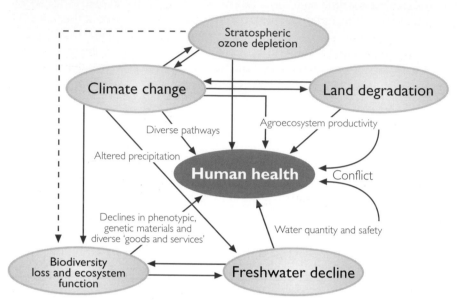

In the meantime, it is instructive to look back over the centuries and begin to understand how climatic changes and events can affect human well-being and health. This is a long and continuing story.

Climate and human health: an ancient struggle

Whoever wishes to investigate medicine properly, should proceed thus: in the first place to consider the seasons of the year, and what effects each of them produces, for they are not all alike, but differ much from themselves in regard to their changes.

Hippocrates, in Airs, Waters, and Places (18)

Recognition that human health can be affected by a wide range of ecological disruptions, consequent upon climate change, is a recent development, reflecting the breadth and sophistication of modern scientific knowledge. Nevertheless, the simpler idea that human health and disease are linked to climate probably predates written history.

The Greek physician Hippocrates (about 400 BC) related epidemics to seasonal weather changes, writing that physicians should have "due regard to the seasons of the year, and the diseases which they produce, and to the states of the wind peculiar to each country and the qualities of its waters" (18). He exhorts them to take note of "the waters which people use, whether they be marshy and soft, or hard and running from elevated and rocky situations, and then if saltish and unfit for cooking," and to observe "the localities of towns, and of the surrounding country, whether they are low or high, hot or cold, wet or dry ... and of the diet and regimen of the inhabitants".

Two thousand years later, Robert Plot, Secretary to the newly-founded Royal Society in England, took weather observations in 1683–84 and noted that if the same observations were made "in many foreign and remote parts at the same time" we would "probably in time thereby learn to be forewarned certainly of divers emergencies (such as heats, colds, deaths, plagues, and other epidemical distempers)".

Between these times, countless climatic disasters befell communities and populations around the world, leading variously to starvation, infectious disease, social collapse and the disappearance of whole populations. One such is the mysterious demise of the Viking settlements in Greenland in the fourteenth and fifteenth centuries, as temperatures in and around Europe began to fall. Established during the Medieval Warm Period during the tenth century AD (see Figure 1.3), these culturally conservative, livestock dependent, settlements could not cope with the progressive deterioration in climate that occurred from the late Middle Ages. Food production declined and food importation became more difficult as sea ice persisted. To compound matters, the native Inuit population in Greenland was pressing southwards, probably in response to the ongoing climate change. The Viking settlements eventually died out or were abandoned in the fourteenth (Western Settlement) and fifteenth centuries (Eastern Settlement) (19).

Historical accounts abound of acute famine episodes occurring in response to climatic fluctuations. Throughout pre-industrial Europe, diets were marginal over many centuries; the mass of people survived on monotonous diets of vegetables, grain gruel and bread. A particularly dramatic example in Europe was the great medieval famine of 1315–17. Climatic conditions were deteriorating and the cold and soggy conditions led to widespread crop failures, food price rises, hunger and death. Social unrest increased, robberies multiplied and bands of desperate peasants swarmed over the countryside. Reports of cannibalism abounded from Ireland to the Baltic. Animal diseases proliferated, contributing to the die-off of over half the sheep and oxen in Europe. This tumultuous event and the Black Death which followed thirty years later, are deemed to have contributed to the weakening and dissolution of feudalism in Europe.

Over these and the ensuing centuries, average daily intakes were less than 2000 calories, falling to around 1800 calories in the poorer regions of Europe. This permanent state of dietary insufficiency led to widespread malnutrition, susceptibility to infectious disease and low life expectancy. The superimposed frequent famines inevitably culled the populations, often drastically. In Tuscany, between the fourteenth and eighteenth centuries there were over 100 years of recorded famine. Meanwhile in China, where the mass rural diet of vegetables and rice accounted for an estimated 98% of caloric intake, between 108 BC and 1910 AD there were famines that involved at least one province in over 90% of years (20).

Food shortages are never due to climate extremes alone; the risk of famine depends also on many social and political factors. For example, a strong El Niño event in 1877 caused failure of the monsoon rains in south and central India (21). However, the intense famine that resulted, which caused somewhere between 6 and 10 million deaths, was only partly due to the drought. There was no shortage of food in India at this time (grain exports to the United Kingdom of Great Britain and Ireland reached an all time high in 1877), but a large proportion of the Indian population was unable to access food reserves, or to find alternative sources when their usual crops failed. There were many reasons for this. Under the British Raj, common lands that previously provided sustenance in times of hardship had been converted to (taxable) private property. Local economies had been impoverished by punitive tariff schemes that favoured imported United Kingdom goods over local products. Aided by the expansion of the railways, community-controlled reserves of food had been replaced by remote stockpiles but there were no moral or regulatory controls over speculation. Because of these and other factors at the end of the nineteenth century many people in India were more vulnerable to adverse effects of drought than ever before.

In the light of this varied (often dramatic) history of the climate-society relationship, it is not surprising that scientists foresee a range of health impacts of a change in global climatic conditions. These will be explored in detail later in this volume. However, the following overview of the potential health impacts of climate change will orient the reader to that later assessment.

Potential health impacts of climate change

Global climate change would affect human health via pathways of varying complexity, scale and directness and with different timing. Similarly, impacts would vary geographically as a function both of environment and topography and of the vulnerability of the local population. Impacts would be both positive and negative (although expert scientific reviews anticipate predominantly negative). This is no surprise since climatic change would disrupt or otherwise alter a large range of natural ecological and physical systems that are an integral part of Earth's life-support system. Via climate change humans are contributing to a change in the conditions of life on Earth.

The main pathways and categories of health impact of climate change are shown in Figure 1.5.

FIGURE 1.5 Pathways by which climate change affects human health, including local modulating influences and the feedback influence of adaptation measures. *Source: adapted from Patz et al., 2000 (22).*

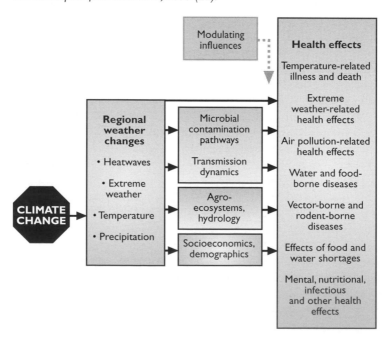

CLIMATE CHANGE AND HUMAN HEALTH

The more direct impacts on health include those due to changes in exposure to weather extremes (heatwaves, winter cold); increases in other extreme weather events (floods, cyclones, storm-surges, droughts); and increased production of certain air pollutants and aeroallergens (spores and moulds). Decreases in winter mortality due to milder winters may compensate for increases in summer mortality due to the increased frequency of heatwaves. In countries with a high level of excess winter mortality, such as the United Kingdom, the beneficial impact may outweigh the detrimental (23, 24). The extent of change in the frequency, intensity and location of extreme weather events due to climate change remains uncertain.

Climate change, acting via less direct mechanisms, would affect the transmission of many infectious diseases (especially water, food and vector-borne diseases) and regional food productivity (especially cereal grains). In the longer term and with considerable variation between populations as a function of geography and vulnerability, these indirect impacts are likely to have greater magnitude than the more direct (25, 26).

For vector-borne infections, the distribution and abundance of vector organisms and intermediate hosts are affected by various physical (temperature, precipitation, humidity, surface water and wind) and biotic factors (vegetation, host species, predators, competitors, parasites and human interventions). Various integrated modelling studies have forecast that an increase in ambient temperature would cause, worldwide, net increases in the geographical distribution of particular vector organisms (e.g. malarial mosquitoes) although some localised decreases also might occur. Further, temperature related changes in the life-cycle dynamics of both the vector species and the pathogenic organisms (flukes, protozoa, bacteria and viruses) would increase the potential transmission of many vector-borne diseases such as malaria (mosquito), dengue fever (mosquito) and leishmaniasis (sand-fly)—although schistosomiasis (water-snail) may undergo a net decrease in response to climate change (27, 28).

Recently, there has been considerable effort in developing mathematical models for making such projections. The models in current use have well recognised limitations—but have provided an important start. For example, from computer multiple modelling studies it seems likely that malaria will significantly extend its geographical range of *potential* transmission and its seasonality during the twenty-first century as average temperatures rise (29).

Allowing for future trends in trade and economic development, modelling studies have been used to estimate the impacts of climate change upon cereal grain yields (which account for two-thirds of world food energy). Globally, a slight downturn appears likely but this would be greater in already food-insecure regions in south Asia, parts of Africa and central America. Such downturns would increase the number of malnourished people by several tens of millions in the world at large—that is, by at least several per cent against a current and projected total, without climate change, of between four and eight hundred million.

By reflecting the increased retention of heat energy in the lower atmosphere, global warming also affects the atmospheric heat budget so as to increase the cooling of the stratosphere (30). Should this cooling persist, the process of ozone depletion could continue even after chlorine and bromine loading (by human emission of ozone-destroying gases) starts to decline. If so, the potential health consequences of stratospheric ozone depletion (increase in incidence of skin cancer in fair-skinned populations; eye lesions such as cataracts; and, perhaps, suppression of immune activity) would become an issue for climate change.

It is likely that climatic change over the past quarter-century has had various incremental impacts on at least some health outcomes. However, the time at which any such health impacts of climate change first become detectable particularly depends upon, firstly, the sensitivity of response (how steep is the rate of increase) and, secondly, whether there is a threshold that results in a "step function". Further, detectability is influenced by the availability of high-quality data and the extent of background variability in the health-related variable under investigation. Detection is a matter of both statistical power and reasonable judgement about attribution. The former depends on numbers of observations and the extent of divergence between observed and expected rates or magnitudes of health outcomes. The latter includes pattern recognition: if a particular infectious disease undergoes changes in occurrence in multiple geographical locations, each in association with local changes in climate, it is more certain to be due to climatic influence than if such a change occurs in just one setting.

The first detectable changes in human health may well be alterations in the geographical range (latitude and altitude) and seasonality of certain vector-borne infectious diseases. Summertime food-borne infections (e.g. salmonellosis) may show longer-lasting annual peaks. There has been debate, as yet unresolved, over whether recent increases of malaria and dengue in highland regions around the world may be due to climate factors or to the several other factors that are known to be significant determinants of transmission. There are several other categories of likely early impact. Hot weather would amplify the production of noxious photochemical smog in urban areas and warmer summers would increase the incidence of food poisoning. By contrast, the public health consequences of the disturbance of natural and managed food-producing ecosystems, rising sea levels and population displacement for reasons of physical hazard, land loss, economic disruption and civil strife, may not become evident for several decades.

Population vulnerability and adaptive responses

Human populations, as with individuals, vary in their vulnerability to certain health outcomes. A population's vulnerability is a joint function of, first, the extent to which a particular health outcome is sensitive to climate change and, second, the population's capacity to adapt to new climatic conditions. The vulnerability of a population depends on factors such as population density, level of economic development, food availability, income level and distribution, local environmental conditions, pre-existing health status and the quality and availability of public health care (32).

Adaptation refers to actions taken to lessen the impact of (anticipated) climate change. There is a hierarchy of control strategies that can help to protect population health. These strategies are categorised as: administrative or legislative, engineering, personal-behavioural. Legislative or regulatory action can be taken by government, requiring compliance by all or designated classes of persons. Alternatively, adaptive action may be encouraged on a voluntary basis, via advocacy, education or economic incentives. The former type of action would normally be taken at a supranational, national or community level; the latter would range from supranational to individual levels. Adaptation strategies will be either reactive, in response to climate impacts, or anticipatory, in order to reduce

BOX 1.1 Health impacts of other types of global environmental change

Global climate change is part of a larger set of global environmental changes. As shown in Figure 1.4, these changes influence one another and often exert interactive impacts when acting in concert.

Stratospheric ozone depletion

Depletion of stratospheric ozone by human-made gases such as chlorofluorocarbons has been occurring over recent decades and is likely to peak around 2020. Ambient ground-level ultraviolet irradiation is estimated to have increased consequently by up to 10% at mid-to-high latitudes over the past two decades. Scenario-based modelling that integrates the processes of emissions accrual, ozone destruction, UVR flux and cancer induction, indicates that European and United States' populations will experience 5–10% excess in skin cancer incidence during the middle decades of the twenty-first century. If climate change and consequent stratospheric cooling delay the recovery of protective ozone, there will be greater numbers of excess skin cancers.

Biodiversity loss and invasive species

Increasing human demand for space, materials and food leads to increasingly rapid extinction of populations and species of plants and animals. An important consequence for humans is the disruption of ecosystems that provide "nature's goods and services". Biodiversity loss also means the loss, before discovery, of many natural chemicals and genes, others of which have conferred enormous medical and health improvement benefits. Myers estimates that five-sixths of tropical vegetative nature's medicinal goods have yet to be recruited for human benefit (31).

Meanwhile, "invasive" species are spreading worldwide into new non-natural environments via intensified human food production, commerce and mobility. The resultant changes in regional species composition have myriad consequences for human health. For example: the choking spread of water hyacinth in eastern Africa's Lake Victoria, introduced from Brazil as a decorative plant, is now a breeding ground for the water snail that transmits schistosomiasis and for the proliferation of diarrhoeal disease organisms.

Impairment of food-producing ecosystems

Increasing pressures of agricultural and livestock production are stressing the world's arable lands and pastures. At the start of the twenty-first century an estimated one-third of the world's previously productive land is seriously damaged: by erosion, compaction, salination, waterlogging and chemicals that destroy organic content. Similar pressures on the world's ocean fisheries have left most of them severely depleted or stressed. Almost certainly an environmentally benign and socially acceptable way of using genetic engineering to increase food yields must be found in order to produce sufficient food for another three billion persons (with higher expectations) over the coming half century.

Allowing for future trends in trade and economic development, modelling studies have estimated that climate change would cause a slight downturn globally of around 2–4% in cereal grain yields (which represent two-thirds of world food energy). The estimated downturn in yield would be considerably greater in the food-insecure regions in South Asia, the Middle East, North Africa and Central America.

Other global environmental changes

Freshwater aquifers in all continents are being depleted of their ancient fossil water supplies. Agricultural and industrial demand amplified by population growth often greatly exceeds the rate of natural recharge. Water-related political and public health crises loom in some regions within decades.

Various semi-volatile organic chemicals (such as polychlorinated biphenyls) are now disseminated worldwide via a sequential distillation process in the cells of the lower atmosphere, thereby transferring chemicals from their usual origins in low to mid latitudes to high, indeed polar, latitudes. Consequently, increasingly high levels are occurring in polar mammals and fish and the traditional human groups that eat them. Clearly chemical pollution is no longer just an issue of local toxicity.

vulnerability. Adaptation can be undertaken at the international/national, community and individual level—that is, at macro, meso and micro-levels.

The reduction of socioeconomic vulnerability remains a priority. The poor (and especially the very young and old) are likely to be at greatest health risk because of their lack of access to material and information resources. Long-term reduction in health inequalities will require income redistribution, full employment, better housing and improved public health infrastructure. There must be improvement in services with a direct impact on health such as primary care, disease control, sanitation and disaster preparedness and relief. The vulnerability of the poor may jeopardise the well-being of more advantaged members of the same population. Examples of spillover effects include spread of infectious diseases from primary foci in poor populations and the opportunity cost of public services committed to dealing with problems related to disadvantage.

Improved environmental management of health-supporting ecosystems (e.g. freshwater resources, agricultural areas) would reduce the adverse health impacts of climate change. A good example is the control of water-borne infections. In many areas increased density of rainfall is likely to lead to more frequent occurrence of significant human infections such as giardiasis and cryptosporidiosis. Traditional public health interventions that focus entirely on personal hygiene and food safety have limited effectiveness. A broader approach would consider the interactions between climate, vegetation, agricultural practices and human activity—and would result in recommendations for the type, time and place of "upstream" public health interventions such as changes in management of water catchment areas.

The maintenance of national public health infrastructure is a crucial element in determining levels of vulnerability and adaptive capacity. The 1990s witnessed the resurgence of several major diseases once thought to have been controlled such as tuberculosis, diphtheria and sexually-transmitted diseases. The major causes were deteriorating public health infrastructure (especially the vaccination programme) as well as socioeconomic instability and population movement (33). Elementary adaptation to climate change can be facilitated by improved monitoring and surveillance systems. Basic indices of population health status (e.g. life expectancy) are available for most countries. However, disease (morbidity) surveillance varies widely depending on locality and the specific disease. To monitor disease incidence/prevalence –which may often provide a sensitive index of impact—low-cost data from primary care facilities could be collected in sentinel populations.

Such top-down approaches should be widely supplemented by adaptation at the community and individual levels. These would include local environmental management, urban design, public education, neighbourhood alert and assistance schemes, and individual behavioural changes. When implementing adaptation technologies care must be taken to prevent adverse secondary impacts (via maladaptation) that is, new health hazards created by the application of technologies. For example, conventional air-conditioning systems can increase the urban heat-island effect and might even exacerbate climate change itself. Water development projects can have significant effects on the local transmission of parasitic diseases including malaria, lymphatic filariasis and schistosomiasis.

Conclusions

Over the ages human societies have degraded or changed local ecosystems and modified regional climates. Without precedent, the aggregate human impact now has attained a global scale, reflecting the recent rapid increase in population size and energy-intensive, high-throughput, mass consumption. The world population is encountering unfamiliar human-induced changes in the lower and middle atmospheres and worldwide depletion of various other natural systems (e.g. soil fertility, aquifers, ocean fisheries and biodiversity in general). Despite early recognition that such changes would affect economic activities, infrastructure and managed ecosystems, there has been less awareness that such large-scale environmental change would weaken the supports for healthy life. Fortunately that is now beginning to change. Indeed, this volume seeks to present a comprehensive discussion of the relationship between global climate change and human population health.

Global climate change is likely to change the frequency of extreme weather events: tropical cyclones may increase as sea surface waters warm; floods may increase as the hydrological cycle intensifies; and heatwaves may increase in mid-continental locations. As discussed in detail in later chapters, a change in the frequency and intensity of heatwaves and cold spells would affect seasonal patterns of morbidity and mortality. The production of various air pollutants and of allergenic spores and pollens would be affected by warmer and wetter conditions. Climate change also is expected to affect health via various indirect pathways, including the patterns of infectious diseases; the yield of food-producing systems on land and at sea; the availability of freshwater; and, by contributing to biodiversity loss, may destabilize and weaken the ecosystem services upon which human society depends.

Adaptations to the health hazard posed by global climate change can be both proactive and reactive, and can occur at the macro, meso and micro-scales; that is, at the population, community and individual levels. Climate change represents a one-off global experiment so there will be limited opportunity to carry out preliminary evaluation of adaptation options. There is therefore a strong case for prudence, both in mitigating climate change and in adapting to its impacts.

This topic is likely to become a major theme in population health research, social policy development and advocacy during this first decade of the twenty-first century. Indeed, consideration of global climatic-environmental hazards to human population health will play a central role in the sustainability transition debate.

References

1   Fagan, B. *Floods, famines and emperors. El Niño and the fate of civilisations.* New York, USA, Basic Books, 1999.

2. McMichael, A.J. *Human frontiers, environments and disease.* Cambridge, UK, Cambridge University Press, 2001.

3. Intergovernmental Panel on Climate Change (IPCC). *Climate Change 2001. Third Assessment Report (Volume I).* Cambridge, UK Cambridge University Press, 2001.

4. World Bank. *World development report. Development and the environment.* Oxford, UK, Oxford University Press, 1992.

5. Folke, C. et al. Renewable resource appropriation. In: *Getting down to earth.* Costanza, R. & Segura, O. eds. Washington, DC, USA, Island Press, 1996.

6. Wackernagel, M. et al. Tracking the ecological overshoot of the human economy. *Proceedings of the national Academy of Sciences,* 99:9266–9271 (2002).

7. Vitousek, P.M. et al. Human domination of Earth's ecosystems. *Science* 277: 494–499 (1997).

8. Intergovernmental Panel on Climate Change (IPCC). *Climate Change 2001: Third Assessment Report (Volume II), Impacts, Vulnerability and Adaptation.* Cambridge: Cambridge University Press, 2001.

9. Intergovernmental Panel on Climate Change (IPCC). Climate change 1995: impacts, adaptations and mitigation of climate change. Contribution of Working Group II. In: *Second Assessment Report of the Intergovernmental Panel on Climate Change.* Watson, R.T. et al. eds. Cambridge, UK, and New York, USA, Cambridge University Press, 1996.

10. Watson, R.T. & McMichael, A.J. Global climate change—the latest assessment: Does global warming warrant a health warning? *Global Change and Human Health* 2: 64–75 (2001).

11. Watson, R.T. et al. *Protecting our planet, securing our future. Linkages among environmental issues and human needs.* UNEP, NASA, World Bank, 1998.

12. McMichael, A.J. *Human frontiers, environments and disease.* Cambridge, UK, Cambridge University Press, 2001.

13. McMichael, A.J. Population, environment, disease, and survival: past patterns, uncertain futures. *Lancet* 359: 1145–1148 (2002).

14. McMichael, A.J. The biosphere, health and sustainable development. *Science* 297: 1093 (2002).

15. McMichael, A.J. et al. The sustainability transition: a new challenge (Editorial). *Bulletin of the World Health Organization* 78: 1067 (2000).

16. McMichael, A.J. et al. eds. *Climate change and human health.* Geneva, Switzerland, World Health Organization, 1996 (WHO/EHG/96.7).

17. McMichael, A.J. et al. *Climate change 1995: scientific–technical analyses of impacts, adaptations, and mitigation of climate change.* Watson, R.T. et al., eds. New York, USA, Cambridge University Press, pp. 561–584, 1996.

18. Hippocrates. Airs, waters and places. An essay on the influence of climate, water supply and situation on health. In: *Hippocratic Writings.* Lloyd G.E.R. ed. London, UK, Penguin, 1978.

19. Pringle, H. Death in Norse Greenland. *Science* 275: 924–926 (1997).

20. Bryson, R.E. & Murray, T.J. *Climates of hunger: mankind and the world's changing weather.* Madison, Wisconsin, USA, University of Wisconsin Press, 1977.

21. Davis, M. *Late Victorian holocausts. The making of the third world.* London, Spectre, 2000.

22. Patz, J. et al. The potential health impacts of climate variability and change for the United States: executive summary of the report of the health sector of the US National Assessment. *Environmental Health Perspectives,* 108: 367–376 (2000).

23. Langford, I.H. & Bentham, G. The potential effects of climate change on winter mortality in England and Wales. *International Journal of Biometeorology* 38: 141–147 (1995).

24. Rooney, C. et al. Excess mortality in England and Wales during the 1995 heatwave. *Journal of Epidemiology and Community Health,* 52: 482–486 (1998).

25. McMichael, A.J. & Githeko, A. Human health. In: *Climate Change 2001: impacts, adaptation, and vulnerability.* Contribution of Working Group II to the Third Assessment Report of the Intergovernmental Panel on Climate Change. McCarthy, J.J. et al. eds. New York, USA, Cambridge University Press, 2001.

26. Epstein, P.R. Climate and health. *Science* 285: 347–348 (1999).

27. Patz, J.A. et al. Global climate change and emerging infectious diseases. *Journal of the American Medical Association* 275: 217–223 (1996).

28. Martens, W.J.M. *Health and climate change: modelling the impacts of global warming and ozone depletion.* London, UK, Earthscan, 1998.

29. Martens, W.J.M. et al. Climate change and future populations at risk of malaria. *Global Environmental Change* 9 Suppl: S89–S107 (1999).

30. Shindell, D.T. et al. Increased polar stratospheric ozone losses and delayed eventual recovery owing to increasing greenhouse gas concentrations. *Nature* 392: 589–592 (1998).

31. Myers, N. Biodiversity's genetic library. *Nature's services. Societal dependence on natural ecosystems.* Washington, DC: Island Press, pp. 255–273, 1997.

32. Woodward, A.J. et al. Protecting human health in a changing world: the role of social and economic development. *Bulletin of the World Health Organization.* 78: 1148–1155 (2000).

33. Lindsay, S. & Martens, W.J.M. 1998. Malaria in the African Highlands: past, present and future. *Bulletin of the World Health Organization* 78, 76:33–45 (2000).

CHAPTER 2

Weather and climate: changing human exposures

K. L. Ebi,[1] L. O. Mearns,[2] B. Nyenzi[3]

Introduction

Research on the potential health effects of weather, climate variability and climate change requires understanding of the exposure of interest. Although often the terms weather and climate are used interchangeably, they actually represent different parts of the same spectrum. Weather is the complex and continuously changing condition of the atmosphere usually considered on a time-scale from minutes to weeks. The atmospheric variables that characterize weather include temperature, precipitation, humidity, pressure, and wind speed and direction. Climate is the average state of the atmosphere, and the associated characteristics of the underlying land or water, in a particular region over a particular time-scale, usually considered over multiple years. Climate variability is the variation around the average climate, including seasonal variations as well as large-scale variations in atmospheric and ocean circulation such as the El Niño/Southern Oscillation (ENSO) or the North Atlantic Oscillation (NAO). Climate change operates over decades or longer time-scales. Research on the health impacts of climate variability and change aims to increase understanding of the potential risks and to identify effective adaptation options.

Understanding the potential health consequences of climate change requires the development of empirical knowledge in three areas (*1*):

1. historical analogue studies to estimate, for specified populations, the risks of climate-sensitive diseases (including understanding the mechanism of effect) and to forecast the potential health effects of comparable exposures either in different geographical regions or in the future;
2. studies seeking early evidence of changes, in either health risk indicators or health status, occurring in response to actual climate change;
3. using existing knowledge and theory to develop empirical-statistical or biophysical models of future health outcomes in relation to defined climate scenarios of change.

The exposures of interest in these studies may lie on different portions of the weather/climate spectrum. This chapter provides basic information to understand weather, climate, climate variability and climate change, and then discusses some analytical methods used to address the unique challenges presented when studying these exposures.

---

[1] World Health Organization Regional Office for Europe, European Centre for Environment and Health, Rome, Italy.
[2] National Center for Atmospheric Research, Boulder, CO, USA.
[3] World Climate Programme, World Meteorological Organization, Geneva, Switzerland.

The climate system and greenhouse gases

Earth's climate is determined by complex interactions among the Sun, oceans, atmosphere, cryosphere, land surface and biosphere (shown schematically in Figure 2.1). These interactions are based on physical laws (conservation of mass, conservation of energy and Newton's second law of motion). The Sun is the principal driving force for weather and climate. The Sun's energy is distributed unevenly on Earth's surface due to the tilt of Earth's axis of rotation. Over the course of a year, the angle of rotation results in equatorial areas receiving more solar energy than those near the poles. As a result, the tropical oceans and land masses absorb a great deal more heat than the other regions of Earth. The atmosphere and oceans act together to redistribute this heat. As the equatorial waters warm air near the ocean surface, it expands, rises (carrying heat and moisture with it) and drifts towards the poles; cooler denser air from the subtropics and the poles moves toward the equator to take its place.

This continual redistribution of heat is modified by the planet's west to east rotation and the Coriolis force associated with the planet's spherical shape, giving rise to the high jet streams and the prevailing westerly trade winds. The winds, in turn, along with Earth's rotation, drive large ocean currents such as the Gulf Stream in the North Atlantic, the Humboldt Current in the South Pacific, and the North and South Equatorial Currents. Ocean currents redistribute warmers waters away from the tropics towards the poles. The ocean and atmosphere exchange heat and water (through evaporation and precipitation), carbon dioxide and other gases. By its mass and high heat capacity, the ocean moderates climate change from season to season and year to year. These complex, changing atmospheric and oceanic patterns help determine weather and climate.

Five layers of atmosphere surround Earth, from surface to outer space. The lowest layer (troposphere) extends from ground level to 8–16 km. The height

FIGURE 2.1 Schematic illustration of the components of the coupled atmosphere/earth /ocean system. *Source: reproduced from reference 2.*

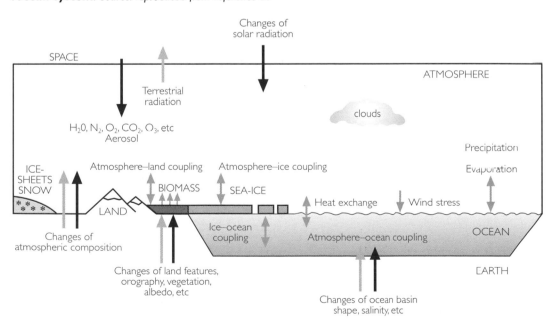

varies with the amount of solar energy reaching Earth; it is lowest at the poles and highest near the equator. On average, air temperature in the troposphere decreases 7 °C for each kilometre increase in altitude, as atmospheric pressure decreases. The troposphere is the level where the weather that affects the surface of Earth develops. The level at which temperature stops decreasing with height is called the tropopause, and temperatures here can be as low as −58 °C. The next layer (stratosphere) extends from the tropopause to about 50 km above the surface, with temperatures slowly increasing to about 4 °C at the top. A high concentration of ozone occurs naturally in the stratosphere at an altitude of about 24 km. Ozone in this region absorbs most of the Sun's ultraviolet rays that would be harmful to life on Earth's surface. Above the stratosphere are three more layers (mesosphere, thermosphere and exosphere) characterized by falling, then rising, temperature patterns.

Overall, the atmosphere reduces the amount of sunlight reaching Earth's surface by about 50%. Greenhouse gases (including water vapour, carbon dioxide, nitrous oxide, methane, halocarbons, and ozone) compose about 2% of the atmosphere. In a clear, cloudless atmosphere they absorb about 17% of the sunlight passing through it (3). Clouds reflect about 30% of the sunlight falling on them and absorb about 15% of the sunlight passing through them. Earth's surface absorbs some sunlight and reradiates it as long-wave (infrared) radiation. Some of this infrared radiation is absorbed by atmospheric greenhouse gases and reradiated back to Earth, thereby warming the surface of Earth by more than would be achieved by incoming solar radiation alone. This atmospheric greenhouse effect is the warming process that raises the average temperature of Earth to its present 15 °C (Figure 2.2). Without this warming, Earth's diurnal temper-

FIGURE 2.2 The greenhouse effect. *Source: reproduced from reference 4.*

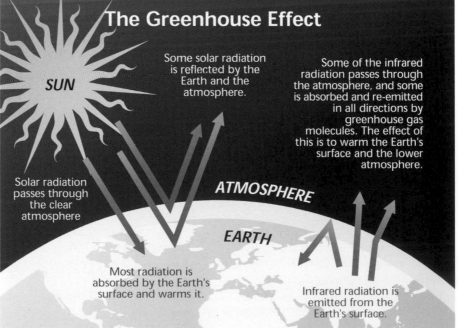

ature range would increase dramatically and the average temperature would be about 33 °C colder (3). Changes in the composition of gases in the atmosphere alter the intensity of the greenhouse effect. This analogy arose because these gases have been likened to the glass of a greenhouse that lets in sunlight but does not allow heat to escape. This is only partially correct—a real greenhouse elevates the temperature not only by the glass absorbing infrared radiation, but also by the enclosed building dramatically reducing convective and advective losses from winds surrounding the building. Yet the misnomer persists.

For Earth as a whole, annual incoming solar radiation is balanced approximately by outgoing infrared radiation. Climate can be affected by any factor that alters the radiation balance or the redistribution of heat energy by the atmosphere or oceans. Perturbations in the climate system that cause local to global climate fluctuations are called forcings. This is short for radiative forcing which can be considered a perturbation in the global radiation (or energy) balance due to internal or external changes in the climate system. Some forcings result from natural events: occasional increases in solar radiation make Earth slightly warmer (positive forcing), volcanic eruptions into the stratosphere release aerosols that reflect more incoming solar radiation causing Earth to cool slightly (negative forcing). Characterization of these forcing agents and their changes over time is required to understand past climate changes in the context of natural variations and to project future climate changes. Other factors, such as orbital fluctuations and impacts from large meteors, also influenced past natural climate change.

Anthropogenic forcing results from the gases and aerosols produced by fossil fuel burning and other greenhouse gas emission sources, and from alterations in Earth's surface from various changes in land use, such as the conversion of forests into agricultural land. Increases in the concentrations of greenhouse gases will increase the amount of heat in the atmosphere. More outgoing terrestrial radiation from the surface will be absorbed, resulting in a positive radiative forcing that tends to warm the lower atmosphere and Earth's surface. The amount of radiative forcing depends on the size of the increase in concentration of each greenhouse gas and its respective radiative properties (5).

The usual unit of measure for climatic forcing agents is the energy perturbation introduced into the climate system (measured in watts per square metre). A common way of representing the consequences of such forcings for the climate system is in the change in average global temperature. The conversion factor from forcing to temperature change is the sensitivity of the climate system (5). This sensitivity is commonly expressed in terms of the global mean temperature change that would be expected after a time sufficient for both atmosphere and ocean to come to equilibrium with the change in climate forcing. Climate feedbacks influence climate sensitivity; the responses of atmospheric water vapour concentration and clouds probably generate the most important feedbacks (6). The nature and extent of these feedbacks give rise to the largest source of uncertainty about climate sensitivity.

When radiative forcing changes (positively or negatively), the climate system responds on various time-scales (5). The longest may last for thousands of years because of time lags in the response of the cryosphere (e.g. sea ice, ice sheets) and deep oceans. Changes over short (weather) time-scales are due to alterations in the global hydrological cycle and short-lived features of the atmosphere such as locations of storm tracks, weather fronts, blocking events and tropical cyclones, which affect regional temperature and precipitation patterns. Greenhouse gases that contribute to forcing include: water vapour, carbon dioxide, nitrous oxide, methane and ozone. Aerosols released in fossil fuel burning also influence climate by reflecting solar radiation.

In addition to adding greenhouse gases and aerosols to the atmosphere, other anthropogenic activities affect climate on local and regional scales. Changes in land use and vegetation can affect climate over a range of spatial scales. Vegetation affects a variety of surface characteristics such as albedo (reflectivity) and roughness (vegetation height), as well as other aspects of the energy balance of the surface through evapotranspiration. Regional temperature and precipitation can be influenced because of changes in vegetation cover. A modelling study by Pielke et al. estimated that loss of vegetation in the South Florida Everglades over the last century decreased rainfall in the region by about 10% (7). Bonan demonstrated that the conversion of forests to cropland in the United States resulted in a regional cooling of about 2 °C (8). There is concern that deforestation induced drought may be occurring in the Amazon and other parts of the tropics (9). However, recent evidence suggests that deforestation and interannual climate fluctuations interact in a non-linear manner such that the response of Amazon rainfall to deforestation also depends on the phase of the El Niño/Southern Oscillation (ENSO) cycle (10). In some transition regions there may be more, not less, precipitation from deforestation. Another land-use impact is the urban heat island wherein cities can be up to 12 °C warmer than surrounding areas due to the extra heat absorbed by asphalt and concrete, and by the relative lack of vegetation to promote evaporative cooling (6).

Water vapour is the major greenhouse gas, contributing a positive forcing ten times greater than that of the other gases. Clouds (condensed water) produce both positive and negative forcing: positive by trapping Earth's outgoing radiation at night, and negative by reflecting sunlight during the day. Understanding how to measure accurately and simulate cloud effects remains one of the most difficult tasks for climate science.

Carbon dioxide currently contributes the largest portion of anthropogenic positive forcing. Atmospheric CO_2 is not destroyed chemically and its removal from the atmosphere occurs through multiple processes that transiently store the carbon in the land and ocean reservoirs, and ultimately in mineral deposits (5). A major removal process depends on the transfer of the carbon content of near-surface waters to the deep ocean, on a century time-scale, with final removal stretching over hundreds of thousands of years. Natural processes currently remove about half the incremental man-made CO_2 added to the atmosphere each year; the balance can remain in the atmosphere for more than 100 years (6). Atmospheric concentrations of CO_2 have increased by 31% since 1750 (5). Current global concentrations average about 370 ppmv (parts per million by volume). This concentration has not been exceeded during the past 420 000 years and probably not during the past 20 million years (3). Measurements begun in the 1950s show that atmospheric CO_2 has been increasing at about 0.5% per year (Figure 2.3). This rate of increase is unprecedented during at least the past 20 000 years (5). About 75% of the anthropogenic CO_2 emissions to the atmosphere during the past 20 years were due to fossil fuel burning (5). Much of the rest were due to land-use change, especially deforestation.

Methane (CH_4) contributes a positive forcing about half that of CO_2 (5). It is released from cultivating rice; raising domestic ruminants (cows, sheep); disposing waste and sewage in landfills; burning biomass; and operating leaking gas pipelines. The atmospheric concentration of methane has increased 151% since 1750 (5). Measurements between the early 1980s and 2000 showed a 10%

FIGURE 2.3 Observed and projected atmospheric CO₂ concentrations from 1000 to 2100.
From ice core data and from direct atmospheric measurements over the past few decades. Projections of CO₂ concentrations for the period 2000–2100 are based on the IS92a scenario (medium), and the highest and lowest of the range of SRES scenarios. *Source: reproduced from reference 11.*

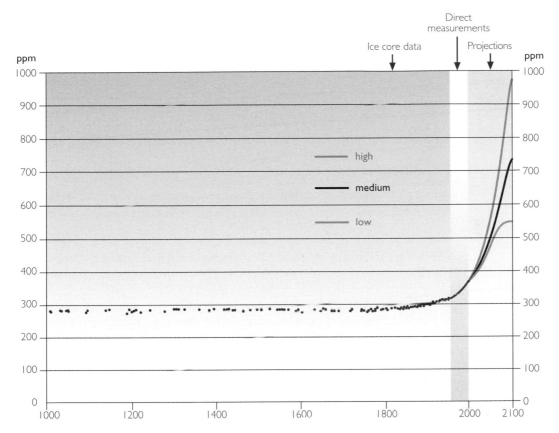

increase in atmospheric CH₄ to 1850 ppb (parts per billion). Although the rate of increase has slowed to near zero in the past two years, present CH₄ concentrations have not been exceeded during the past 420 000 years. CH₄ remains in the atmosphere about 10 years. The primary removal mechanism is by chemical reaction in the stratosphere with hydroxyl ions to produce carbon dioxide and water vapour.

Other greenhouse gases include nitrous oxide and ozone. Nitrous oxide is emitted by both natural and anthropogenic sources, and removed from the atmosphere by chemical reactions. The atmospheric concentration of nitrous oxide has increased steadily since the Industrial Revolution and is now about 16% larger than in 1750 (5). Nitrous oxide has a long atmospheric lifetime.

Ozone (O₃) is not emitted directly but formed from photochemical processes involving both natural and anthropogenic species. Ozone remains in the atmosphere for weeks to months. Its role in climate forcing depends on altitude: in the upper troposphere it contributes a small positive forcing, while in the stratosphere it caused negative forcing over the past two decades (5). Based on limited observations, global tropospheric ozone has increased by about 35% since pre-industrial times.

TABLE 2.1 Examples of greenhouse gases that are affected by human activities.

	CO₂ (Carbon Dioxide)	CH₄ (Methane)	N₂O (Nitrous Oxide)	CFC-11 (chlorofluoro-carbon-11)	HFC-23 (Hydrofluoro-carbon-23)	CF₄ (Perfluoromethane)
Pre-industrial concentration	~280 ppm	~700 ppb	~270 ppb	Zero	Zero	40 ppt
Concentration in 1998	365 ppm	1745 ppb	314 ppb	268 ppt	14 ppt	80 ppt
Rate of Concentration change[b]	1.5 ppm/yr[a]	7.0 ppb/yr[a]	0.8 ppb/yr	−1.4 ppt/yr	0.55 ppt/yr	1 ppt/yr
Atmospheric lifetime	5–200 yr[c]	12 yr[d]	114 yr[d]	45 yr	260 yr	>50,000 yr

[a] Rate has fluctuated between 0.9 ppm/yr and 2.8 ppm/yr for CO_2 and between 0 and 13 ppb/yr for CH_4 over the period 1990 to 1999.

[b] Rate is calculated over the period 1990 to 1999.

[c] No single lifetime can be defined for CO_2 because of the different rates of uptake by different removal processes.

[d] This lifetime has been defined as an "adjustment time" that takes into account the indirect effect of the gas on its own residence time.

Source: reproduced from reference 5.

Aerosols are microscopic particles or droplets in air, their major anthropogenic sources are fossil fuel and biomass burning. They can reflect solar radiation and can alter cloud properties and lifetimes. Depending on their size and chemistry, aerosols contribute either positive or negative forcing. For example, sulphate particles scatter sunlight and cause cooling. Soot (black carbon particles) can warm the climate system by absorbing solar radiation. Aerosols have a lifetime of days to weeks and so respond fairly quickly to changes in emissions. They are less well measured than greenhouse gases.

Table 2.1 provides examples of several greenhouse gases and summarizes their 1790 and 1998 concentrations; rate of change over the period 1990–1999; and atmospheric lifetime. The atmospheric lifetime is highly relevant to policy-makers because emissions of gases with long lifetimes is a quasi-irreversible commitment to sustained positive forcing over decades, centuries or millennia (3).

Weather, climate and climate variability

The terms weather and climate often are used interchangeably, but they actually represent different parts of the same spectrum. Weather is the day-to-day changing atmospheric conditions. Climate is the average state of the atmosphere and the underlying land or water in a particular region over a particular time-scale. Put more simply, climate is what you expect and weather is what you get. Climate variability is the variation around the mean climate; this includes seasonal variations and irregular events such as the El Niño/Southern Oscillation. These differences amongst weather, climate and climate variability have not been applied consistently across studies of potential health impacts, which can lead to confusion and/or misinterpretation.

Elements of daily weather operate on a variety of scales. Well-defined patterns dominate the distribution of atmospheric pressure and winds across Earth. These large-scale patterns are called the general circulation. Smaller patterns are found on the synoptic scale, on the order of hundreds or thousands of square kilometres. Synoptic scale features (e.g. cyclones, troughs and ridges) persist for a period of days to as much as a couple of weeks. Other elements of daily weather operate at the mesoscale, which is on the order of tens of square kilometres, and for periods as brief as half an hour. The smallest scale at which heat and moisture transfers occur is the microscale, such as across the surface of a single leaf.

Climate is typically described by the summary statistics of a set of atmospheric and surface variables such as: temperature, precipitation, wind, humidity, cloudiness, soil moisture, sea surface temperature, and the concentration and thickness of sea ice. The official average value of a meteorological element for a specific location over 30 years is defined as a climate normal (12). Included are data from weather stations meeting quality standards prescribed by the World Meteorological Organization. Climate normals are used to compare current conditions and are calculated every 10 years.

Climatologists use climatic normals as a basis of comparison for climate during the following decade. Comparison of normals between 30-year periods may lead to erroneous conclusions about climatic change due to changes over the decades in station location, instrumentation used, methods of weather observations and how the various normals were computed (12). The differences between normals due to these primarily anthropogenically-induced changes may be larger than those due to a true change in climate.

The climate normal for the 1990s was the period 1961–1990. This was the baseline for the analyses of climatic trends summarized by the IPCC Third Assessment Report. In January 2002, the climate normal period changed to 1971–2000. This change in the climate normal means a change in the baseline of comparison; different conclusions may result when comparisons are made using different baselines.

A climate normal is simply an average and therefore does not completely characterize a particular climate. Some measure of the variability of the climate also is desirable. This is especially true for precipitation in dry climates, and with temperatures in continental locations that frequently experience large swings from cold to warm air masses. Typical measures of variability include the standard deviation and interquartile range. Some measures of the extremes of the climate are useful also.

A variety of organizations and individuals summarize weather over various temporal and spatial scales to create a picture of the average meteorological conditions in a region. There are well-known spatial latitudinal and altitudinal temperature gradients. For example, under typical conditions in mountainous terrain, the average surface air or soil temperature decreases by about 6.5 °C for every 1000 m increase in elevation, and along an equator to pole gradient a distance of 1000 km corresponds to an average surface temperature change of about 5 °C (6). Superimposed on these large-scale gradients are more complex regional and local patterns.

Temporal climate variations are most obviously recognized in normal diurnal and seasonal variations. The amplitude of the diurnal temperature cycle at most locations is typically in the range of 5–15 °C (3). The amplitude of seasonal variability is generally larger than that of the diurnal cycle at high latitudes and smaller at low latitudes. Years of research on seasonal to interannual variations have uncovered several recurring pressure and wind patterns that are termed modes of climate variability (6).

The El Niño/Southern Oscillation (ENSO) cycle is one of Earth's dominant modes of climate variability. ENSO is the strongest natural fluctuation of climate on interannual time-scales, with global weather consequences (13, 14). An El Niño event occurs approximately every two to seven years. Originally the term applied only to a warm ocean current that ran southwards along the coast of Peru about Christmas time. Subsequently an atmospheric component, the Southern Oscillation, was found to be connected with El Niño events. The atmosphere and ocean interact to create the ENSO cycle: there is a complex interplay between the strength of surface winds that blow westward along the equator and subsurface currents and temperatures (13). The ocean and atmospheric conditions in the tropical Pacific fluctuate somewhat irregularly between El Niño and La Niña events (which consist of cooling in the tropical Pacific) (15). The most intense phase of each event usually lasts about one year.

Worldwide changes in temperature and precipitation result from changes in sea surface temperature during the ENSO cycle (14, 16). During El Niño events, abnormally heavy rainfall occurs along part of the west coast of South America, while drought conditions often occur in parts of Australia, Malaysia, Indonesia, Micronesia, Africa, north-east Brazil and Central America (13). These changes can have a strong effect on the health of individuals and populations because of associated droughts, floods, heatwaves and changes that can disrupt food production (16). Predictions of ENSO associated regional anomalies (deviations or departures from the normal) are generally given in probabilistic terms because the likelihood of occurrence of any projected anomaly varies from one region to another, and with the strength and specific configuration of the equatorial Pacific sea surface temperature anomalies (12).

ENSO is not the only mode of climate variability. The Pacific Decadal Oscillation (PDO) and the North Atlantic–Artic Oscillation (NAO–AO) are well established as influences on regional climate. The NAO is a large-scale oscillation in atmospheric pressure between the subtropical high near the Azores and the sub-polar low near Iceland (17). The latter appears to have a particularly large decadal signal (18). The PDO signal may fluctuate over several decades.

A note about terminology used by meteorologists and climatologists is relevant. The terms forecast and prediction each refer to statements about future events: predictions are statements that relate to the results of a single numerical model; forecasts are statements that relate to a synthesis of a number of predictions (6). Forecasts and predictions are currently most relevant to future (i.e. near-term) weather conditions and seasonal climate conditions. Estimates of long-term climate change usually are discussed in terms of projections, which are less certain than predictions or forecasts. Projections (of future climate) are based on estimates of possible future changes with no specific probability attached to them.

Climate change

Climate change operates over decades or longer. Changes in climate occur as a result of both internal variability within the climate system and external factors (both natural and anthropogenic). The climate record clearly shows that climate is always changing (Figure 2.4). One feature of the record is that climate over the past 10 000 years has been both warm and relatively stable (5).

Past changes could not be observed directly, but are inferred through a variety of proxy records such as ice cores and tree rings. Such records can be used to make inferences about climate and atmospheric composition extending back as far as 400 000 years. These data indicate that the range of natural climate variability is in excess of several degrees Celsius on local and regional spatial scales over periods as short as a decade (5). Precipitation also has varied widely.

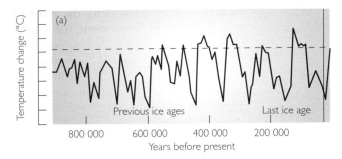

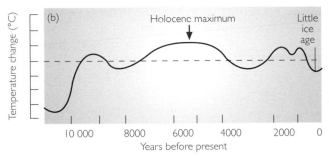

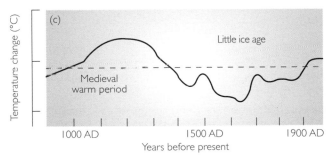

FIGURE 2.4 Schematic diagrams of global temperature variations since the Pleistocene on three time-scales: (a) last million years (b) last 10000 years (c) last 1000 years. The dotted line nominally represents conditions near the beginning of the century. *Source: reproduced from reference 19.*

On century to millennial scales, climate changes such as the European 'little ice age' from the fourteenth to eighteenth centuries occur (20). Over the past approximately million years, the global climate record is characterized by larger glacial-interglacial transitions, with multiple periodicities of roughly 20000, 40000 and 100000 years (6). These are correlated with the effects of Earth-Sun orbital variations. The amplitudes of these transitions are on the order of 5–10 °C and are accompanied by large extensions and retreats of polar and glacial ice.

In 1861, instrumental records began recording temperature, precipitation and other weather elements. Figure 2.5 shows the annual global temperature (average of near surface air temperature over land and of sea surface temperatures) expressed as anomalies or departures from the 1961 to 1990 baseline. Over the twentieth century, the global average surface temperature increased about 0.6 °C ± 0.2 °C, the 1990s being the warmest decade and 1998 the warmest year in the Northern Hemisphere (5). The high global temperatures associated with the 1997–1998 El Niño event are apparent, even taking into account recent warming trends. The increase in temperature over the twentieth century is likely to have been the largest of any century during the past 1000 years (Figure 2.6) (5). The warmth of the 1990s was outside the 95% confidence interval of temperature uncertainty, defined by historical variation, during even the warmest periods of the last millennium (3).

FIGURE 2.5 Combined annual land-surface, air, and sea surface temperature anomalies (°C) from 1861 to 2000, relative to 1961 to 1990. *Source: produced from data from reference 21.*

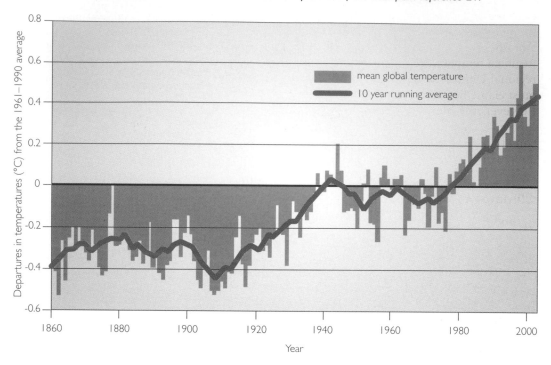

FIGURE 2.6 Millennial Northern Hemisphere (NH) temperature reconstruction from AD 1000 to 1999. Measurements before the 1850s are based on tree rings, corals, ice cores, and historical records. Later records are from instrumental data. The dark line is a smoother version of the series and the shaded area represents two standard error limits. *Source: reproduced from reference 5.*

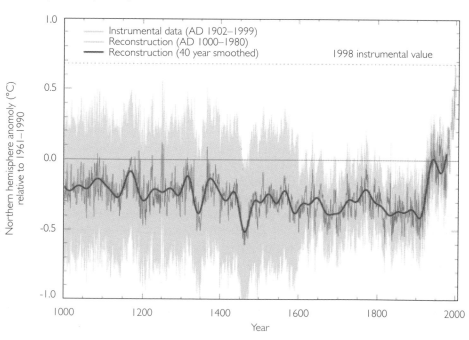

The regional patterns of warming that occurred in the early twentieth century differ from those of the latter part (5). The largest increases in temperature occurred over the mid and high latitudes of the continents in the Northern Hemisphere. Regional temperature patterns are related, in part, to various phases of atmospheric-oceanic oscillations, such as the North Atlantic–Artic Oscillation (5). Regional temperature patterns over a few decades can be influenced strongly by regional climate variability, causing a departure from global trends. More time must elapse before the importance of recent temperature trends can be assessed. However, the Northern Hemisphere temperatures of the 1990s were warmer than any other time in the past six to ten centuries (5). Less is known about the conditions that prevailed in the Southern Hemisphere prior to 1861 because limited data are available.

Climate variability and change over the twentieth century

In the Third Assessment Report of the Intergovernmental Panel on Climate Change, Working Group I summarized climatic changes that occurred over the twentieth century. A concerted effort was made to express the uncertainty about climate trends in a consistent and meaningful fashion. Thus, confidence in their judgements was expressed as: virtually certain (>99% chance that a result is true); very likely (90–99% chance); likely (66–90% chance); medium likely (33–66% chance); unlikely (10–33% chance); and very unlikely (1–10% chance) (5). This terminology is used in the summary below.

On average, between 1950 and 1993, night time daily minimum air temperatures over land increased by about 0.2 °C per decade, although this did not happen everywhere (5). This increase may be due to a likely increase in cloud cover of about 2% since the beginning of the twentieth century (5). This increase was about twice the rate of increase in daytime daily maximum air temperatures (0.1 °C per decade) and lengthened the freeze-free season in many mid and high latitude regions.

Along with these temperature changes, snow cover and ice extent decreased. Snow cover has very likely decreased about 10% since the late 1960s, and spring and summer sea ice extent decreased about 10–15% since the 1950s. There is now ample evidence to support a major retreat of alpine and continental glaciers in response to twentieth century warming. However, in a few maritime regions, increases in precipitation overshadowed increases in temperature in the past two decades, and glaciers re-advanced (5). Sea ice is important because it reflects more incoming solar radiation than the sea surface and insulates the sea from heat loss. Therefore, reduction of sea ice causes positive climate forcing at high latitudes.

Data show that global average sea level rose between 0.1 and 0.2 m during the twentieth century. Based on tide gauge data, the rate of global mean sea level rise was in the range of 1.0 to 2.0 mm per year, compared to an average rate of about 0.1 to 0.2 mm per year over the last 3000 years (5). This does not mean that sea level is rising in all areas: the retreat of glacial ice in the past several thousand years has lead to a rebound of land in some areas. Sea level has been rising for several reasons. First, ocean water expands as it warms. On the basis of observations and model results, thermal expansion is one of the major contributors to historical sea level changes (5). Thermal expansion is expected to be the largest contributor to sea level rise over the next 100 years. As deep ocean temperatures change slowly, thermal expansion is expected to continue for many centuries after stabilization of greenhouse gases. Second, after thermal expansion, the melting of mountain glaciers and ice caps is expected to make the next largest contribution to sea level rise over the next 100 years. These glaciers and ice caps are much smaller than the large ice sheets of Greenland and Antarctica, and are more sensitive to climate change. Third, processes unrelated to climate change influence sea level; these processes could have regional effects on sea level, such as coastal subsidence in river delta regions.

Other changes include the following:

- it is very likely that precipitation increased by 0.5–1.0% per decade over most mid and high latitudes of the Northern Hemisphere continents, and likely that rainfall increased 0.2–0.3% per decade over tropical land areas. Also it is likely that rainfall decreased over much of the Northern Hemisphere subtropical land areas. Comparable systematic changes were not detected over the Southern Hemisphere;
- it is likely that there was a 2–4% increase in the frequency of heavy precipitation events in mid and high latitudes of the Northern Hemisphere over the latter half of the twentieth century;
- since 1950, it is very likely that there was a reduction in the frequency of extremely low temperatures, with a smaller increase in the frequency of extreme high temperatures;
- El Niño events were more frequent, persistent and intense since the mid-1970s, compared with the previous 100 years;
- in parts of Asia and Africa, the frequency and intensity of droughts increased in recent decades.

However, not all aspects of climate changed during the last century (5). A few areas of the globe cooled in recent decades, mainly over some parts of the Southern Hemisphere oceans and parts of Antarctica. No significant trends of Antarctic sea ice extent are apparent since 1978. No systematic changes in the frequency of tornadoes or other severe storms are evident.

One area of concern is the possibility of a sudden, large change in the climate system in response to accumulated climatic forcing. The paleoclimate record contains examples of such changes, at least on regional scales.

Special Report on Emission Scenarios

The projection of future climate change first requires projection of future emissions of greenhouse gases and aerosols, for example, the future fossil fuel and land-use sources of CO_2 and other gases and aerosols. How much of the carbon from future use of fossil fuels will increase atmospheric CO_2 will depend on what fractions are taken up by land and the oceans. Future climate change depends also on climate sensitivity.

For the Third Assessment Report, the IPCC developed a series of scenarios that include a broad range of assumptions about future economic and technological development to encompass the uncertainty about the structure of society in 2100. These scenarios are called collectively the SRES, from the Special Report on Emission Scenarios (22). An earlier baseline, or business as usual, scenario (or IS92a) assumed rapid growth rates such that annual greenhouse gas emissions continue to accelerate; this scenario was developed for the Second Assessment Report. The SRES scenarios produce a range of emission projections that are both larger and smaller in 2100 than the IS92a scenario. The SRES scenarios are grouped into four narrative storylines. The storylines can be categorized basically in a 2x2 table, with the axes global versus regional focus, and a world focused more on consumerism versus a world focused more on conservation. The basic storylines are A1 (world markets), B1 (global sustainability), A2 (provincial enterprise) and B2 (local stewardship). Each storyline contains underlying assumptions about population growth, economic development, life style choices, technological change and energy alternatives. Each leads to different patterns and concentrations of emissions of greenhouse gases. In some storylines, the large growth in emissions could lead to degradation of the global environment in ways beyond climate change (5). No attempt was made to assign probabilities to the SRES scenarios; they are designed to illustrate a wide range of possible emissions outcomes. Much of the summary climate change information provided below is based on results from climate models that used the SRES scenarios.

Anthropogenic climate change

Several key questions are asked about climate change:

- was there detectable climate change during the twentieth century?
- if so, how much warming already experienced was likely to be due to human activities?
- how much additional warming is likely to occur if we increase the atmospheric levels of greenhouse gases?
- what will be the likely impacts?

To distinguish anthropogenic climate changes from natural variation requires that the anthropogenic signal be identified against the noise of natural climate variability. The third question is important because the climate system has a great deal of inertia—changes to the atmosphere today may continue to affect the climate for decades or even centuries. Similarly, the consequences of efforts to reduce the magnitude of future change may not become apparent for decades to centuries. The question of impacts is addressed in other chapters in this book.

Two of the tasks of the IPCC's Third Assessment Report (TAR) were to determine whether there has been a detectable signal of climate change (in a statistical sense) and if so, to determine if any of the change could be attributed confidently to anthropogenic causes. One conclusion was that best agreement between observations and model simulations over the past 140 years was found when both natural factors and anthropogenic forcings were included in the models (Figure 2.7) (5). Further, the IPCC authors concluded that most of the warming observed over the past 50 years is attributable to human activities and that human influences will continue to change atmospheric composition throughout the twenty-first century (5). The IPCC authors concluded that emissions of CO_2 due to fossil fuel burning are virtually certain to be the dominant contributor to the trends in atmospheric CO_2 concentration throughout the twenty-first century.

By 2100, atmospheric concentrations of CO_2 are projected to be between 490 and 1260 ppm (75–350% above the concentration of 280 ppm in 1750) (5). Based on climate model results using the SRES scenarios, the IPCC projected that the global mean temperature of Earth would increase by the end of the twenty-first century by between 1.4 and 5.8 °C. Global precipitation also would increase. This projected rate of warming is much larger than the observed changes during the

FIGURE 2.7 Global mean surface temperature anomalies relative to the 1880 to 1920 mean from the instrumental record, compared with ensembles of four simulations with a coupled ocean-atmosphere climate model. The line shows the instrumental data while the shaded area shows the range of outputs from individual model simulations. Data are annual mean values. *Source: reproduced from reference 5.*

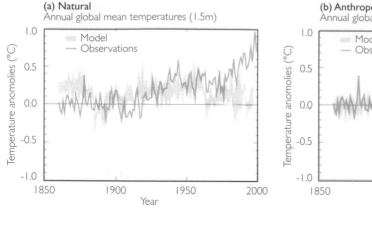

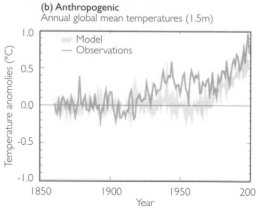

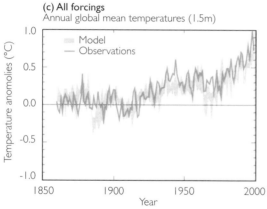

twentieth century and is very likely to be without precedent during at least the last 10 000 years (*5*). These projections strongly depend on climate sensitivity. Recent publications, using different approaches and different models, conclude that these IPCC estimates are likely to be conservative (*23–25*). Andronova and Schlesinger concluded that there is a 54% likelihood that true climate sensitivity lies outside the IPCC range and that global average temperature increases could be higher or lower than those projected by the IPCC (*23*). Knutti et al. concluded that there is 40% probability that the warming could exceed that projected by the IPCC, and only 5% probability that it will be lower (*24*). In addition, these studies suggest that much of the warming over the next few decades will be due to past greenhouse gas emissions and thus relatively insensitive to mitigation efforts. Only beyond mid-century could mitigation efforts begin to affect global mean temperatures.

Average temperature increases are projected to be greatest in the northern regions of North America, and northern and central Asia. Precipitation also is projected to increase, particularly over the northern mid to high latitudes and

FIGURE 2.8 Distribution of July daily maximum temperatures in Chicago, today and in 2095 under one climate change scenario (GFDL). *Source: reproduced from reference 27.*

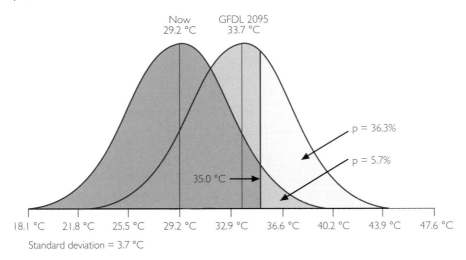

Antarctica in winter. Of particular note is that the shift in the mean of meteorological variables, such as temperature and precipitation, will result in a shift in extremes (Figure 2.8). For example, in Chicago, Illinois, currently about 6% of days in July and August are above 35 °C; under one climate scenario by 2095 that could rise to 36%. Global climate change is not likely to be spatially uniform and is expected to include changes in temperature and the hydrologic cycle. Associated health effects also will vary spatially. Higher evaporation rates will accelerate the drying of soils following rain events resulting in drier average conditions in some regions (6). Larger year-to-year variations in precipitation are very likely over most areas where an increase in mean precipitation is projected, including an increase in heavy rain events (3). Changes in other features of the climate system also could occur, for example the frequency and intensity of tropical and mid latitude storms. Global climate change may also influence the behaviour of ENSO or other modes of climate variability (26). Chapter 5 summarizes the health effects associated with El Niño events.

Between 1990 and 2100 global mean sea level is projected to rise 0.09–0.88 m (5, 28). This will be due primarily to thermal expansion of the oceans and loss of mass from glaciers and ice caps. Sea levels are projected to continue rising for hundreds of years after stabilization of greenhouse gas concentrations due to the long time-scales on which the deep ocean adjusts to climate change.

Table 2.2 summarizes the confidence in observed changes in extremes of weather and climate during the latter part of the twentieth century and in projected changes during the twenty-first century.

Climate modelling

Projections of future climatic conditions are produced using climate system models. Atmosphere-Ocean general circulation models (AOGCMs) are mathematical expressions of the thermodynamics; fluid motions; chemical reactions; and radiative transfer of the complete climate system that are as comprehensive as allowed by computational feasibility and scientific understanding of their for-

TABLE 2.2 Estimates of confidence in observed and projected changes in extreme weather and climate events. The table depicts an assessment of confidence in observed changes in extremes of weather and climate during the latter half of the 20th century (left column) and in projected changes during the 21st century (right column).[a] This assessment relies on observational and modelling studies, as well as physical plausibility of future projections across all commonly used scenarios and is based on expert judgment. *Adapted from reference 5.*

Confidence in observed changes (latter half of the 20th century)	Changes in Phenomenon	Confidence in projected changes (during the 21st century)
Likely	Higher maximum temperatures and more hot days over nearly all land areas	Very likely
Very likely	Higher minimum temperatures, fewer cold days and frost days over nearly all land areas	Very likely
Very likely	Reduced diurnal temperature range over most land areas	Very likely
Likely, over many areas	Increase of heat index[a] over land areas	Very likely, over most areas
Likely, over many Northern Hemisphere mid-to high latitude land areas	More intense precipitation events[b]	Very likely, over many areas
Likely, in a few areas	Increased summer continental drying and associated risk of drought	Likely, over most mid-latitude continental interiors (Lack of consistent projections in other areas)
Not observed in the few analyses available	Increase in tropical cyclone peak wind intensities[c]	Likely, over some areas
Insufficient data for assessment	Increase in tropical cyclone mean and peak precipitation intensities	Likely, over some areas

[a] Heat index: A combination of temperature and humidity that measures effects on human comfort.
[b] For other areas there are either insufficient data or conflicting analyses.
[c] Past and future changes in tropical cyclone location and frequency are uncertain.

mulation (6). The models couple known laws of physics with prescribed initial and boundary conditions of the atmosphere to compute the evolving state of the global atmosphere, ocean, land surface and sea ice in response to external natural and anthropogenic forcings. Boundary conditions are external factors that influence Earth's climate system, such as the intensity of sunlight, the composition of the atmosphere, etc. The climate system adjusts when one or more of these external factors change: for example, global average temperatures would be expected to increase with an increase in solar output. The ultimate aim is to model as much as possible of the climate system, especially the complex feedbacks among the various components.

A number of models are in operation in various research institutes and universities worldwide. Although the models are based on the same laws of physics, each has different ways of dealing with processes that cannot be represented explicitly by physical laws, such as formation of clouds and precipitation. Variations in these parameterizations lead to different regional projections of climate change, particularly for precipitation.

AOGCMs cannot simulate all aspects of climate and there are large uncertainties associated with clouds, yet there is increasing confidence that they can provide useful projections of future climate. This is due to their improved ability to simulate the important interactions between ocean and atmosphere for past and current climate on a range of temporal and spatial scales (5). In particular, simulations that include estimates of natural and anthropogenic forcing reproduce the large-scale changes in surface temperature over the twentieth century

(5). Current AOGCMs simulate well some of the key modes of climate variability such as the North Atlantic Oscillation and ENSO.

AOGCMs generally have horizontal spatial resolutions of about 250 km for their atmospheric component. This coarse spatial scale creates problems for successfully simulating possible regional climate change and its impacts. For example, two AOGCMs were used in the United States' National Assessment of the Potential Impacts of Climate Variability and Change. In one (from the UK Hadley Centre) the state of Florida was too small to be resolved, and in the other (the Canadian climate model), the Great Lakes were not represented. It is not possible to project regional climate patterns with confidence if significant geographical features are missing.

With limitations of spatial scale and other factors, AOGCMs still have difficulty portraying accurately precipitation patterns in mountainous regions and resolving important synoptic weather features (such as Mesoscale Convective Systems) that strongly influence precipitation patterns and amounts in many agricultural regions. The typical biases in reproduction of observed regional scale climate by AOGCMS are in the range of about ±4 °C for temperature, and −40 to +80% for precipitation (29). However, larger biases do occur. It is assumed that the ability to reproduce faithfully the current climate is a necessary condition for simulating future climate in a meaningful way (30).

Some techniques can ameliorate the problem of spatial scale. These regionalization techniques include statistical downscaling; regional climate modelling; and application of high resolution and variable resolution atmospheric models (AGCMS). All of these result in higher resolution simulations and, usually, better representation of regional climate (31). Statistical downscaling and regional modelling are the most popular techniques that have been used to provide improved regional climate representation for use in impacts studies (30–35). The guiding principle of both techniques is to use output from the coarse resolution AOGCMs to produce more detailed regional information.

In regional modelling, lateral and initial boundary conditions from an AOGCM are used to drive regional climate models (usually derived from mesoscale weather forecasting models) over a particular region of interest. AOGCMs provide the large-scale responses of the climate system while the regional model provides regional scale details. Regional models, which are run at higher resolutions (e.g. 30–50 km) are much more successful at simulating accurately regional climate, particularly in regions with complex topography, coastlines, or land use patterns. This ability to reproduce regional climate is limited, however, by the quality of the boundary conditions from the global model. Regional models cannot overcome large errors in the AOGCMs. Climate change experiments with regional models have been performed over many regions, including North America, Europe, Australia, China and India (36).

In statistical downscaling, the cross-scale relationship (i.e. large scale to regional/local scale) is expressed as a function between large-scale variables (predictors) and regional or local scale climate variables (predictands). Usually the large and local scale variables are different. It is important that the predictor variables be of relevance to the local variables. The technique of statistical downscaling relies on the assumption that the statistical relationships developed under observed climate conditions are valid under future climate conditions and that the predictors fully represent the climate change signal. The literature on statistical downscaling is quite large and the technique has been applied over most regions of the world (30–31).

These regionalization techniques sometimes produce climate changes, particularly of precipitation, that differ substantially from those of the global model that provides the large scale information for the techniques. There remains uncertainty regarding which projection of climate change (from the global model or from the regionalization technique) is more likely correct. However, at least in the case of regional modelling, there is good evidence to suggest that in regions of complex topography, the regional model is more apt to provide a more realistic response to increased greenhouse gases. Applications of regionalized scenarios to impacts models (e.g. hydrologic and agricultural models) usually produce impacts that are different from those obtained using the corresponding coarse scale scenario (35).

Despite the regional climate limitations of AOGCMs, they remain the major source of information on possible climatic changes that can be useful for projections of possible health impacts of climate change. However, scenarios developed using regionalization techniques will become increasingly available. It is also the case that the typical spatial resolution of AOGCMs will continue to improve in the coming years.

The rate of climate change also is of particular importance for understanding potential impacts, especially from the point of view of possible adaptations of human and natural systems. Some periods of past climate change occurred very rapidly. For example, during the last major ice age shifts of temperatures of up to 5 °C occurred in fewer than 50 years (20). Moreover, the recent rate of global temperature change in the last (20th) century is the greatest century scale change in the past millennium. The range of climate change indicated by the projected changes from climate models represents rates of change from roughly 0.14 to 0.58 °C per decade. It is expected that natural ecosystems in particular could have difficulty in adapting to the higher rates of change. There are also issues of very sudden changes as well as climate surprises (37). These include such events as the possible collapse of the West Antarctic Ice Sheet and the shutting down of the thermohaline circulation of the ocean in the North Atlantic. Such punctuated events would have dramatic effects on sea level rise for the former phenomenon and temperature for the latter. However, neither event is considered to have any significant likelihood in the next 100 years (26).

Exposure assessment

This section begins the discussion, continued throughout the remainder of the book, on how weather, climate variability and change can influence population health. There are descriptions of some of the methods and tools that can be used to assess exposure, along with illustrative examples. Further discussion of methods to assess relevant exposures to weather and climate can be found in Ebi and Patz (1).

Studying the natural complexities of weather and climate variability in relation to health outcomes offers unique challenges. Weather and climate can be considered over various spatial and temporal scales, with different scales of relevance to different health outcomes. For example, one categorization of temporal scales is into episodes, short-term weather variability and longer-term variation. The consequences of a single event, from a heatwave to an El Niño event, may be useful analogues for similar events. However, a single event may not be representative of all events; it might be weaker or stronger, or may be shorter or last longer than typical events.

Although informative for future directions in research and adaptation, the predictive value of analogue studies may be limited because future events may differ from historical events and because the extent of vulnerability of a population changes over time. The 1995 heatwave caused considerably more loss of life in midwestern states in the United States than a similar heatwave in 1999, in part because of programmes established in the interim (38). There are similar issues for geographical analogues, such as using the current experience of heatwaves in a more southern region to predict what might happen in the future in a more northern region. Regions differ on a number of important factors, including living standards and behaviours. Therefore, scenario-based modelling approaches are needed to project what might happen under different climate conditions.

As noted in other chapters, one of the difficulties faced by researchers studying the health impacts of climate variability and change is the often limited availability of both weather and health outcome data on the same temporal and geographical scale.

Exposure assessment begins by incorporating a definition of the exposure of interest into the study hypothesis. To use heatwaves as an example, a heatwave needs to be defined and methods for assessment determined. Heatwaves may be defined by temperature alone or a combination of temperature with other weather variables. There are various definitions of heatwaves. In the Netherlands, for example, the Royal Meteorological Institute defines a heatwave as a period of at least five days, each of which has a maximum temperature of at least 25 °C, including at least three days with a maximum temperature of at least 30 °C (39). It should be noted that the adverse health effects observed during and following a heatwave do not depend on weather alone: the physiological, behavioural and other adaptations of the population exposed to the heatwave are additional determinants of outcomes.

As well as defining the exposure of interest, decisions need to be made on the appropriate lag period between exposure and effect, and how long health outcomes may be increased after an exposure. Lag periods ranging from a few days to a year have been used, depending on the presumed underlying mechanism of effect. Deaths in the 1995 Chicago heatwave were highest two days after temperatures peaked (40). In a study of viral pneumonia, a seven-day lag was used (41); a study of water-borne disease outbreaks used lags of one and two months (42); and studies of El Niño and malaria epidemics used one-year (43–45).

The following examples describe a variety of approaches used to summarize exposures to weather and climate. Informative exposure assessment is required for development of quantitative estimates of current vulnerability to climate-sensitive diseases using empirical epidemiological approaches.

An example of an episode analysis is a study that took advantage of the 1997/8 El Niño extreme event to assess the effects of unseasonable conditions on diarrhoeal disease in Peru (46). Checkley et al. used harmonic regression (to account for seasonality) and autoregressive-moving average models to show an increased risk of these diseases following the El Niño event.

Synoptic climatological approaches are one method used to summarize short-term weather. For example, McGregor investigated the association between weather and winter ischaemic heart disease deaths (47). A principle component analysis followed by a cluster analysis of meteorological data for seven weather variables was used to determine winter air masses. Increases in ischaemic heart disease mortality appeared to be associated with concurrent meteorological conditions and with antecedent and rapidly changing conditions.

Time series analyses are used frequently to analyse exposures associated with short-term variability of climate. Time series analyses can take account of cyclical patterns, such as seasonal patterns, when evaluating longitudinal trends in disease rates in one geographically defined population. Seasonal patterns may be due to the seasonality of climate or to other factors, such as the school year. Two generally used approaches to time series analyses are generalized additive models (GAM) and generalized estimating equations (GEE). The generalized additive model entails the application of a series of semi-parametric Poisson models that use smoothing functions to capture long-term patterns and seasonal trends from data. The generalized estimating equation approach is similar to GAM in the use of a Poisson regression model to estimate health events in relation to weather data. However, no *a priori* smoothing is performed for the time series. Instead, the GEE model allows for the removal of long-term patterns in the data by adjusting for overdispersion and autocorrelation. Autocorrelation needs to be controlled for in time series data of weather measurements because today's weather is correlated with weather on the previous and subsequent days. Overdispersion may be present in count data (health outcomes) that are assumed to follow a Poisson distribution.

Time series analysis was used in a study that described and compared the associations between certain weather variables and hospitalizations for viral pneumonia, including influenza, during normal weather periods and El Niño events in three regions of California (41). Temperature variables, precipitation and sea surface temperature were analysed. Sea surface temperature was included as a marker for weather variables not included in the analysis, such as cloud cover. The cut points for the weather variables were approximately one standard deviation from the mean. A seven-day lag period was used. Specific changes in temperature or precipitation alone could not describe the hospitalization patterns found across the three regions. Also, developing a model based on either the inland or one of the coastal regions would not have been predictive for the other regions. These results underscore the difficulties in trying to model the potential health effects of climate variability.

Another example is a study of the association between extreme precipitation and water-borne disease outbreaks in the United States (42). The goal of the analysis was to determine whether outbreaks clustered around extreme precipitation events as opposed to geographical clustering. Curriero et al. defined extreme precipitation events with Z-score thresholds: scores greater than 0.84, 1.28 and 1.65 corresponded to total monthly precipitation in the highest 20%, 10% and 5%, respectively. A Monte Carlo version of Fisher's exact test was used to test for statistical significance of associations. The authors repeatedly generated sets of outbreaks in a random fashion, tabulating the percentage of these artificial outbreaks with extreme levels of precipitation at each step. This process produced a distribution of coincident percentages under the assumption of no association that was then compared with the observed percentage to calculate a p-value. Analyses were conducted at the watershed level, including outbreaks due to both ground and surface water contamination, and were further stratified by season and hydrologic region. Of the 548 outbreaks, 51% were preceded by a precipitation event above the 90%ile and 68% above the 80%ile (p < 0.001). Outbreaks due to surface water contamination were associated with extreme precipitation during the month of the outbreak, while outbreaks due to ground water contamination had the strongest association with extreme precipitation two months prior to the outbreak.

A study by Hay et al. demonstrates one approach for looking at longer-term variability (48). The authors used spectral analysis to investigate periodicity in both climate and epidemiological time series data of dengue haemorrhagic fever (DHF) in Bangkok, Thailand. DHF exhibits strong seasonality, with peak incidences in Bangkok occurring during the months of July, August and September. This seasonality has been attributed to temperature variations. Spectral analysis (or Fourier analysis) uses stationary sinusoidal functions to deconstruct time series into separate periodic components. A broad band of two to four year periodic components was identified as well as a large seasonal periodicity. One limitation of this method is that it is applicable only for stationary time series in which the periodic components do not change.

Another approach, used to study the association between El Niño events and malaria outbreaks, analysed historical malaria epidemics using an El Niño/Southern Oscillation index (43–46). This is discussed in more detail in chapter five.

Other statistical methods for analysing epidemiological studies of the health impacts of weather and climate are being developed. Improved methods still are needed to fit time series, as are methods to handle bivariate data (one series of counts and one of continuous data). However, these methods are not enough. Convergence of expertise, methods and databases from multiple disciplines is required to understand and prepare for a different future climate and its potential ecological, social and population health impacts. Capacity building to improve human and ecological data quality, and the development of innovative interdisciplinary methods, remain high priorities when facing the challenges of assessing actual and potential risks from global climate change.

Conclusions

Main findings

1. The IPCC Third Assessment Report concluded that the best agreement between climate observations and model simulations over the past 140 years was found when both natural factors and anthropogenic forcings were included in the models (5). Further, most of the warming observed over the past 50 years is attributable to human activities.
2. Human influences will continue to change atmospheric composition throughout the twenty-first century. Global average temperature is projected to rise by 1.4 to 5.8 °C over the period 1990–2100 (5). Global climate change will not likely be spatially uniform, and is expected to include changes in temperature and in the hydrologic cycle.
3. Studying the natural complexities of weather and climate variability in relation to health outcomes offers unique challenges. Weather and climate can be summarized over various spatial and temporal scales. The appropriate scale of analysis will depend on the study hypothesis. Each study hypothesis needs to define the exposure of interest and the lag period between exposure and effect.
4. The predictive value of analogue studies may be limited because future events may be different than historical events, and because the extent of vulnerability of a population changes over time. For these and other reasons, scenario-based modelling is needed to project what might happen under different climate conditions.

5. Analysis methods need to take account of the changing climate baseline. An additional consideration is that the shift in the mean of distributions of meteorological variables is likely to change the extremes of the distribution.

Research gaps

1. Innovative approaches to analysing weather/climate in the context of human health are needed. Many standard epidemiology approaches and methods are inadequate because these exposures operate on a population level. Methods used in other disciplines need to be modified and new methods developed to enhance the ability to study and project potential health impacts in a future that may have a markedly different climate.
2. Long-term data sets with weather and health outcome data on the same spatial and temporal scales are required. Currently it is not possible to answer key questions such as the contribution of climate variability and change to the spread of malaria in African highlands, because the appropriate health, weather and other data (i.e. land use change) are not being collected in the same locations on the same scales. There is currently little coordination across disciplines and institutions; these links need to be established and maintained.
3. Improved understanding is needed of how to incorporate outputs from multiple AOGCMs into health studies to highlight better the range of uncertainties associated with projected future health impacts. Including several climate scenarios can illustrate the range of possible future changes, thus allowing decision-makers to identify populations that may be particularly vulnerable to adverse health impacts and to use this information when prioritizing strategies, policies and measures to enhance the adaptive capacity of future generations.

References

1. Ebi, K.L. & Patz, J.A. Epidemiological and impacts assessment methods. In: *Global environmental change and human health*. Martens, P. & McMichael, A.J. Cambridge, UK: Cambridge University Press, 2002.
2. Cubash, U. & Cess, R.D. Processes and modelling. In *Climate Change: the IPCC Scientific Assessment*. Houghton, J.T. et al. eds. Cambridge, UK: Cambridge University Press, p. 75, 1990.
3. Burroughs, W.J. Changing weather. In: *Weather*. Burroughs, W.J. et al. Sydney, Australia: The Nature Company Guides / Time-Life Books, 1996.
4. United States Environmental Protection Agency (EPA). Greenhouse effect schematic (2001) http://www.epa.gov/air/aqtrnd95/globwarm.html
5. Albritton, D.L. & Meira Filho, L.G. coordinating lead authors. Technical Summary. In: *Climate change 2001: the scientific basis*. Contribution of Working Group I to the Third Assessment Report of the Intergovernmental Panel on Climate Change. Cambridge, UK and New York, USA: Cambridge University Press, 2001.
6. Committee on Climate, Ecosystems, Infectious Diseases, and Human Health. *Under the weather: climate, ecosystems, and infectious disease.* National Research Council Board on Atmospheric Sciences and Climate, Division of Earth and Life Sciences. Washington, DC, USA: National Academy Press, 2001.
7. Pielke Sr., R.A. et al. The influence of anthropogenic landscape changes on weather in south Florida. *Monthly Weather Review* 127: 1663–1674 (1999).
8. Bonan, G.B. Frost followed the plow: impacts of deforestation on the climate of the United States. *Ecological Applications* 9: 1305–1315 (1999).

9. Gash. J.H.C. et al. *Amazonian deforestation and climate.* Chichester, UK: Wiley, 1996.

10. Arritt, R.W. et al. Interaction of natural and anthropogenic factors in Amazon rainfall. *Presentation to the American Meteorological Society 2002 meeting*; abstract J8.2.

11. Watson, R.T. and the core writing team. *Climate change 2001: synthesis report. Summary for policymakers.* A Report of the Intergovernmental Panel on Climate Change. Geneva, Switzerland: IPCC Secretariat, c/o World Meteorological Organization, p. 33, 2001.

12. http://www.ncdc.noaa.gov/ol/climate/normals/usnormals.html

13. http://www.pmel.noaa.gov/tao/elNiño /el-Niño -story.html

14. Philander, S.G. *El Niño, La Niña, and the Southern Oscillation.* San Diego, California, USA: Academic Press, Inc. Harcourt Brace Jovanovich, 1990.

15. Glantz, M.H. *La Niña and its impacts: facts and speculation.* Tokyo, Japan: United Nations University Press, 2002.

16. Glantz, M.H. *Currents of change: El Niño's impact on climate and society.* Cambridge, UK: Cambridge University Press, 1996.

17. http://www.ldeo.columbia.edu/NAO/

18. Hurrel, J. Decadal trends in the North Atlantic Oscillation regional temperatures and precipitation. *Science* 269: 676–679 (1995).

19. Folland, C.K. Karl, T. & Ya. Vinnikov, K. Observed climate variations and change. In: *Climate change: the IPCC scientific assessment.* Houghton, J.T. et al. eds. Cambridge, UK: Cambridge University Press, p. 202, 1990.

20. Lamb, H.H. *Climate, history and the modern world.* 2nd Edition. London, UK: Routledge, 1995.

21. Climate Research Unit, University of East Anglia. Climate Monitor Online. (2003) http://www.cru.uea.ac.uk/cru/climon/data/

22. Nakicenovic, N. et al. *IPCC special report on emission scenarios.* Cambridge, UK: Cambridge University Press, 2000.

23. Andronova, N.G. & Schlesinger, M.E. Objective estimation of the probability density function for climate sensitivity. *Journal of Geophysical Research* 106: D19, 22605–22612 (2002).

24. Knutti, R. et al. Constraints on radiative forcing and future climate change from observations and climate model ensembles. *Nature* 416: 719–723 (2002).

25. Stott, P.A. & Kettleborough, J.A. Origins and estimates of uncertainty in predictions of twenty-first century temperature rise. *Nature* 416: 723–726 (2002).

26. Cubasch, U. et al. Projections of future climate change. In: *Climate change 2001: the scientific basis.* Houghton, J.T. et al. eds. Cambridge, UK: Cambridge University Press, chapter 9, 2001.

27. National Climatic Data Center (NCDC). *Probabilities of temperature extremes in the USA*, Version 1 CD-ROM. (2001)

28. Church, J.A. et al. Changes in sea level. In: *Climate change 2001: the scientific basis.* Houghton, J.T. et al. eds. Cambridge: Cambridge University Press, chapter 11, 2001.

29. Giorgi, F. & Mearns, L.O. Calculation of average, uncertainty range, and reliability of regional climate changes from AOGCM simulations via the "Reliability Ensemble Averaging" (REA) method. *Journal of Climate* 15: 1141–1158 (2002).

30. Wilby, R.L. & Wigley, T.M.L. Downscaling general circulation model output: a review of methods and limitations. *Progress in Physical Geography* 21: 530–548 (1997).

31. Giorgi, F. et al. Regional climate information—evaluation and projections. In: *Climate change 2001: the scientific basis.* Houghton, J.T. et al. eds. Cambridge, UK: Cambridge University Press, chapter 10, 2001.

32. Murphy, J.M. Predictions of climate change over Europe using statistical and dynamical techniques for downscaling local climate. *Journal of Climate* 12: 2256–2284 (1999).

33. McGregor, J.L. Regional climate modeling. *Meteorology and Atmospheric Physics* 63: 105–117 (1997).

34. Giorgi, F. & Mearns, L.O. Regional climate modeling revisited. An introduction to the special issue. *Journal of Geophysical Research* 104: 6335–6352 (1999).

35. Mearns, L.O. et al. *Climate scenario development. In: Climate change 2001: the scientific basis.* Houghton, J.T. et al. eds. Cambridge, UK: Cambridge University Press, chapter 13, 2001.

36. Hassel, D. & Jones, R.G. *Simulating climatic change of the southern Asian monsoon using a nested regional climate model* (HadRM2), HCTN 8. Bracknell, UK: Hadley Centre for Climate Prediction and Research, 1999.

37. Streets. D.G. & Glantz, M.H. Exploring the concept of climate surprise. *Global Environmental Change* 10: 97–107 (2000).

38. Palecki, M.A. et al. The nature and impacts of the July 1999 heat wave in the midwestern U.S.: learning from the lessons of 1995. *Bulletin of the American Meteorological Society* 82: 1353–1367 (2001).

39. Huynen, M.M.T.E. et al. The impact of cold spells and heatwaves on mortality rates in the Dutch population. *Environmental Health Perspectives* 109: 463–70 (2001).

40. Whitman, S. et al. Mortality in Chicago attributed to the July 1995 heat wave. *American Journal of Public Health* 87: 1515–1518 (1997).

41. Ebi, K.L. et al. Association of normal weather periods and El Niño events with viral pneumonia hospitalizations in females, California 1983–1998. *American Journal of Public Health* 91: 1200–1208 (2001).

42. Curriero, F.C. et al. Analysis of the association between extreme precipitation and waterborne disease outbreaks in the United States, 1948–1994. *American Journal of Public Health* 91: 1194–1199 (2001).

43. Bouma, M.J. & van der Kaay, H.J. El Niño Southern Oscillation and the historic malaria epidemics on the Indian subcontinent and Sri Lanka: an early warning system for future epidemics? *Tropical Medicine and International Health* 1: 86–96 (1996).

44. Bouma, M.J. & Dye, C. Cycles of malaria associated with El Niño in Venezuela. Journal of the American Medical Association 278: 1772–1774 (1997).

45. Bouma, M.J. et al. Predicting high-risk years for malaria in Colombia using parameters of El Niño Southern Oscillation. *Tropical Medicine and International Health* 2: 1122–1127 (1997).

46. Checkley, W. et al. Effects of the El Niño and ambient temperature on hospital admissions for diarrhoeal diseases in Peruvian children. *Lancet* 355: 442–450 (2000).

47. McGregor, G.R. Winter ischaemic heart disease deaths in Birmingham, United Kingdom: a synoptic climatological analysis. *Climate Research* 13: 17–31 (1999).

48. Hay, S.I. et al. Etiology of interepidemic periods of mosquito-borne disease. *Proceedings of the National Academy of Sciences of the United States of America* 2000; 97(16): 9335–9339 (2000).

CHAPTER 3

International consensus on the science of climate and health: the IPCC Third Assessment Report

A. K. Githeko,[1] A. Woodward[2]

Introduction

In the early 1990s, there was very little awareness of the risks posed to the health of human populations by global climate change. In part this reflected epidemiologists' limited conventional approach to environmental health. The environment was viewed predominantly as a repository for specific human-made pollutants: in air, water, soil and food, each of which was suspected of posing particular risks to human communities and individual consumers. This was compounded by a general lack of understanding of how the disruption of biophysical and ecological systems might pose threats to the longer-term health of populations. There was also little awareness among physical and natural scientists that changes in their particular objects of study—climatic conditions, biodiversity stocks, ecosystem productivity, and so on—were of potential importance to human health. Indeed, this was clearly reflected in the content of the first major report of the United Nation's Intergovernmental Panel on Climate Change (IPCC), published in 1991. That report devoted just several paragraphs to the possibility that global climate change might affect human health.

Things have changed. The Second Assessment Report of the IPCC (1996) contained a full chapter assessing the potential risks to human population health. The Third Assessment Report (1) did likewise, including discussion of the emergence of actual health impacts as well as the larger issue of potential health effects. The Third Assessment Report also included a review of health impacts in regional populations around the world.

The IPCC

Recognizing that global climate change posed a range of potentially serious, often new, hazards to human societies, the World Meteorological Organization (WMO) and the United Nations Environment Programme (UNEP) established the Intergovernmental Panel on Climate Change (IPCC) in 1988. The role of the IPCC is to assess published scientific literature on how human-induced changes to the gaseous composition of the lower atmosphere, caused by an increase in the emission of greenhouse gases, are likely to influence world climatic patterns; how this in turn would affect a range of systems and processes important to human societies (including human health); and what range of economic and social response options exists.

[1] Kenya Medical Research Institute, Kisumu, Kenya.
[2] Wellington School of Medicine, University of Otago, Wellington, New Zealand.

The IPCC does not carry out research, neither does it monitor climate-related data or other relevant parameters. Assessments are based on peer-reviewed and other accessible published scientific and technical literature.

A prime source of this chapter is the recent work of Working Group II of the IPCC (WGII), as this was set up as the internationally authoritative scientific review body in climate change and its impacts. Its work has been authorized by national governments through the processes of the United Nations' system; membership has been extensive, international and representative of world scientific skills and opinions, and its review processes have been open to external review by other scientific peers and government scientific advisors.

The scale of the IPCC endeavour is reflected in the size of the reports produced by its working parties and the number of scientists who take part. For instance, the most recent report from WGII is almost a thousand pages long; the list of authors and reviewers includes more than 650 names from 74 countries. Given the number of scientists involved, the formal review processes required and the need to reach consensus, the IPCC is by nature a conservative body.

WGII builds on the reports of Working Group I, in which climatologists describe the evidence that climate is changing due to human intervention, and construct and test the computer models on which future projections of climate are based. Such climate models are based on physical understanding of the climate system and build upon first principles of dynamic systems. Climatic changes due to a specified, plausible, rise in greenhouse gas concentrations can be forecast with greater confidence at global and regional scales than can changes occurring at a local level. How quickly greenhouse concentrations actually rise in future will depend on many factors, including future trends in fertility, economic development, resource consumption levels and technological choices. As a guide to policy-makers, scientists therefore must devise plausible scenarios that include these features of the future world.

It has been usual to view this topic area as comprising three kinds of health impacts. First, those that are relatively direct and foreseeable. Second, the effects that arise via indirect processes of environmental change and ecological disruption, occurring in response to climate change. Third, diverse health consequences—traumatic, infectious, nutritional, psychological and other—that occur in demoralized and displaced populations in the wake of climate-induced economic dislocation, environmental decline and conflict situations.

Our understanding of the impacts of climate change and variability on human health has increased considerably in recent years. However, research in this area faces three main difficulties:

1. It is difficult to describe clearly the main environmental and biological influences on health, while at the same time including important interactions with ecological and social processes. There must be a balance between complexity and simplicity.
2. There are many sources of scientific and contextual uncertainty. The IPCC has sought a satisfactory way to describe the level of confidence that can be assigned to each statement about a particular health impact (see Box 3.2).
3. Climate change is one of several global environmental changes that affect human health. Various large-scale environmental changes now simultaneously impinge on human population health, often interactively (2). An obvious example is the transmission of vector-borne infectious diseases. These are affected by: climatic conditions; population movement; forest clearance and land-use patterns; freshwater surface configurations; human population density; and the population density of insectivorous predators (3).

The effects of climate on the transmission biology of human diseases

Climate change involves a change in both the mean meteorological values and variability of these values. The anticipated change in mean climatic conditions is expected to be a slow process, occurring over many decades. Climate variability, however, occurs on a time-scale from weeks or months (e.g. storms and floods) to years (e.g. the ENSO cycle, oscillating with an approximately 5-year periodicity).

The health impacts of climate variability are, in general, likely to be more pronounced over the near term than are those of climate change. For example, large

---

Box 3.2 Dealing with uncertainties

Since the First and the Second IPCC Assessment Reports (1991, 1996), substantial advances have been made in understanding the impacts on health of climate change. Furthermore, the IPCC scientist-authors have attempted to assign a degree of confidence to each of the conclusions in the Third Assessment Report (2001).

In the Summary for Policy-makers for Working Group II, the IPCC (2001) used the following terms to indicate levels of confidence (based upon the collective judgment of the authors):

- *very high (95% or greater)*
- *high (67–<95%)*
- *medium (33–<67%)*
- *low (5–<33%)*
- *very low (less than 5%)*

In some other evaluative judgements within the Third Assessment Report, the IPCC used a qualitative scale to convey the level of scientific understanding. The scale comprised these categories: *well established, established—but-incomplete, competing explanations, and speculative.*

anomalies in temperature and rainfall in a particular season could cause a number of vector-borne and water-borne epidemics, thereafter the weather could return to normal. Extremes of heat can cause heat exhaustion, cardiovascular disease (heart attacks and strokes) while cold spells can lead to hypothermia and increase morbidity and mortality from cardiovascular disease. Storms, tropical cyclones and extreme rainfall can cause immediate death and injuries, as well as increased risk of water-borne diseases in the medium-term and psychological stress on affected communities in the long-term.

Slow changes in climatic conditions may allow human populations time to adapt. For example, people or communities may develop new ways of coping with, or attenuating, rising residential temperatures. In contrast, abrupt climate changes due to anomalous seasonal climate variability do not allow such opportunities.

The complexities of interactions between environment and host are best shown by the example of vector-borne diseases. The success of pathogens and vectors is determined partly by their reproductive rate. Malaria-carrying mosquito populations can increase tremendously within a very short time. Equally the *Plasmodium* parasite species proliferates rapidly in both mosquito and human hosts. In contrast, tsetse flies have a low reproductive rate and their populations take much longer to increase under favourable conditions. Hence, infectious diseases transmitted by the tsetse fly (including human sleeping sickness) respond less rapidly to variations in climate than do many mosquito-borne infections. Vectors' ability to transmit disease is also affected by feeding frequency. Hard ticks (such as the vectors of Lyme disease) feed more frequently and for shorter periods than soft ticks. Hard ticks therefore tend to be much more efficient vectors of human diseases. Overall, high vector and pathogen reproductive capacity; preference for humans as a source of blood meals; low life cycle complexity; and high sensitivity to temperature changes result in an infectious disease that has high sensitivity to climate variability.

While climate and environmental factors often initiate changes in the rate of disease (e.g. triggering an epidemic) health service interventions often play a major role in containing the spread of disease. Therefore, in disease outbreaks it is often unclear whether the outcome is a result of either altered climatic and environmental conditions or intervention failures. This is an example of the general problem, it is known that climate has an effect on infections and other health problems but it is difficult to tell *how much* disease and injury can be attributed to this factor.

Mathematical models provide one important means to answer the "what if?" question about the future effects of climate change on infectious disease occurrence. Both biologically based and statistical-empirical models have been used in recent years. More sophisticated integrated models are being developed to take into account the effects of other determinants such as economics and human behaviour. Historical examples of the health correlates of climate variability, such as the El Niño phenomenon, also provide insights into possible future climate and health scenarios.

The traditional role of surveillance in epidemiological assessment of diseases may not stand up to the speed with which epidemics evolve under climate change. Quite often it is difficult to tell whether a rise in the number of cases of malaria is simply normal seasonal variation or the beginning of a large-scale epidemic. At first the number of cases grows slowly, but may rapidly move into a phase of exponential growth, in which case the health care system may be overwhelmed. Hence the value of disease forecasting methods that can estimate the size of a developing epidemic depending on the level of climate anomaly.

IPCC Third Assessment Report

Bearing in mind the general caveat that they are necessarily operating within a penumbra of uncertainty, scientists have estimated the likely range of future health impacts of climate change on human health. For the moment, the most comprehensive and widely reviewed estimates come from the work of the IPCC and the remainder of this chapter therefore provides an overview of that assessment. Unless otherwise specified, all references are to the contribution of WGII to the Third Assessment Report.

The Third Assessment Report included sectoral and regional analyses of published literature related to impacts of climate change. The Report considered the weight of evidence supporting its conclusions and attributed levels of confidence to the conclusions (these can be found in the technical summary of the IPCC-TAR: see Box 3.2 above).

The health chapter in the WGII report included a discussion of specific diseases and regions that have been impacted upon by climate variability, vulnerable populations and their adaptation options and capacity. The overall conclusion was that global climate change will have diverse impacts on human health— some positive, most negative. Changes in the frequencies of extreme heat and cold, of floods and droughts, and the profile of local air pollution and aeroallergens would affect population health directly. Other indirect health impacts would result from the effects of climate change on ecological and social systems. These impacts would include changes in occurrence of infectious diseases, local food production and under-nutrition, and various health consequences of population displacement and economic disruption.

As yet, there is little firm evidence that changes in population health status have occurred in response to observed trends in climate over recent decades. A recurring difficulty in identifying such impacts is that the causation of most human disorders is multi-factorial, and the background socioeconomic, demographic and environmental contexts change over time, so that conclusively proving (or disproving) a link with climate change is highly problematic.

Direct effects on health

Heatwaves and other extreme events

Human populations have, over time, acclimatized and adapted to local climates and also are able to cope with a range of weather changes. However, within populations, there is a range of individual sensitivity to extreme weather events. If heatwaves increase in frequency and intensity, the risk of death and serious illness would increase principally in the older age groups, those with pre-existing cardio-respiratory diseases, and the urban poor. The effects of an increase in heatwaves often would be exacerbated by increased humidity and urban air pollution. The greatest increases in thermal stress are forecast for mid to high latitude cities, especially in populations with unadapted architecture and limited air conditioning.

Modelling of heatwave impacts in urban populations, allowing for acclimatization, suggests that many United States' cities would experience, on average, several hundred extra deaths each summer (4). Although climate change may have considerable impact on thermal stress-related mortality in cities in developing countries, there has been little research in such populations. Warmer winters and fewer cold spells will decrease cold-related mortality in many temperate countries. In some instances in the temperate zones, reduced winter deaths probably would outnumber increased summer deaths (5).

Any increase in frequency of extreme events such as storms, floods, droughts and cyclones would harm human health through a variety of pathways. These natural hazards can cause direct loss of life and injury and affect health indirectly through loss of shelter; population displacement; contamination of water supplies; loss of food production (leading to hunger and malnutrition); increased risk of infectious disease epidemics (including diarrhoeal and respiratory diseases); and damage to infrastructure for provision of health services. If cyclones were to increase regionally, there might be devastating impacts particularly in densely settled populations with inadequate resources. Over recent years climate-related disasters have caused hundreds of thousands of deaths in countries such as China, Bangladesh, Venezuela and Mozambique.

Air pollution

Weather conditions can influence the transportation of air-borne pollutants, pollen production and levels of fossil fuel pollutants resulting from household heating and energy demands. Climate change may increase the concentration of ground level ozone but the magnitude of the effect is uncertain (6). For other pollutants, the effects of climate change and/or weather are even less well known.

Climate change is expected to increase the risks of forest and rangeland fires and associated smoke hazards. Major fires in 1997 in south-east Asia and the Americas were associated with increases in respiratory and eye symptoms (7). In Malaysia, a two to three fold increase in outpatient visits for respiratory disease and 14% decrease in lung function in school children was reported.

Aeroallergens

Experimental research has shown that doubling CO_2 levels from about 300 to 600 ppm induces a four-fold increase in the production of ragweed pollen (8, 9). Pollen counts from birch trees (the main cause of allergies in northern Europe) rise with increasing temperature (10).

Indirect effects on health

Food production and supply

Climate change will have mixed effects on food production globally. Most of the research to date has focused on cereal grain production—an important indicator of total food production, since it accounts for around 70% of global food energy. The probability of reduced food yields is, in general, greatest in developing countries where it is estimated that approximately 790 million people currently are undernourished (11). Populations in isolated areas with poor access to markets will be particularly vulnerable to local decreases or disruptions of food supply.

Vector-borne infectious diseases

Recent studies of disease variations associated with inter-annual climate variability (such as those related to the El Niño cycle) have provided much useful evidence of the sensitivity to climate of many disease processes. This is particularly so for mosquito-borne diseases. The combination of knowledge from such empirical research; the resultant theoretical understanding of biological and ecological processes; and the output of scenario-based modelling; leads to several conclusions about the future effects of climate change on human populations.

Higher temperatures, changes in precipitation and climate variability would alter the geographical range and seasonality of transmission of many vector-borne diseases. Mostly, range and seasonality would be extended; in some cases reduced. Currently 40% of the world population lives in areas in which endemic malaria occurs (12). In areas with limited or deteriorating public health infrastructure, increased temperatures will tend to expand the geographical range of malaria transmission to higher altitudes and latitudes. Higher temperatures in combination with conducive patterns of rainfall and surface water will extend the transmission season in some locations. Changes in climate mean conditions and variability would affect many other vector-borne infections (such as dengue, leishmaniasis, Lyme disease, and tick-borne encephalitis) at the margins of their current distributions. For some vector-borne diseases in some locations, climate change will decrease the likelihood of transmission via a reduction in rainfall, or temperatures that are too high for transmission.

A range of mathematical models, based on observed climatic effects on the population biology of pathogens and vectors, indicate that climate change scenarios over the coming century would cause a small net increase in the proportion of the world population living in regions of potential transmission of malaria and dengue (13, 14, 15). An alternative modelling approach, based on a direct correlation of the observed distribution of disease distribution against a range of climate variables, suggests that there will be little change in malaria distributions, as areas that become permissible for transmission are balanced by others that become unsuitable for at least one climatic factor. Neither approach attempts to incorporate the effects of socioeconomic factors or control programmes on the distribution of current or future disease.

Water-borne infectious diseases

There are complex relationships between human health and water quality, water quantity, sanitation and hygiene. Increases in water stress are projected under climate change (see chapter 4, IPCC –TAR WG II), but it is difficult to translate these changes into risk of water-related diseases.

Heavy rainfall events can transport terrestrial microbiological agents into drinking-water sources resulting in outbreaks of crytosporidiosis, giardiasis, amoebiasis, typhoid and other infections (19, 20, 21, 22). Recent evidence indicates that copepod zooplankton provide a marine reservoir for the cholera pathogen and thereby facilitate its long-term persistence and disseminated spread to human consumers via the marine food-web (23). Epidemiological evidence has pointed to a widespread environmental cause for recent outbreaks of cholera, rather than a point source contamination as seen in Peru in 1991 and East Africa in 1997/98. Strong links are found between cholera infections, bathing and drinking water from east African lakes (24). Cholera epidemics also are associated with positive surface temperature anomalies in coastal and inland lake waters (23).

Global warming is expected to lead to changes in the marine environment that alter risks of bio-toxin poisoning from human consumption of fish and shellfish. For example, bio-toxins associated with warm waters, such as ciguatera in

TABLE 3.1 Main vector-borne diseases: populations at risk and burden of diseases. *Based on data from reference 1, with updated DALY estimates from reference (16).*

Disease	Vector	Population at risk	Number currently infected or new cases per year	Disability adjusted life years lost[a]	Present distribution
Malaria	Mosquito	2400 million (40% world population)	272 925 000	42 280 000	Tropics/subtropics
Schistosomiasis	Water snail	500–600 million	120 million	1 760 000	Tropics/subtropics
Lymphatic filariasis	Mosquito	1000 million	120 million	5 644 000	Tropics/subtropics
African trypanosomiasis (Sleeping sickness)	Tsetse fly	55 million	300 000–500 000	1 598 000	Tropical Africa
Leishmaniasis	Sand Fly	350 million	1.5–2 million	2 357 000	Asia, Africa, Southern Europe, Americas
Onchocerciasis River blindness	Black fly	120 million	18 million	987 000	Africa, Latin America, Yemen
American trypanosomiasis (Chagas' disease)	Triatomine bug	100 million	16–18 million	649 000	Central and South America
Dengue	Mosquito	3000 million	Tens of millions	653 000[b]	All tropical countries
Yellow fever	Mosquito	468 million in Africa	200 000	Not available	Tropical South America and Africa
Japanese encephalitis	Mosquito	300 million	50 000	767 000	Asia

[a] The Disability–Adjusted Life Year (DALY) is a measure of population health deficit that combines chronic illness or disability and premature death (17). Numbers are rounded up to nearest 100 000.
[b] Other analyses suggest this value could be as high as 1 800 000 (18).

tropical waters, could extend their range to higher latitudes. Higher sea surface temperatures also would increase the occurrence of algal blooms that may affect human health directly, and which are also ecologically and economically damaging.

Changes in surface water quality and quantity are likely to affect the incidence of diarrhoeal diseases (25). This group of diseases includes conditions caused by bacteria such as cholera and typhoid as well as parasitic diseases such as amoebiasis, giardiasis and cryptosporidium. Infections with cholera and typhoid bacteria are dependent on the concentration of the pathogens in water or food. Currently the World Health Organization (WHO) estimates more than one billion people worldwide to be without access to safe drinking water, and that every year approximately 1.7 million die prematurely because they do not have access to safe drinking water and sanitation (16). Climate can increase directly the amount of pathogen in the water through increasing the biotic reservoir of the infectious agent (cholera) or by decreasing the amount of water in a river or a pond and thus raising concentration of the bacteria (typhoid). Floods can cause contamination of public water supplies with both bacteria and parasites as surface discharge flows into rivers and reservoirs, while drought can increase the concentration of pathogens in the limited water supplies. A reduction in the availability of clean water increases the risk of drinking contaminated supplies and also reduces the amount of water available for personal hygiene thus leading to skin infections.

Effects of social and economic disruptions

In some settings, the impacts of climate change may cause severe social disruptions, local economic decline and population displacement that would affect human health (26). Of particular concern is the impact of a rising sea level (estimated, with a wide band of uncertainty, at around 0.5 m over the coming century) on island and coastal populations currently living not far above the shoreline. Population displacement resulting from sea level rise, natural disasters or environmental degradation is likely to lead to substantial health problems, both physical and mental.

Assessments of health impacts by IPCC region

Africa

Africa has a number of climate-sensitive diseases, the most prominent being malaria, meningitis and cholera.

Malaria epidemics in the past 15 years have been reported mainly in the highlands of east Africa, Rwanda and Zimbabwe, associated with inter-annual climate variability (such as the occurrence of El Niño events). Following flooding in the arid regions of Somalia and Kenya, malaria outbreaks were reported during the 1997/98 El Niño event. Meanwhile, in the Sahel region malaria transmission has declined in the past 30 years due to long term drought.

From 1931 (when Rift Valley Fever was first described) until the end of the 1970s, the disease was considered to be a relatively benign zoonosis that developed periodically in domestic animals (especially sheep) following heavy rains (27). Thereafter extensive research on mosquito vectors of Rift Valley Fever in Kenya (mainly *Aedes* and *Culex* species) clearly has linked the risk of outbreaks with flooding (28). Following the 1997/8 El Niño event in East Africa, a Rift Valley Fever outbreak in Somalia and northern Kenya killed up to 80% of livestock and affected their owners (29). In West Africa the disease is linked to epizootic diseases with increased risks during the wet season. The IPCC (2001) concluded that increased precipitation as a consequence of climate change will increase the risk of infections of this kind to livestock and people.

Currently the seventh cholera pandemic is active across Asia, Africa and South America. During the 1997/98 El Niño, the rise in sea-surface temperature and excessive flooding (29) provided two conducive factors for cholera epidemics which were observed in Djibouti, Somalia, Kenya, Mozambique and the United Republic of Tanzania, all of which border the Indian Ocean. Cholera epidemics also have been observed in areas surrounding the Great Lakes in the Great Rift Valley region. A significant association between bathing, drinking water from Lake Tanganyika and the risk of infection with cholera has been found (24). It is likely that warming in these African lakes may cause conditions that increase the risk of cholera transmission.

Major epidemics of bacterial meningococcal infection

TABLE 3.2. Summary of the number of countries in Africa that reported disease outbreaks to WHO from January 1997 to June 1999 (1).

Disease	1997	1998	1999 January–July
Malaria	0	2	2
Rift Valley Fever	0	4	1
Yellow Fever	1	1	0
Meningitis (bacterial)	3	2	4
Plague	2	1	2
Cholera	8	10	7
Dengue	0	0	0

usually occur every five to ten years within the African meningitis belt, and typically start in the middle of the dry season and end a few months later with the onset of the rains (30). Between February and April 1996, the disease affected thousands of people in parts of northern Nigeria, many of whom died (31). This epidemic spread from the traditional meningitis belt to Kenya, Uganda, Rwanda, Zambia and the United Republic of Tanzania (32). One of the environmental factors that predispose to infection and epidemics is low humidity (33). To date this disease has been limited to the semi-arid areas of Africa, suggesting that future distribution could expand due to increased warming and reduced precipitation.

Plague is a flea-borne disease and the major reservoirs of infection are rodents such as the common rat. Rodent populations fluctuate widely with the availability of food which in turn depends on rainfall. Exceptionally heavy rainfall can increase food abundance; as a consequence the population of rodents and fleas may multiply rapidly. During severe droughts, rodents may leave their wild habitats in search of food in human houses and this can also increase the risk of plague transmission. Plague outbreaks in Africa have in the last few years been reported in Mozambique, Namibia, Malawi, Zambia and Uganda (See Table 3.2 above).

Asia

In Asia, as in Africa, the main health concerns under climate change and variability are malaria and cholera, but thermal stress and air-pollution related illnesses also are important. Malaria still is one of the most important vector-borne diseases in India, Bangladesh, Sri Lanka, Thailand, Malaysia, Cambodia, the Lao People's Democratic Republic, Viet Nam, Indonesia, Papua New Guinea and parts of China. Vector resistance to insecticides, and parasites' to chloroquine, compound the problem of malaria control. The IPCC concluded that changes in environmental temperature and precipitation could expand the geographical range of malaria in the temperate and arid parts of Asia.

Water-borne diseases such as cholera, and various diarrhoeal diseases such as giardiasis, salmonellosis and cryptosporidiosis, occur commonly with contamination of drinking water in many south Asian countries. These diseases could become more frequent in many parts of south Asia in a warmer climate.

The direct effects of heat are important public health issues in this region. The heat index (derived from daily mean temperature, and humidity) is closely related to the occurrence of heat stroke in males aged 65 years and above residing in Tokyo. In the city of Nanjing, China, a marked increase in the number of heat stroke patients and mortality was observed when the maximum daily temperature exceeded 36°C for 17 days during July 1988. Similar events were observed when the temperature exceeded 31°C in Tokyo, Japan.

Australia and New Zealand

In Christchurch, New Zealand, an increase of 1°C above 21.5°C was associated with a 1.3% increase in all-cause mortality. There were more than the expected numbers of deaths in winter also, although this was not statistically significant. Since 1800, deaths specifically ascribed to climate hazards have averaged about 50 per year in Australia (34), of which 60% are estimated to be caused by heat-waves, 20% by tropical cyclones and floods. Climate change would increase the number of heatwaves in Australia but the future frequency of storms and floods is less certain.

In Australia the number of notified cases of arbovirus infections (caused by insect-borne virus) appears to have increased in recent years. Exotic species such as *Aedes albopictus* and *Aedes camptorhynchus*, competent vectors of (respectively) the dengue and Ross River viruses, have been detected in New Zealand. Outbreaks of Ross River virus disease and Murray Valley encephalitis in south-eastern Australia tend to follow heavy rainfall events. In south-western Australia the major vector for Ross River virus is the salt-water breeding mosquito *Ae. camptorhynchus,* and variations in sea level have been associated with outbreaks. Climate scenarios suggest that conditions in some parts of Australia and New Zealand will become more favourable for the transmission of several vector-borne diseases. However, whether this potential risk will translate into an increase in cases of disease will depend on other factors such as the maintenance and expansion of the public health surveillance and response system.

Ozone and other photochemical oxidants are a concern as air pollutants in several major Australian cities and in Auckland, New Zealand (35). In Brisbane, Australia, current levels of ozone and particulates have been associated with increased hospital admission rates (36). Warm weather promotes formation of these pollutants, although other factors such as wind speed and cloud cover are also important, if more difficult to anticipate.

Europe

The major impacts of climate change and variability on health in Europe are mainly via thermal stress and air pollution, vector and food-borne diseases, water-related diseases and flood effects.

In many European cities total daily mortality rises as summer temperatures increase. Heatwaves in July 1976 and July–August 1995 were accompanied by a 15% increase in mortality in Greater London and particularly from cardio-respiratory diseases at older age (37, 38). A major heatwave in July 1987 in Athens was associated with 2000 excess deaths (39, 40). Warmer winters, however, would result in reduced cold-related mortalities. It has been estimated that 9000 deaths per year could be avoided by 2025 in England and Wales under a 2.5 °C increase in average winter temperature (41).

With deteriorating health systems, the recent resurgence of malaria in south-eastern Europe could be amplified by a warmer climate. Small numbers of locally transmitted cases currently occur in the Mediterranean region (42). However, existing public health resources and reduction of breeding habitats for *Anopheles* mosquitoes make it unlikely that malaria will re-emerge on a large scale in western Europe, whatever changes take place in the climate. There has been no dengue transmission in Europe in recent times, but the appearance of the vector *Aedes albopictus* in Italy and Albania is a matter of concern.

The two common forms of leishmaniasis: visceral and cutaneous, are transmitted to humans and dogs in all the Mediterranean countries by phlebotomine flies (43). This disease is associated with dry habitats. Higher temperatures are likely to shift northwards the range of the disease.

Lyme disease and tick-borne encephalitis (TBE) are transmitted by hard ticks such as *Ixodes ricinus* and *I. persulcatus* found in the temperate regions of Europe. Recent observations in Sweden suggest that the incidence of TBE has increased following milder winters in combination with extended spring and autumn in two successive years (44). There is also some evidence that the northern limit of the tick's distribution in Sweden has shifted northwards as a result of a higher frequency of milder winters (44, 45, 46), although this relationship remains contentious (47). Climate change may extend the tick-borne disease transmission season and also its range towards the north, but disrupt transmission in more southerly regions (48).

Some countries in eastern Europe with restricted access to water at home could be affected by any climate-related decrease in supplies. For instance, an increase in the frequency and intensity of extreme precipitation could increase the risk of transmission of cryptosporidiosis.

The distribution of carriers of food-borne diseases such as flies, cockroaches and rodents could change due to climate change. In the United Kingdom of Great Britain and Northern Ireland, a study of food-borne illness found a strong relationship between incidence and temperature in the month preceding the illness (49).

Leptospirosis, a disease associated with flooding, is a major concern in some parts of Europe. Outbreaks of the disease have been reported following floods in Ukraine and the Czech Republic in 1997 (50, 51) and Portugal in 1967 (52). As well as the direct injuries and infections resulting from flooding, psychological distress including cases of suicide has been associated with the event.

Latin America

Like other regions located in the tropics, some parts of Latin America are home to many tropical infectious diseases such as malaria, dengue, leishmaniasis, yellow fever, Chagas' disease and cholera. The regional assessment of health impacts in the region indicated that the main concerns are heat stress, malaria, dengue, cholera and other water-borne diseases. The region also has been particularly affected by extreme weather events, notably those associated with El Niño.

Although Latin America is home to many vector-borne diseases, few cases of climate driven diseases outbreaks were reported in the Third Assessment Report of the IPCC. However, it was noted that near the equator in Iquitos, Peru, the seasonality of malaria transmission is driven by small temperature fluctuations of 1–2°C (53). Changes of this magnitude can be expected to occur with global warming and this may drive disease transmission to higher altitudes and lower latitudes in Latin America. However, in some parts of the region, increases in temperature could reduce malaria transmission as has been observed in the southern part of Honduras.

In semi-arid zones in Mexico, rainfall has been observed to cause outbreaks of bubonic plague (54), probably as a result of an increase in the rodent reservoirs. Rodents escaping floods in Colombia are suspected to have been the primary cause of leptospirosis outbreaks. The effects of water-borne diseases are well documented in this region. Between 1991 and 1996 cholera affected 21 counties in Peru resulting in almost 200 000 cases and 11 700 deaths. Climate variability was linked to later outbreaks in Peru and Ecuador during the 1997/98 El Niño event. Besides cholera in Peru, other diarrhoeal diseases such as *Salmonella typhi* have been linked to environmental change, climate and sanitary conditions.

Considering all causes of disease and injury, it is apparent that the El Niño weather phenomenon has particularly strong effects on health in Latin America.

In 1983, during a particularly strong event, total mortality increased by 40% and infant mortality by 103% in Peru (55).

North America

The direct impacts of climate change and variability in this region include heat stress, injury and mortality due to convective storms, floods, hurricanes, tornadoes and ice storms.

Photochemical smog and fine particulate matter are important environmental health issues in this region. It is unclear precisely what effect climate change will have on urban air quality in North America, but higher temperatures increase the risk of significant photochemical smog. In 1997, approximately 107 million people in the United States lived in counties that did not meet air quality for at least one regulated pollutant (56), while more than half of Canadians live in areas where ground level ozone may reach unacceptable levels during summer months (57). In the United States floods are the most frequent, and the leading cause of death from, natural disasters. The mean annual loss of life has been estimated to be 147 deaths (56). In Canada the Red River flood of 1997 displaced more than 25 000 people (58).

The areas of the United States most vulnerable to heat-related illnesses appear to be the north-east and mid-west. Recent examples of heat-related deaths include 118 persons in Philadelphia in 1993 (59), 91 persons in Milwaukee and 726 in Chicago in 1995 (60). In Canada, the urbanized area of south-eastern Ontario and Quebec could be affected very negatively by warmer temperatures as shown by modelling studies (61). However, warmer winters could result in fewer cold related deaths.

The frequency of natural disasters may be increasing in the United States. There were less than 20 natural disasters reported annually in the 1950s and 1960s, but more than 40 per year in the 1990s (62). Ice storms can have very large negative impacts, as demonstrated by the January 1998 event—this left 45 dead and nearly 5 million people without electricity in winter in Ontario, Quebec and New York (63).

Several vector-borne diseases are endemic in North America. The World Health Organization declared the United States free of malaria in 1970, while in Canada the disease disappeared at the end of the nineteenth century. However, other climate sensitive vector-borne diseases in the United States include Lyme disease, Rocky Mountain spotted fever, St Louis encephalitis (SLE), western equine encephalitis (WEE) and snowshoe hare virus (SHV). Dengue transmission occurs in Mexico and to a much lesser extent in Texas. Hantavirus is now the major rodent-borne disease, and cryptosporidiosis and giardiasis the most common water-borne diseases.

Lyme disease is the most common vector-borne condition in the United States, approximately 10 000 cases being reported in 1994. However, in Canada where the tick vector has been reported, no disease transmission has been detected. It is expected that, with climate change, Lyme disease and Rocky Mountain spotted fever could spread to Canada.

Hantavirus infection (in humans a severe disease with a mortality rate around 40%) was associated with unusually prolonged rainfall in the 1991–92 El Niño (64), and by 1999, 231 cases had been reported in the United States (56). A number of cases subsequently have been reported in Canadian provinces of British Colombia, Alberta and Saskatchewan. This disease is sensitive to changes in climate and ecology, but because of difficulties in predicting local rainfall it is difficult to predict the likely changes in the prevalence of hantavirus pulmonary syndrome in the United States and Canada (57).

Of the water-borne diseases, giardia cysts are fairly common in treated water in Canada (18.2%) and very frequent in raw sewage samples (73%) (65). The pathogen cryptosporidium also is widely distributed and capable of causing large-scale outbreaks. For example, in 1993, more that 400 000 cases (including 54 deaths) from a cryptosporidium outbreak were reported in Milwaukee, Wisconsin (66). A positive correlation between rainfall, concentration in river water and human diseases has been noted for both cryptosporidiosis and giardiasis.

Polar regions

At the time of the Third Assessment Report of the IPCC no studies were available on the impacts of climate change on human health in the polar region. A number of studies have been conducted subsequently and a summary report, the Arctic Climate Impact Assessment, is being prepared by the Arctic Council (an intergovernmental forum of countries making up the Arctic region), and the International Arctic Science Committee (a non-governmental organization for research and co-operation in the region).

Small island states

Many tropical islands report outbreaks of vector-borne and water-borne infectious diseases that are attributed, in part, to changes in temperature and rainfall regimes. In some regions, e.g. the Pacific, it has been noted that extreme weather events appear to be occurring at a greater frequency than elsewhere (67). As a consequence, physical injuries arising from these events can be expected to increase. Some of the small island states such as the Bahamas, Kiribati, the Marshall Islands and the Maldives are a mere 2–4 m above sea level, which predisposes them to inundation with seawater and consequent salinization of fresh water supplies and flooding from sea level rise.

Post-TAR assessments

The IPCC TAR assessment ended in 2001, however other regional and country specific assessments have since begun. In Europe and a number of developing countries, a post TAR assessment on adaptation strategies is being carried out 2001–2004. The results of these assessments will be reviewed in the fourth assessment of the IPCC.

Conclusions

The IPCC found that global climate change will affect human health in many ways. Overall, negative effects are expected to outweigh positive impacts. Important influences on health will include changes in the frequency and intensity of extremes of heat, cold, droughts, floods, hurricanes, tornadoes and other forms of extreme weather. Climate change also will impinge on health by disrupting ecological and social systems, resulting in changes in infectious disease transmission, food production, air pollution, population displacement and other forms of social disruption.

Climate change impacts on health that were judged to be of high confidence included increased heat-related mortality and morbidity; decreased cold-related mortality in temperate countries; greater frequency of infectious disease epidemics following floods and storms; and substantial health effects following population displacement from sea level rise and increased storm activity.

For each of the potential impacts of climate change it is possible to identify groups that will be particularly vulnerable to disease and injury. For instance, those most at risk of suffering harm from thermal extremes will include socially isolated city dwellers, the elderly and the poor. Populations living at the present margins of malaria and dengue, and without effective primary health care, will be most susceptible if these diseases expand their range in a warmer world.

The IPCC report shows that understanding of the links between climate, climate change and human health has increased considerably in the last ten years. However, there are still many uncertainties. There are many gaps in knowledge of exposure, vulnerability and adaptability of physical, ecological and social systems to climate change. The next chapter explores the challenges faced by researchers as they strive to fill these gaps and provide a stronger information base for responses to climate change.

References

1. Intergovernmental Panel on Climate Change (IPCC). *Climate change 2001: third assessment report, impacts, adaptations and vulnerability of climate change*. McCarthy, J.J. et al. eds. Cambridge, UK, Cambridge University Press, 2001.
2. Watson, R.T. et al. eds. *The regional impacts of climate change*. An assessment of vulnerability: a special report of IPCC Working Group II . Cambridge, UK, Cambridge University Press, pp. 517, 1998.
3. Gubler, D.J. Dengue and dengue haemorrhagic fever. *Clinical Microbiology Review* 11: 480–496 (1998).
4. Kalkstein, L.S. & Greene, J.S. An evaluation of climate/mortality relationships in large U.S. cities and the possible impacts of a climate change. *Environmental Health Perspectives* 105: 84–93 (1997).
5. Donaldson, G.C. et al. Heat and cold related mortality and morbidity and climate change. In: *Health effects of climate change in the UK*. London, UK, Department of Health, 2001.
6. Patz, J.A. et al. The potential health impacts of climate variability and change for the United States. Executive summary of the report of the health sector of the U.S. National Assessment. *Journal of Environmental Health* 64: 20–28 (2001).
7. World Health Organization (WHO). *Health guidelines for episodic vegetation fire events*. Geneva, Switzerland, World Health Organization, 1999 (WHO/EHG/99.7).
8. Ziska, L.H. & Caulfield, F.A. The potential influence of raising atmospheric carbon dioxide (CO_2) on public health: pollen production of common ragweed as a test case. *World Resources Review* 12: 449–457 (2000a).
9. Ziska, L.H. & Caulfield F.A. Pollen production of common ragweed (Ambrosia artemisiifolia) a known allergy–inducing species: implications for public health. *Australian Journal of Plant Physiology* 27: 893–898 (2000b).
10. Ahlholm, J.U. et al. Genetic and environmental factors affecting the allergenicity of birch (Betula pubescens ssp. czerepanovii [Orl.] Hamet-ahti) pollen. *Clinical and Experimental Allergy* 28: 1384–1388 (1998).
11. Food and Agricultural Organization of the United Nations (FAO). *The state of food insecurity in the world 1999*. Rome, Italy, FAO, pp. 32. 1999.
12. World Health Organization (WHO). *Malaria* -WHO Fact Sheet No. 94 Geneva, Switzerland, World Health Organization, 1998.

13. Martens, W.J.M. et al. Potential impact of global climate change on malaria risk. *Environmental Health Perspectives* 103: 458–464 (1995).

14. Martens, W.J.M. et al. Climate change and vector-borne diseases: a global modeling perspective. *Global Environmental Change* 5: 195–209 (1995).

15. Jetten, T.H. & Focks, D.A. Potential changes in the distribution of dengue transmission under climate warming. *American Journal of Tropical Medicine and Hygiene* 57:285–287 (1997).

16. World Health Organization (WHO). *The world health report 2002*. Geneva, Switzerland, World Health Organization, 2002.

17. Murray, C.J.L. & Lopez, A.D. eds. *Global burden of disease*: global burden of diseases and injury series: Vol. 1. Boston, USA, Harvard School of Public Health, Harvard University, 1996.

18. Gubler, D.J. & Meltzer, M. The impact of dengue/dengue haemorrhagic fever in the developing world. In: *Advances in virus research* Vol. 53. Maramorosch, K. et al. eds. San Diego, CA, USA, Academic Press, pp. 35–70, 2000.

19. Lisle, J.T. & Rose, J.B. Cryptosporidium contamination of water in the US and UK: a mini-review. *Aqua* 44: 103–117 (1995).

20. Atherholt, T.B. et al. Effects of rainfall on giardia and cryptosporidium. *Journal of the American Water Works Association* 90: 66–80 (1998).

21. Rose, J.B.S. et al. Climate and waterborne outbreaks in the US: a preliminary descriptive analysis. *Journal of the American Water Works Association* 92: 77–86 (2000).

22. Curriero, F.C. et al. The association between extreme precipitation and waterborne disease outbreaks in the United States, 1948–1994. *American Journal of Public Health* 91: 1194–1199 (2001).

23. Colwell, R.R. Global warming and infectious diseases. *Science* 274: 2025–2031 (1996).

24. Birmingham, M.E. et al. Epidemic cholera in Burundi: patterns in the Great Rift Valley Lake region. *Lancet* 349: 981–985 (1997).

25. Lipp, E.K. & Rose, J.B. The role of seafood in foodborne diseases in the United States of America. *Revue Scientific et Technicale. Office Internationale des Epizootics*, 16: 620–640 (1997).

26. Prothero, R.M. Forced movements of populations and health hazards in tropical Africa. *International Journal of Epidemiology* 23: 657–663 (1994).

27. Lefevre, P.C. Current status of Rift Valley fever. What lessons to deduce from the epidemic of 1977 and 1987. *Medicine Tropicale* 57: (3 Suppl) 61–64 (1997).

28. Linthicum, K.J. Application of polar-orbiting, meteorological satellite data to detect flooding of Rift Valley Fever virus vector mosquito habitats in Kenya. *Medical and Veterinary Entomology* 4: 433–438 (1990).

29. World Health Organization (WHO)—*Weekly Epidemiological Records*: 20, 15th May, (1998).

30. Greenwood, B.M. et al. Meningococcal disease and season in sub-Saharan Africa. *Lancet* 1: 1339 –1342 (1984).

31. Angyo, I.A. & Okpeh, E.S. Clinical predictor of epidemic outcome in meningococcal infections in Jos, Nigeria . *East African Medical Journal* 74: 423–426 (1997).

32. Hart, C.A. & Cuevas L.E. Meningococcal disease in Africa. *Annal of Tropical Medicine & Parasitology* 91: 777–785 (1997).

33. Tikhomirov, E. et al. Meningococcal disease: public health burden and control. *World Health Statistics* 50: 170–177 (1997).

34. Pittock, A.B. Coral reef and environmental change: adaptation to what? *American Zoologist* 39: 10–29 (1999).

35. Woodward, A. et al. Tropospheric ozone: respiratory effects and Australian air quality goals. *Journal of Epidemiology and Community Health* 49: 401–407 (1995).

36. Petroeschevsky, A. et al. Associations between outdoor air pollution and hospital admissions in Brisbane, Australia. *Archives of Environmental Health* 56: 37–52 (2001).

37. McMichael, A.J. & Kovats, S. Assessment of the impact on mortality in England and Wales of the heat wave and associated air pollution episode of 1976. *Report to the Department of Health,* London, U.K, London School of Tropical Medicine (1998).

38. Rooney, C. et al. Excess mortality in England and Wales, and in Greater London, during the 1995 heatwave. *Journal of Epidemiology and Community Health* 52: 482–486 (1998).

39. Katsouyanni, K. et al. The 1987 Athens heatwave. *Lancet* 8610: 573 (1988).

40. Katsouyanni, K. et al. Evidence for interaction between air pollution and high temperature in the causation of excess mortality. *Architecture and Environmental Health* 48: 235–242 (1993).

41. Martens, P. Climate change, thermal stress and mortality changes. *Social Science and Medicine* 46: 331–334 (1997).

42. Balderi, M. et al. Malaria in Maremma, Italy. *Lancet* 351: 1246–1247 (1998).

43. Dedet, J.P. et al. Leishmaniasis and human immunodeficiency virus infections. *Presse Medicale* 24: 1036–1040 (1995).

44. Talleklint, L. & Jaenson, T.G. Increasing geographical distribution and density of Ixodes ricinus (Acari: Ixodidae) in central and northern Sweden. *Journal of Medical Entomology* 35: 521–526 (1998).

45. Lindgren, E. et al. Impact of climatic change on the northern latitude limit and population density of the disease-transmitting European tick Ixodes ricinus. *Environmental Health Perspectives* 108: 119–123 (2000).

46. Lindgren, E. & Gustafson, R. Tick-borne encephalitis in Sweden and climate change. *Lancet* 358. 16–18 (2001).

47. Randolph, S. Tick-borne encephalitis in Europe. *Lancet* 358: 1731–1732 (2001).

48. Randolph, S.E. & Rogers, D.J. Fragile transmission cycles of tick-borne encephalitis virus may be disrupted by predicted climate change. *Proceedings of the Royal Society of London Series B-Biological Sciences* 267: 1741–1744 (2000).

49. Bentham, G. & Langford, I.H. Climate change and the incidence of food poisoning in England and Wales. *International Journal of Biometeorology* 39: 81–86 (1995).

50. Kríz, B. Infectious diseases: consequences of the massive 1997 summer flood in the Czech Republic. (1998). (EHRO 020502/12 Working group paper)

51. Kríz, B. et al. Monitoring of the epidemiological situation in flooded areas of the Czech Republic in 1997. Proceedings of the Conference DDD'98 11–12 May 1998 Podebrady, Czech Republic (1998).

52. Simoes, J. et al. Some aspects of the Weil's disease epidemiology based on a recent epidemic after a flood in Lisbon (1967). *Anais da Escola Nacional de Saúde Pública e de Medicina Tropical (Lisb)* 3: 19–32 (1969).

53. Patz, J.A. Climate change and health: new research challenges. *Health and Environment Digest* 12: 49–53 (1998).

54. Parmenter, R.R. et al. Incidence of plague associated with increased winter-spring precipitation in New Mexico. *American Journal of Tropical Medicine and Hygiene.* 61: 814–821 (1999).

55. Toledo-Tito, J. Impacto en la Salud del Fenomenno de El Niño 1982–83 en le Peru. In: *Proceedings of the Health Impacts of the El Niño Phenomenon, Central American Workshop* held in San Jose Costa Rica 3–5 November 1997. Washington, DC, USA, Pan American Health Organization and World Health Organization, (in Spanish), 1999.

56. Patz, J.A. et al. The potential health impacts of climate variability and change for the United States. Executive summary of the report of the health sector of the U.S. National Assessment. *Journal of Environmental Health* 64: 20–28 (2000).

57. Duncan, K. et al. In: *The Canadian country study: climate impacts and adaptations.* Koshida, G. & Avis, W. eds. Toronto, ON, Canada, Environment Canada, pp. 501–590 (1998).

58. Manitoba Water Commission. *An independent review of action taken during the 1997 Red River flood.* Winnipeg, MB, Canada, Minister of Natural Resources, 1998.

59. Centres for Disease Control and Prevention. Heat-related deaths: United States 1993. *Morbidity and Mortality Weekly Report* 42: 558–560 (1993).

60. Centres for Disease Control and Prevention. Heat-related deaths: Chicago July 1995. *Morbidity and Mortality Weekly Report* 44: 577–597 (1995).

61. Kalkstein, L. & Smoyer, K.E. The impact of climate on Canadian mortality: present relationships and future scenarios. *Canadian Climate Centre Report 93–7*, Downview ON, Canada, Atmospheric Environment Service. pp. 50 (1993).

62. Miller, A. et al. *What's fair? Consumer and Climate Change. Redefining Progress.* San Francisco, CA, USA, pp. 67 (2000).

63. Centres for Disease Control and Prevention. Community needs assessment and morbidity surveillance following an ice storm—Maine, January 1998. *Morbidity and Mortality Weekly Report* 47: 351–354 (1998).

64. Glass, G.E. et al. Using remotely sensed data to identify areas at risk for hantavirus pulmonary syndrome. *Emerging Infectious Diseases* 6: 238–247 (2000).

65. Wallis, P.M. et al. Prevalence of giardia cysts and cryptosporidium oocysts and characterization of giardia spp. Isolated from drinking water in Canada. *Applied Environmental Microbiology* 62: 2789–2797 (1996).

66. MacKenzie, W.R. et al. A massive outbreak in Milwaukee of cryptosporidium infection transmitted through the public water supply. *New England Journal of Medicine* 331:1529–1530 (1994).

67. Timmermann, A. & Oberhuber, J. Increased El Niño frequency in climate model forced future greenhouse-warming. *Nature* 398, 634–698 (1997).

Looking to the future: challenges for scientists studying climate change and health

A. Woodward,[1] J.D. Scheraga[2,3]

Introduction

Chapter 3 describes ways in which climate change may affect human health and summarizes the findings of the Third Assessment Report of the IPCC (*1*). This chapter looks ahead and considers the challenges awaiting researchers who seek to advance knowledge of this area beyond what is contained in reports from the IPCC and other bodies. This begins with an outline of important ways in which climate change is different from other environmental health problems and explores the implications for researchers.

The biggest challenge is scale. Both the geographical spread of climate-related health problems and the much elongated time spans that often apply, are largely unfamiliar to public health researchers. Research on climate change typically is conducted on three time-scales:

1. relatively short periods between altered climate (expressed as weather) and the effects on health.
2. intermediate time periods that include recurring, inter-annual events like El Niño and La Niña.
3. longer intervals (decades or centuries) between the release of greenhouse gases and subsequent change in the climate. This category of research is most troublesome to standard epidemiological methods.

Researchers in the public health sciences are accustomed to studying geographically localized problems that have a relatively rapid onset and impact directly on human health. There are exceptions (e.g. the global spread of AIDS and tobacco-related diseases) but, typically, health problems (and control strategies) are defined by boundaries at a finer scale: neighbourhood, town or province. The standards that researchers bring to the evaluation of evidence frequently are born out of an experimental research tradition. In this vein the natural unit of observation tends to be the individual rather than the group and when thinking about causes the emphasis lies on specific agents acting downstream in the causal process.

Weather and climate variability do not fit well the conventional research model, partly because there is no easily identified unexposed control group and little variation in exposures between individuals in a geographical region. Consequently, studies of the effects on health of weather and climate variability need to use ecological designs (in which the study unit is a population). Following a period when population-based studies were somewhat out of favour, epidemi-

[1] Wellington School of Medicine, University of Otago, Wellington, New Zealand.
[2] Global Change Program, US Environmental Protection Agency, Washington, DC, USA.
[3] The views expressed are the author's own and do not reflect official USEPA policy.

ologists are re-visiting their use and exploring ways in which evidence from ecological investigations can be combined with information collected at the individual level (2). While the exposure is common to a geographical area, there are frequently variations in coping capacity that cause considerable differences in outcomes. For example, excess mortality in the 1995 Chicago heatwave varied almost one hundred-fold between neighbourhoods as a result of factors such as housing quality and community cohesion (3).

Other important differences from traditional environmental exposures are the directness of the association between exposure and disease, and the degree to which interactions and feedbacks could occur. There are many pathways—some more direct than others—through which climate change could affect human health (see chapter 2): the effects of temperature extremes on health are direct, while the effects of changes in temperature and cloud cover on air pollution-related diseases involve several intermediate steps. Similarly, ecosystem change will be one mediator of the potential effects of changes in temperature and precipitation on vector-borne diseases. As another example, climate change may increase the amount of time taken for stratospheric ozone levels to return to pre-industrial concentrations. This delay could be decades or longer than expected under the Montreal Protocol (1). During this time, increased exposure to UV radiation is expected to continue to increase rates of skin cancer, cataracts and other diseases (see chapter 8).

Hypotheses of the effects of climate on health cannot be tested in experimental studies because climate cannot be assigned at the whim (random or otherwise) of the investigator. It may be possible to study some early effects using standard observational methods. However, the effects of future climate variability and change can be estimated only by analogue studies, using current weather or climate variability (such as El Niño events—chapter 5) that mimics in some way what might be expected under climate change, or by models. Such models cannot predict what will happen, but instead sketch out what would occur *if* certain conditions were fulfilled. Some—at least in theory—could assign probabilities, such as the chance of sea level rise along the coast of the United States of America over t he next 100 years. Ideally, models include scenarios of future societal, economic and technological conditions, since the impact of climate depends very heavily on these factors. It is important for these models to capture the effects of humans as an added stressor on the environment, and their ability to respond to change. Climate/health models should be informed but need not be constrained by historical data. For example, it is possible to construct a simulation model based upon assumed conditions and processes (e.g. with thresholds and non-linearities) different to those experienced historically.

The less than ideal fit between the problems and available study methods presents a challenge; this chapter describes some of the ways in which researchers are responding. Developments include new ways of estimating the impacts of future threats (such as scenario based assessments). Methods applied elsewhere to the study of complex non-linear systems are being translated to the health sphere (e.g. modelling of infectious disease). A more sophisticated approach to uncertainty assessment includes not only statistical sources of error (arising from sampling processes) but also the uncertainty that results from judgements that must be made to bridge knowledge gaps. Whereas in the past the variability in response between study units tended to be regarded as noise around the exposure-outcome signal, now this variability is seen as important in its own right. For example, to learn about possible mechanisms of adaptation to extreme

FIGURE 4.1 Tasks for public health science.

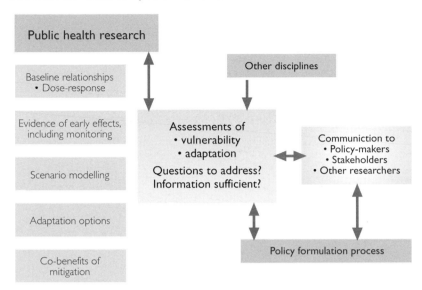

weather events, researchers in the United States have investigated the reasons for wide variations in the change in mortality for a given increase in daily temperature in cities (4).

The main tasks of public health science in assessing the potential health effects of climate variability and change (Figure 4.1) include:

- establishing baseline relationships between weather and health (introduced in chapter 2)
- seeking evidence for early effects of climate change (discussed in detail in chapter 10)
- developing scenario-based models (referred to in a number of chapters)
- evaluating adaptation options (discussed in detail in chapter 11)
- estimating the coincidental benefits and costs of mitigation and adaptation.

Consideration of the links between science and policy development must be incorporated in each of these steps. More precisely, the question is how science can best inform decision-makers in a timely and useful fashion.

Tasks for public health scientists

Establishing baseline relationships

For centuries the relation between weather and health has attracted attention. However, interest in the topic has increased following the first signs that human activity may be influencing the world's climate. There are many unresolved questions about the sensitivity of particular health outcomes to weather, climate variability and climate-induced changes in environmental conditions critical to health. Yet a good deal is known about the effects of weather on aspects of human health, as shown in the following example.

The major pathogens responsible for acute gastroenteritis multiply more rapidly in warmer conditions so it would be expected that higher temperatures would be associated with greater risk of illness, all else being equal. This appears

FIGURE 4.2 Relationship between mean temperature and monthly reports of *Salmonella* cases in New Zealand.

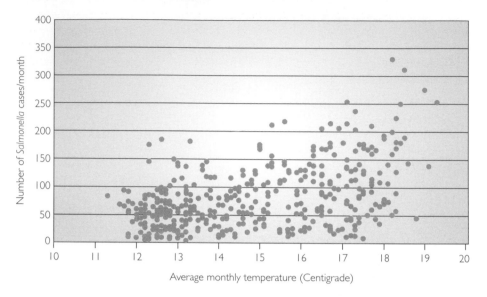

to be the case: over the last 30 years the number of notifications of salmonella infections in New Zealand has been clearly related to the average temperature during the same month. (Figure 4.2).

To investigate the relation between ambient temperatures and the rate of enteric infections, Bentham and colleagues collected British data on all cases of food poisoning in a ten year period (1982–1991) (5). Statistical models of the relationship between the monthly incidence of food poisoning and temperatures were developed. The numbers of reported cases were compared with temperatures of both that and the previous month, on the basis of the known biology. An association was found with both temperature measures although the previous month showed the stronger effect. This suggests that temperature may act through effects on storage, preparation and hygiene close to the point of consumption. The even stronger relationship with the temperature of the previous month suggests the possible importance of conditions earlier in the food production process. It is important to emphasize that correlation is not causality; more work is needed to understand the mechanisms by which environmental change affects disease risk.

The challenges faced by researchers investigating base-line relationships are similar to those faced in public health sciences generally. They include difficulties with accurately measuring outcomes, obtaining meaningful assessments of the exposure and dealing with large numbers of potential confounding factors. In the case of the studies reported here, month-by-month variations in notifications of infectious disease are not likely to be due to reporting artefact. Changes in the accuracy of diagnosis and completeness of reporting are a greater concern when long-term time trends are analysed.

Seeking evidence for early health effects of climate change

The changes in the world's climate in the past 50 years or so are remarkable when compared with patterns over the past 10 000 years. The rate of change of

CLIMATE CHANGE AND HUMAN HEALTH

atmospheric concentrations of greenhouse gases is outside human experience and recently the rise in average temperatures has exceeded what is considered the bounds of natural variability (chapter 2). It would seem reasonable therefore to look for evidence of early health effects of climate change: changes in health outcomes that become apparent soon after the onset of climate change. The Intergovernmental Panel on Climate Change (chapter 3) considered the following lines of evidence:

- observed variations in disease incidence and range in association with short term climate variability;
- long-term disease trends and the factors that may be responsible;
- projections based on first principle relations between temperature and the development of disease;
- empirical disease models based on the current geographical distribution of climate-sensitive diseases.

There are many studies of physical and biological systems reporting a variety of phenomena linked to long-term, decade-scale change in temperature and rainfall. These include events such as permafrost melting, ice sheets breaking up and glaciers retreating. Biological changes associated with climate alteration include lengthening of growing seasons, earlier flowering, changes in egg laying and poleward shift in distribution of a number of insects (6). In one example of this kind of work (and showing climate change to have positive as well as negative effects) Nicholls charted changes in Australian wheat yields from 1952 to 1996 (7). Over this period the average yield increased from 1.1 to almost 1.6 tonnes per hectare. In the same years the diurnal temperature range fell by almost 0.5 °C, largely as a consequence of warmer overnight temperatures (relevant because wheat is sensitive to frost). Several techniques were used to attempt to control for non-climatic influences such as changes in crop types and land use patterns. Especially persuasive was the strong correlation between year-to-year variation in temperature range and wheat yield. On the basis of these data Nicholls estimates that climate trends in the second half of the twentieth century (including natural and human-induced climate change) were responsible for 30–50% of the observed increase in wheat yields.

There are few reports of the effects of climate change on human health to match the range of observations on physical and ecological effects. Lindgren has reported geographical changes in tick-borne encephalitis in Sweden that match shifts in both the vector and long-term climate. She observed also a northward extension of the tick population following a trend of warmer winters (8). In Africa some investigators have reported changes in the occurrence of malaria (including rising altitude limits) that they think unlikely to be due to changes in land-use patterns, increasing drug resistance or diminishing effectiveness of health services (9). A good deal has been written about associations of weather and health, but apart from the examples given here there are few studies documenting effects that might be attributed to year on year change in climate.

Why the relative paucity of evidence of early health impacts? Our species is not immune to changes in climate—there are plenty of studies reporting acute effects of extreme weather such as heatwaves, floods and storms, and short-term variations in climate such as El Niño (10, 11). Less work is being done in this area than in the natural sciences—that is, the denominator (all climate/health research) is smaller. More importantly, research on free-living human populations includes a layer of complexity that does not apply to investigators working with butterflies, ticks or wheat. Not only are there non-climatic confounding factors but also there is the phenomenon of social adaptation. Humans are unrivalled in their capacity to adapt to changing environmental circumstances. European butterflies have no choice but to head north when it heats up; humans have many coping strategies at their disposal from planting shade trees, changing hours of work or making use of cooler times of the day, to installing air-conditioning.

Time trends in disaster losses show how effectively humans can protect themselves against climate. In the United States in the 1900s, hurricane damages in constant dollar terms increased from about US$ 5 billion in the 1930s to more than US$ 30 billion in the 1990s. This increase was due principally to massively increased exposure to financial losses (resulting from new settlements on flood plains and in coastal regions) rather than any change in the frequency or severity of storms. Over the same period the number of deaths due to hurricanes was reduced ten-fold. In the United States the effect of improved communications, early warning systems, transport, building standards and other elements of civil defence has been to insulate the population to a large extent from the effects of climate extremes (*12*).

The challenge for public health scientists is to pick the settings, populations and health outcomes where there is the best chance of firstly, detecting changes and secondly, attributing some portion of these to climate change. This means seeking research opportunities where the necessary information can be found: because climate change spreads over decades, time series data on outcomes and confounders over a similar period are required. From first principles, impacts are likely to be seen most clearly where the exposure-outcome gradient is steepest and adaptive capacity weakest. Attribution is most straightforward when there are few competing explanations for observed associations and when these links can be clearly specified.

With such conditions in mind, what are the best bets for studies of early effects? Vector-borne diseases may be relatively sensitive indicators because transmission involves intermediate organisms, such as mosquitoes, open to environmental influences. Intestinal infections (food poisoning) show very strong seasonal patterns (suggesting a powerful effect of climate variability) and have been routinely reported for many years (although the data are known to be incomplete). Deaths, injuries and illnesses caused by extreme events (such as heatwaves, cold spells, floods and storms) satisfy the condition of few competing explanations, but in many populations it may be difficult to distinguish the climate change signal from much stronger mitigating effects of social and economic development.

Developing scenario-based models (future effects)

With climate change, researchers are attempting to quantify future effects of future weather exposures and, more fundamentally, trying to understand and delineate the magnitude of future risks. The risks arise from a wide range of diseases and injuries, some readily quantified (e.g. deaths due to storms and floods) others more difficult to capture statistically (the health consequences of food insecurity). Whether the potential for climate-related disease is translated into actual occurrence of death and illness depends on both how quickly the climate changes and how successfully humans adapt to new conditions. Interacting exposures may make it difficult to forecast dose-response relations from historical data: a study of 29 European cities found the effect of particulate air pollution on daily mortality was almost three times stronger in the warmest cities than in the coldest (13). This suggests that at least some part of the effects of rising temperatures will be dependent on future trends in urban air pollution. Multidisciplinary research is required to make progress in this field: modelling changes in future socioeconomic status and examination of its implications for susceptibility will require health scientists, economists and experts from other disciplines to work together.

One response to this challenge has been the development of integrated assessments, a term that has been variously defined. Under one definition, "integrated assessment is an interdisciplinary process of combining, interpreting, and communicating knowledge from diverse scientific disciplines in such a way that the whole set of cause-effect interactions of a problem can be evaluated from a synoptic perspective with two characteristics: it should have added value compared to single disciplinary oriented assessment; and it should provide useful information to decision-makers" (14). The essential points in this definition are synthesis (combining information from more than one discipline) and application to decision-making. The integration may be horizontal (between disciplines) or vertical (bringing together assessments at different levels of complexity within the one discipline). An example of the former is an assessment of the impact of climate change on the prevalence of hunger, which was based on climate forecasts, plant science, demographic scenarios and economic models of food trade (15). A vertical assessment is exemplified by the work done by UV researchers to estimate impacts of stratospheric ozone depletion on skin cancer rates (16). Such approaches have proven useful in evaluating potential impacts of climate change other than effects on human health, and are beginning to be applied to a range of climate change and health issues (17). The goal of integrated assessment is to provide insights that cannot be gained from traditional, single disciplinary research (see chapter 12 discussion about prioritization of research agenda). An example of integrative assessment is the inclusion of pathogen transmission dynamics; contextual elements such as changing land use; and demographic forces such as population movement into an evaluation of the potential impacts of climate change on infectious diseases (17). Integrated assessment also allows evaluation of how feedback mechanisms and adaptation measures could change the system response. As with any type of risk assessment, the choices about which variables to include (or exclude) may make a big difference to the conclusions.

Modelling is one of several methods employed to conduct an integrated assessment. An array of component models of varying degrees of complexity (from simpler parametric to more detailed process-oriented models) is developed, each with mathematical representations of cause-effect relationships. These are linked to show the interrelationships and feedback mechanisms among the key components. The resulting framework aids the identification and prioritization of scientific uncertainties. Sensitivity analyses can be conducted to understand better the sensitivity of the system to changes in each relationship (17). Although modelling is useful for risk categorization, it can imply more precision than is appropriate where full data are absent, and relevant available data may be excluded or overemphasized. In addition, the technical limitations of the model might defeat the ultimate value of the analysis (18).

If integrated models are to advance understanding of phenomena such as climate change, they must simulate to some degree the interrelationships and feedbacks that occur in complex systems. If the results are to be widely accepted and applied, these models must be congruent with the points of view and understandings of those who are ultimately the users of the information. Studies indicate that integrated assessments are more likely to be accepted when they incorporate a variety of methods (demonstrating that similar answers are obtained via different routes), include multiple objectives (since users are likely to be concerned with a range of outcomes) and when the users receive immediate feedback on the implications of changing key parameters (*19*, *20*).

Evaluating adaptation options

Adaptation means taking steps to reduce the potential damage that occurs when the environment changes. As a policy option this approach has a number of attractions. First, it offers opportunities for win-win strategies (building adaptive capacity to risks under current conditions may be beneficial, regardless of future climate change). Second, it recognises that—to some extent—the world is committed to climate change. Even if immediate substantial reductions were made in greenhouse emissions, over the next 50 years the planet would continue to warm and sea levels continue to rise for hundreds of years due to the time it takes for basic global systems to reach a new equilibrium.

The reason why some systems cope better than others when stressed has attracted the interest of researchers in many disciplines. Geneticists have been intrigued by the particular sensitivity of some populations to introduced infections, and some have proposed that variability (in genotypes) might be an important protective factor (*21*). Ecologists have explored the attributes of ecosystems that are relatively robust under pressure. Complexity or interconnectedness and diversity are two elements that appear to enhance adaptability. Diverse plant communities, for example, are more effective than simple communities in fixing atmospheric carbon (*22*) and more stable in the face of severe environmental stress such as drought (*23*). The distribution of risk also is important. Amory Lovins has described a resilient energy system as "one that has many relatively small dispersed elements, each having a low cost of failure. These substitutable components are interconnected, not at a central hub but by many short, robust links" (*24*). This is analogous to the system of veins that distributes nutrients through the leaves of a tree, or the electronic links that make up the Internet. It might be a model for secure food supplies or robust global health care.

Public health researchers have applied many of these ideas when investigating why some populations are more vulnerable than others to disease and injury. Studies of disaster preparedness have emphasized the importance of diverse community networks, both formal and informal (*25*). Populations dependent on a few staple foods are at greater risk of famine than those with a variety of foods (*26*). Studies of historical climate-related disasters have shown that political structures and economic arrangements are very important in modifying the effect of droughts and other extreme events on health outcomes (*27).*

When applied to climate change, vulnerability may be defined as the degree to which a system is susceptible to, or unable to cope with, the adverse effects of climate change including variability and extremes. Components include the:

- extent to which health, or the natural or social systems on which health outcomes depend, are sensitive to changes in weather and climate (i.e.

TABLE 4.1 Examples of factors affecting vulnerability.

Level	Influence on vulnerability	Description
Individual	Disease status	Those with pre-existing cardiovascular disease, for example, may be more vulnerable to direct effects such as heatwaves
	Socioeconomic factors	Poor in general are more vulnerable
	Demographic factors	Elderly are more vulnerable to heatwaves, infants to diarrhoeal diseases
Community	Integrity of water and sanitation systems and their capacity to resist extreme events	
	Local food supplies and distribution systems	
	Access to information	Lack of early warnings of extreme events
	Local disease vector distribution and control programmes	
Geographical	Exposure to extreme events	Influence of El Niño cycle or occurrence of extreme weather events more common in some parts of the world
	Altitude	Low-lying coastal populations more vulnerable to the effects of sea level rise
	Proximity to high-risk disease areas	Populations bordering current distributions of vector-borne disease may be particularly vulnerable to changes in distribution
	Rurality	Rural residents often have less access to adequate health care; urban residents more vulnerable to air pollution and heat island effects
	Ecological integrity	Environmentally degraded and deforested areas more vulnerable to extreme weather events.

Source: reproduced from reference 1.

exposure-response relationship—drinking water contamination associated with heavy rainfall);

- exposure to the weather or climate-related hazard (includes character, magnitude and rate of climate variation);
- adaptive capacity—the ability of institutions, systems and individuals to adjust to potential damages, take advantage of opportunities, or cope with the consequences (for example: watershed protection policies, or effective public warning systems for boil-water alerts and beach closings) (*1*).

Individual, community and geographical factors all contribute to capacity to adapt to change in climate (Table 4.1). These include the level of material resources, effectiveness of governance and civil institutions, quality of public health infrastructure, access to relevant local information on extreme weather threats, many other socioeconomic factors, and pre-existing level of disease (*28*).

Scheraga and Grambsch propose several principles for policy-makers considering options for adaptation (*29*). First, the principle of heterogeneity: the effects of climate change will vary by region, between different demographic groups and over time. This means that adaptive responses need to be specific to a particular setting. Second, the effects of climate change will not occur in isolation from other social and environmental stressors so adaptation must take account of coincidental factors such as population growth and environmental degradation. Third, the costs and effectiveness of adaptive options must be weighed when setting priorities (current efforts to cope with climate variability may provide a guide to this). Lastly, it is important to bear in mind that attempts at adaptation can do more harm than good: poorly designed coastal defences may increase vulnerability to storms and tidal surges if they engender a false sense of security and promote settlement in marginal coastal areas.

Estimating ancillary benefits and costs

Decisions on climate change are driven largely by the anticipated consequences in the medium and long-term future. However, steps taken to reduce emissions of greenhouse gas (mitigation) or to lessen the impact of climate change on health (adaptation) may have immediate effects. A greater emphasis on public transport at the expense of personal motor vehicles may not only reduce emissions of CO_2, but also improve public health in the short-term by reducing air pollution and traffic accidents. The magnitude and timing of ancillary costs and benefits are very important for policy-makers. Political decision-making occurs within a short time horizon but the costs of climate change actions often must be borne before there is any discernible effect on climate or the health consequences. As a result, it may be helpful to include outcomes that can be attributed to climate change interventions and occur soon after the introduction of such interventions. Such changes may be positive or negative. Policies that restrict access to motor cars but do not ensure the availability of alternative forms of transport, might do more harm than good in health terms (for example, making it more difficult for people to get to health care facilities).

To illustrate this point: improving energy efficiency and increasing the use of low pollution energy sources would not only reduce greenhouse emissions, but also improve air quality for much of the world's population. An analysis was carried out to estimate the local health benefits of adopting greenhouse mitigation policies in four major cities (Santiago, São Paulo, Mexico City and New York) (30). In this study it was assumed that climate mitigation policies would lead to a 10% reduction in levels of fine particles and ozone in urban areas. Upper and lower bounds of particle-related mortality were estimated, as were a number of other health impacts (hospital admissions, asthma attacks, lost workdays due to acute illness). The conclusion of the study was that "policies aimed at mitigating greenhouse gas emissions can provide a broad range of more immediate air pollution benefits to public health".

In general terms, the challenge for researchers is to conceive study designs that capture information on coincidental benefits (and costs). Co-benefits apply not only to measures that reduce greenhouse gases but also to adaptation policies. These may yield co-benefits in the form of reductions of other (non-climate-related) environmental problems. They may also provide co-benefits in the form of reductions in greenhouse gases.

Informing policy

How can research into the effects of climate change contribute most usefully to policy? Early risk assessment/risk management models describe a linear, unidirectional process in which science precedes (and is remote from) issues of values, trade-offs and ethics (31). However, close examination of what happens in practice shows that it is difficult to draw a clear distinction between scientific risk assessment and social risk management. Two examples in the literature are a study of the health effects of acid rain (32) and Conrad Brunk's analysis of three assessments of the pesticide, Alachlor. In his case study Brunk pointed out numerous decision points where the risk assessors' values were paramount (*33*). It was differences in these value-driven choices (such as whether to assume "adhering to best practice" or "plausible real behaviour" exposure scenarios) that explained the three orders of magnitude difference between the risks calculated by Monsanto and Health and Welfare Canada, not disagreements over scientific methods or the basic data.

The titles of recent United States' National Research Council (NRC) publications on risk reflect the shift in thinking that has occurred away from the linear objective science—subjective policy model. "Science and judgement" (*34*), was followed by "Informing decisions in a democratic society" (*35*), and "Towards environmental justice" (*36*). Scientific assessments are not remote technical exercises, they are part of the messy, problematic and negotiated world of social decision-making. The NRC report "The science of regional and global change: putting knowledge to work," states:

> "*Assessment and policy analysis are essential to understand the overall impact of changes in human behavior and natural processes, to link research agendas with decision needs, and to monitor the results of policy actions. Effective assessment aims to integrate the concepts, methods, and results of the physical, biological, and social sciences into a decision support framework*" (*37*).

There are important implications in accepting such a view. One is that scientific assessments should not assume a common understanding of the problem faced, let alone a single view of the solution. Scheraga and Furlow argue that for an assessment to be informative, the assessors must know the particular issues and questions of interest to stakeholders (e.g. public health officials) (*38*). Stakeholders should be engaged from the outset of the assessment process and involved in the analytical process throughout the assessment. Openness and inclusiveness enable different participants to bring a diversity of views and information that may benefit the assessment process: including all interested parties makes the assessment process more transparent and credible. Another writer has argued that an important part of scientists' work is to try to achieve "widespread agreement on what questions are being asked, why they are important, what counts as answers to them and what the social use of these answers might be" (*32*). Aron puts it this way: "the process of integrated assessment must grapple with questions of values up front" (*19*). Without this clarity, policy-makers may find it difficult to interpret insights from assessments and may struggle to appreciate the reasons for disagreement between different researchers.

The purpose of assessment is to inform decision-makers, *not* to make policy choices. The ways in which information is presented may strongly affect subsequent choices, but scientists cannot determine what is the best policy choice. Determining what is best is a societal decision, involving societal values. The next section examines the interplay of values and scientific assessment in relation to climate change, and other factors that contribute to uncertainty.

Recognizing and responding to uncertainty

The effect of climate change on human health depends on a sequence of events that produces a "cascade of uncertainty" (*39*). There is no doubt that human activity has altered the composition of Earth's atmosphere. The resultant effects on the world's climate are less certain but, as already noted, most of the warming that has occurred in the last 50 years probably is human-induced (*1*). It is more difficult to forecast what will happen in the future—complicating factors include future trends in greenhouse emissions, biophysical feedback forces and threshold phenomena in the climate system that by their very nature are difficult to anticipate. Impacts on human health will depend not only on the nature and rate of climate change at the local level, but also on the ability of ecological systems to buffer climate variability. Yet as discussed in chapter 12, assessments can provide insights even if predictions cannot be made, conducting bounding exercises to estimate the potential magnitude of particular impacts and the importance of their effects.

Sources of uncertainty can be examined by taking as a particular case study the relation between climate change and mosquito-borne diseases (discussed in more detail in chapter 6). Mosquito-borne diseases are a major cause of human ill-health worldwide—each year there are hundreds of millions of cases of malaria and dengue, the two most common infections transmitted in this way. New and emerging infections also are a potential threat, as shown by the surprisingly rapid spread of West Nile virus in North America.

Mosquitoes cannot regulate their internal temperatures and therefore are exquisitely sensitive to external temperatures and moisture levels. Specifically, temperature influences the size of vector populations (via rates of growth and reproduction), infectivity (resulting from the effects on mosquito longevity and pathogen incubation) and geographical range. As a result, any comprehensive global study of the impacts of climate change must include possible effects on the global pattern of malaria, dengue and other mosquito-borne diseases. Climate change will be associated with changes in temperature, precipitation and possibly soil moisture, all factors that may affect disease prevalence. Yet mosquitoes are not only temperature dependent they are also very adaptable: these insects can adjust to adverse temperature and moisture by exploiting microenvironments such as containers and drains (and as a result can over-winter in places that are theoretically too cold for mosquitoes). Many other factors are important in transmission dynamics. For example, dengue fever is greatly influenced by house structure, human behaviour and general socioeconomic conditions. This is illustrated very well by the marked difference in the incidence of the disease above and below the United States–Mexico border: in the period 1980–1996, 43 cases were recorded in Texas compared to 50 333 in the three contiguous border states in Mexico (*40*).

There are few instances in which effects of long-term climate change on human health have been observed directly. Where there have been substantial changes in rates of mosquito-borne disease in recent times, there is little evidence that climate has played a major part (*41, 42*). This means that assessments of the impact of future climate change rely on expert judgement, informed by analogue studies, deduction from basic principles and modelling of health outcomes related to climate inputs.

The IPCC's Third Assessment Report concluded that rising temperatures and changing rainfall would have mixed effects on the potential for infections such as malaria and dengue worldwide (chapter 3). It reported that in areas with limited public health resources, warming in conjunction with adequate rainfall would likely cause certain mosquito-borne infections to move to higher altitudes (medium to high confidence) and higher latitudes (medium to low confidence). The IPCC concluded also that transmission seasons would be extended in some endemic locations (medium to high confidence).

CLIMATE CHANGE AND HUMAN HEALTH

There are other views than those of the IPCC. Some argue that simple climate change models provide no useful information about future disease rates because factors other than climate are bound to be more important (*43, 44*). Others conclude that there is sufficient evidence that global warming already is extending the geographical range of significant mosquito-borne diseases (*45*).

Why should scientific assessments that agree largely on evidence and methods come to rather different conclusions? One possible explanation is that the assessments are attempting to answer different questions. For example, the IPCC conclusion is couched explicitly in terms of disease potential (what would happen *if* certain conditions were to apply). Others, such as Reiter, have focused more strongly on predictions of actual disease incidence (not what *might*, but what *will* happen) (*43*).

The "what might happen" view is more wide-ranging, consistent with the position that science has a legitimate function in tackling "what if?" questions. According to this position, one objective of climate change research is not to propose testable long-range hypotheses rather to provide indicative forecasts to guide pre-emptive policy-making. Strictly speaking, no scientific study (not even the most tightly controlled experiment) can predict the consequences of a particular course of action and all research operates in the realm of what might happen. What is more, extrapolations from the research environment to the real world setting inevitably are hedged by conditions: it may not be explicit, but the conclusion always takes the form "*if* the setting in the laboratory was to be replicated, *then* the following outcomes would be expected . . ."

This is not to suggest that there should be only one way of carrying out climate research. One useful approach is the traditional hypothesis testing approach. Another is the "what if?" analysis—useful for risk management decisions as well as contributing to weight-of-evidence arguments. Both approaches have value but they may lead to different types of insights and information.

Assessments of climate change impacts are based, necessarily, on assumptions about the state of a future world. Any summary statement about the likely effects of a changed climate assumes a certain level of susceptibility, whether the status quo or what might be predicted from current trends, formal modelling, scenario analyses, or a worst-case scenario. An optimistic scenario would be a configuration of high disease control capacity and low population susceptibility (as a result of socioeconomic improvements, for example). Another assessment might include a range of possible futures, some of which are more disease-prone, such as settings in which public health services deteriorate, economic productivity declines and social order unravels.

Scale may be another factor that contributes to differences between scientists, adding to uncertainty for policy-makers. Factors that produce the most noticeable changes over a short time may not be the same as those that cause long-term changes in disease rates. Similar considerations apply to spatial distributions: malaria has been pantropical for centuries, essentially a function of climate. As noted already, recent retreats and advances of the disease are not due primarily to climate, but on a global scale they are movements on the margins. McMichael has applied this idea to the pool of disease, emphasizing the difference between factors that agitate the surface (short-term, localized variations) and the conditions that determine the depth of the pool (*46*). The latter class of causes also has been described as the driving force of disease incidence (*47*). The relative importance of different causes will depend partly on the scale that is chosen: what time and space-defining windows should be applied depend on the problem in hand.

An investigation of an outbreak of disease will naturally focus on the causes of short-term and localized variations in incidence, but an assessment of long-term future trends in disease is more likely to capture influences that occur upstream in the sequence of causes. If applying the outbreak frame of reference, then climate change is invisible because it is too big to fit into the study picture. This does not mean that climate change is irrelevant, rather that other approaches, on larger scales, are needed to comprehend this particular problem.

General issues concerning uncertainty

The IPCC raises the question (without giving an answer) of whether "science for policy" is different from "science itself". Researchers concerned with producing science for the purpose of informing decision-makers have additional challenges. They need to consider the specific questions asked by particular decision-makers and the time frame in which the questions must be answered. Their research must focus on answering those questions—as best they can—in a timely fashion. They also must characterize all relevant uncertainties and the implications of these for the decisions under consideration. Whether or not the science is able to inform policy, decisions *will* be made under uncertainty. Fundamentally, an informed decision is preferred to an uninformed decision. Are there ways of retaining quantitative estimates, but tempering them with a full description of uncertainty bounds? Possible approaches included in the IPCC report are standardized graphical displays (*1*) and verbal summaries of confidence categories.

The approach taken depends on the information needs of the decision-makers (e.g. public health officials), how the information most usefully can be conveyed to them and the questions they want answered by the assessors. The issue is not so much whether there are different kinds of science or different standards of proof. Really it is a question of the types of methods that exist to inform decision-makers who must make decisions despite the existence of uncertainties. An alternative, perhaps more constructive and useful approach, would be to ask the question: "What would one have to believe is true for the following to happen?" By asking this, there is no judgment on the relative likelihood of an outcome. Rather, the decision-maker is provided with insights into the necessary conditions for an outcome to occur and can then decide the likelihood of those conditions for the particular situation being addressed—and the level of risk.

Studies of perceptions of risk show that values and social positions have an important bearing on the way individuals view environmental hazards (*48*). For this reason it is important to engage stakeholders in the problem formulation phase of an assessment process (to help shape the questions that will be the focus of the assessment), the assessment itself and communication of the results.

Van Asselt and Rotmans liken risk assessment to a group of hikers crossing an unfamiliar landscape (*49*). Although the hikers start together and face the same terrain, it would not be surprising if they choose different routes and, as a result, come to different destinations. The choices made along the way (to cross a dangerous river or take a lengthy diversion, follow footpaths or forge new routes, continue when the weather is threatening or take shelter) depend on past experiences, preferences, interests and preconceptions about the nature of the land ahead.

In a similar manner, no one begins an assessment of the effects of climate change with a completely open mind. Scientists bring to the task expectations, attitudes and values that influence the questions asked, help to make sense of the data and inevitably, shape the meaning given to results. Important dimensions of difference include presumptions about nature (which might be viewed as capricious, benign, forgiving, fragile or any combination of these qualities), the limits of human capacities and priorities given to core values (such as equity and individual liberties). Even between disciplines there may be major differences. A survey of experts in the field found that natural scientists' estimates of the total damage caused by climate change tended to be much greater than those of economists. It was suggested "economists know little about the intricate web of natural ecosystems, whereas scientists know equally little about the incredible adaptability of human economies" (50). This is why it is so important to engage different disciplines in the assessment process—and have them converse with one another.

Do multi-disciplinary assessments such as those carried out by the IPCC, underestimate uncertainty? In theory, they might. Collaborations like the IPCC have important strengths: no one discipline holds all the knowledge required to deal with complex environmental problems like climate change. Multi-disciplinary assessments may capture more uncertainties because one discipline may identify matters never thought of by another discipline. For example, an ecologist might identify health outcome sensitive considerations that a public health expert would not. This might increase, not decrease, uncertainties. Conversely, a group with many disciplines may find it more difficult to provide a complete account of uncertainties than is possible when all the authors speak the same technical language. Contingencies tend to be overlooked when scientific debates move from specialized forums to public settings and a similar process has been observed with scientific panels in the past. Wynne states: "Doubts and uncertainties of core specialists are diminished by the overlaps and interpenetrations with adjacent disciplines ... the net result is a more secure collective belief in the policy knowledge, or the technology, than one might have obtained from any of the separate contributing disciplines" (51).

Conclusions

Researchers seek to understand ways in which weather and climate may affect human health and (where they exist) to estimate the size, timing, and character of such effects. For public health scientists and officials the primary goal is to prevent any increases in disease associated with changing weather and climate, while recognizing that the scarce resources available to the public health community may have to be diverted to other higher-priority public health problems. Both scientists and policy-makers are interested in the magnitude of potential effects and their distribution—which regions and which populations most likely will be affected? Why? When? How can vulnerability be reduced and adaptive capacity increased?

The research community has provided some answers to these important questions (including those summarized by the IPCC) but still there is much to be done. It is not just that understanding of the science is never complete. In this instance policy choices will need to be made before all relevant scientific information is available. It would be convenient if a decision on re-setting the global thermostat could be delayed until the health costs and benefits of a warmer world were known. However, climate systems cannot be turned on and off like an air conditioning unit. The considerable lag period between greenhouse emissions and climate impacts, and the large inertia in the climate system's response to perturbations, means that policy-makers must make decisions many years before the full effects are apparent. This means extra pressures on researchers to provide robust scientific advice on climate change and health, as early as possible.

References

1. McCarthy, J.J. et al. *Climate change 2001: impacts, adaptation, and vulnerability.* Contribution of Working Group II to the Third Assessment Report of the Intergovernmental Panel on Climate Change. Cambridge, UK, Cambridge University Press, 2001.
2. Blakely, T. & Woodward, A. Ecological effects in multi-level studies. *Journal of Epidemiology and Community Health* 54: 367–374 (2000).
3. Klinenberg, E. *Heat wave. A social autopsy of disaster in Chicago.* Chicago, USA, University of Chicago Press, 2002.
4. Chestnut, L.G. et al. Analysis of differences in hot-weather-related mortality across 44 United States metropolitan areas. *Environmental Science and Policy* 1: 59–70 (1998).
5. Bentham, G. & Langford, I. H. Climate change and the incidence of food poisoning in England and Wales. *International Journal of Biometeorology* 39(2): 81–86 (1995).
6. Walther, G. et al. Ecological responses to recent climate change. *Nature* 416: 389–395 (2002).
7. Nicholls, N. Increased Australian wheat yield due to recent climate trends. *Nature* 387: 484–485 (1997).
8. Lindgren, E. & Gustafson, R. Tick-borne encephalitis in Sweden and climate change. *Lancet* 358 (9275): 16–87 (2001).
9. Patz, J.A. & Lindsay, S.W. New challenges, new tools: the impact of climate change on infectious diseases. *Current Opinion in Microbiology* 2(4): 445–451 (1999).
10. Saez, M. et al. Relationship between weather temperature and mortality: a time series analysis approach in Barcelona. *International Journal of Epidemiology* 24: 576–582 (1995).
11. Bouma, M.J. et al. Global assessment of El Niño's disaster burden. *Lancet* 350 (9089): 1435–1438 (1997).
12. Kunkel, K.E. et al. Temporal fluctuations in weather and climate extremes that cause economic and human health impacts: a review. *Bulletin American Meteorological Society* 80(6): 1077–1098 (1999).
13. Katsouyanni, K. et al. Confounding and effect modification in the short-term effects of ambient particles on total mortality: results from 29 European cities within the APHEA2 project. *Epidemiology* 12(5): 521–531 (2001).
14. Rotmans, J. & Dowlatabadi, H. Integrated assessment modeling. In: *Human Choice and Climate Change, Vol. 3: The tools for policy analysis.* Rayner, S. & Malone, E. eds. Columbus, Ohio, USA, Batelle Press, 1998.
15. Parry, M. et al. Millions at risk. Defining critical climate change targets and threats. *Global Environmental Change* 11: 1–3 (2001).
16. Slaper, H. et al. Estimates of ozone depletion and skin cancer incidence to examine the Vienna Convention achievements. *Nature* 384 (6606): 256–258 (1996).
17. Chan, N.Y. et al. An integrated assessment framework for climate change and infectious diseases. *Environmental Health Perspectives* 107: 329–337 (1999).
18. Bernard, S. & Ebi, K.L. Comments on the process and product of the health impacts assessment component of the national assessment of the potential consequences of climate variability and change for the United States. *Environmental Health Perspectives* 109: supplement 2: 177–184 (2001).
19. Aron, J.L. et.al. Integrated assessment. In: *Ecosystem change and public health. A global perspective.* Aron, J.L. & Patz, J. eds. Baltimore, USA, John Hopkins Press: pp116–162, 2001.

20. Hobbs, B.F. & Meier, P. *Energy decisions and the environment: a guide to the use of multicriteria decision methods.* Norwell, Massachusetts, USA, Kluwer Academic Press, 2000.

21. Black, F.L. Why did they die? *Science* 258: 1739–1740 (1992).

22. Naeem, S. et al. Declining biodiversity can alter the performance of ecosystems. *Nature* 368: 734–737 (1994).

23. Tilman, D. & Downing, J.A. Biodiversity and stability in grasslands. *Nature* 367: 363–365 (1994).

24. Lovins, A. Energy, people and industrialisation. In: *Resources, environment and population.* Davis, K. & Bernstam, M.S. eds. New York, USA, Oxford University Press: 95–124, 1991.

25. Nigg, J.M. *Disaster recovery as a social process.* Wellington, New Zealand, The New Zealand Earthquake Commission, 1995.

26. Watts, J. Eight million threatened with food crisis in North Korea. *Lancet* 356: 1993 (2000).

27. Sen, A. *Poverty and famines: an essay on entitlement and deprivation.* Oxford, UK, Clarendon Press, 1980.

28. Woodward, A. et al. Climate change and human health in the Asia Pacific: who will be most vulnerable? *Climate Research* 11: 31–38 (1998).

29. Scheraga, J.D. & Grambsch, A. Risks, opportunities, and adaptation to climate change. *Climate Research* 10: 85–95 (1998).

30. Cifuentes, L. et al. Assessing the health benefits of urban air pollution reductions associated with climate change mitigation (2000–2020): Santiago, Sao Pãulo, Mexico City, and New York City. *Environmental Health Perspectives* 109, supplement 3: 419–425 (2001).

31. National Research Council. *Risk assessment in the federal government: managing the process.* Washington, USA, National Academy Press, 1983.

32. Herrick, C. & Jamieson, D. The social construction of acid rain: some implications for science/policy assessment. *Global Environmental Change* 5(2): 105–112 (1995).

33. Brunk, C.G. et al. *Value assumptions in risk assessment. A case study of the Alachlor controversy.* Waterloo, Ontario, Canada, Wilfrid Laurier University Press, 1991.

34. National Research Council. *Science and judgement in risk assessment.* Washington, USA, National Academy Press, 1994.

35. National Research Council. *Understanding risk: informing decisions in a democratic society.* Washington, USA, National Academy Press, 1996.

36. National Research Council. *Toward environmental justice: research, education and health policy.* Washington, USA, National Academy Press, 1999.

37. National Research Council. *The science of regional and global change: putting knowledge to work.* Washington, USA, National Academy Press, 2001.

38. Scheraga, J.D. & Furlow, J.J. From assessment to policy: lessons learned from the US National Assessment. *Human and Ecological Risk Assessment* 7(5) (2001).

39. Schneider, S.H. et al. Imaginable surprise in global change science. *Journal of Risk Research* 1(2): 165–185 (1998).

40. Patz, J.A. & McGeehin, M.A. The potential health impacts of climate variability and change for the United States: executive summary of the report of the health sector of the U.S. national assessment. *Environmental Health Perspectives* 108(4): 367–376 (2000).

41. Kovats, R.S. et al. Early effects of climate change: do they include changes in vector-borne disease? *Philosophical Transactions of the Royal Society of London Series B* 356: 1–12 (2001).

42. Reiter, P. Climate change and mosquito-borne disease. *Environmental Health Perspectives* 109, supplement 1: 141–161 (2001).

43. Reiter, P. Malaria and global warming in perspective? *Emerging Infectious Diseases* 6(4): 438–439 (2000).

44. Masood, E. Biting back. *New Scientist* 167 (2257): 41–42 (2000).

45. Epstein, P.R. Is global warming harmful to health? *Scientific American* 283(2): 50–57 (2000).
46. McMichael, A.J. The health of persons, populations and planets: epidemiology comes full circle. *Epidemiology* 6: 633–636 (1995).
47. Corvalán, C.F. et al. Health, environment and sustainable development: identifying links and indicators to promote action. *Epidemiology* 10(5): 656–660 (1999).
48. Flynn, J. et al. Gender, race and perception of environmental health risks. *Risk Analysis* 14: 1101–1108 (1994).
49. van Asselt, M.B.A. & Rotmans, J. Uncertainty in perspective. *Global Environmental Change* 6: 121–157 (1996).
50. Nordhaus, W.D. Expert opinion on climatic change. *American Scientist* 82 (January–February): 45–51 (1994).
51. Godard, O. *Integrating scientific expertise into regulatory decision-making.* Florence, Italy, European University Institute, 1996.

CHAPTER 5

Impacts on health of climate extremes

S. Hales,[1] S.J. Edwards,[2] R.S. Kovats[2]

Introduction

Extreme climate events are expected to become more frequent as a result of climate change. Climate extremes can have devastating effects on human societies. History records widespread disasters, famines and disease outbreaks triggered by droughts and floods. These complex, large-scale disruptions exert their worst effects in poor countries but even the richest industrial societies are not immune. Extreme weather events are, by definition, rare stochastic events. There are two categories (1):

- extremes based on simple climate statistics, such as very low or very high temperatures;
- more complex, event driven extremes: droughts, floods, or hurricanes—these do not necessarily occur every year at a given location.

With climate change, even if the statistical distribution of simple extreme events remains the same, a shift in the mean will result in a non-linear change in the frequency of extreme events. The detection of change in simple climate extremes is more likely than the detection of changes in event-driven extremes.

Climate variability can be expressed at various temporal scales (by day, season and year) and is an inherent characteristic of climate, whether or not the climate system is subject to change. Much attention has focused on the influence of El Niño-Southern Oscillation (ENSO) on weather patterns in many parts of the world. In sensitive regions, ENSO events may cause significant inter-annual perturbations in temperature and/or rainfall within a loose 2–7 year cycle. However, it is important that such perturbations are not confused with climate change. In reality, these fluctuations introduce more noise into the long-term trends, making it more difficult to detect the climate change signal.

The effect of climate change on the frequency and/or amplitude of El Niño is uncertain. However, even with little or no change in amplitude, climate change is likely to lead to greater extremes of drying and heavy rainfall and increase the risk of droughts and floods that occur with El Niño in many regions.

A range of physical, ecological and social mechanisms can explain an association between extremes of climate and disease (Figure 5.1, Table 5.1 and Table 5.2). Social mechanisms may be very important but are difficult to quantify: for example, droughts and floods often cause population displacement. Outbreaks of infectious disease are common in refugee populations due to inadequate public

---

[1] Wellington School of Medicine, University of Otago, Wellington, New Zealand.
[2] London School of Hygiene and Tropical Medicine, London, England.

FIGURE 5.1 ENSO and disease. ENSO events cause physical effects such as droughts and floods (blue circle). Where these overlap and interact with suitable ecological and socioeconomic conditions (within dotted lines) they may cause disease outbreaks (dark shaded area).

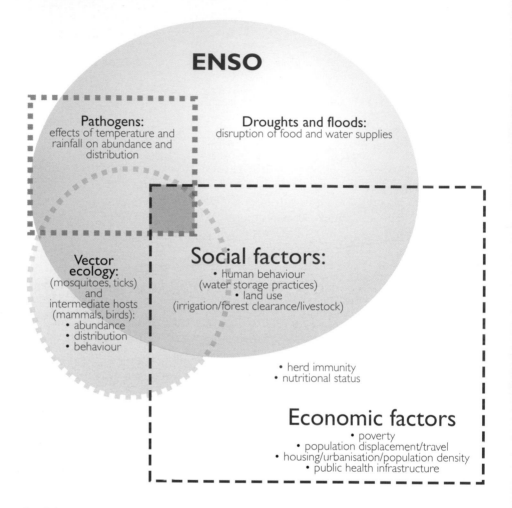

health infrastructure, poor water and sanitation, overcrowding and lack of shelter. Climate also can affect infectious diseases that are spread via contaminated water or food. Water-related diseases are a particular problem in poor countries and communities, where water supplies and sanitation often are inadequate. Outbreaks of cholera, typhoid and diarrhoeal diseases can occur after flooding if the floodwaters become contaminated with human or animal waste, while drought reduces the water available for washing and sanitation and also tends to increase the risk of disease.

There is a web of interactions between ecosystems, climate and human societies, which influences the occurrence of infections. For example, the resurgence of communicable diseases in the past few decades is thought to have resulted primarily from social factors including population growth, urbanization, changes in land use and agricultural practices, deforestation, international travel and breakdown in public health infrastructure (3). From the opposite perspective, major communicable diseases such as malaria also can severely limit social development (4).

TABLE 5.1 Mechanisms by which above-average rainfall can affect health.

Event	Type	Description	Potential health impact
Heavy precipitation event	meteorological	"extreme event"	increased mosquito abundance or decreased (if breeding sites are washed away)
Flood	hydrological	river/stream over tops its banks	changes in mosquito abundance contamination of surface water
Flood	social	property or crops damaged	changes in mosquito abundance contamination of water with faecal matter and rat urine (leptospirosis).
Flood	catastrophic flood /"disaster"	Flood leading to >10 killed, and/or 200 affected, and/or government call for external assistance.	changes in mosquito abundance contamination of water with faecal matter and rat urine and increased risk of respiratory and diarrhoeal disease deaths (drowning) injuries health effects associated with population displacement loss of food supply psychosocial impacts

Source: reproduced from reference (2).

TABLE 5.2 Mechanisms by which below-average rainfall can affect health.

Event	Type	Description	Potential health impact
Drought	meteorological	evaporation exceeds water absorption, soil moisture decreases. Several indices have been developed based on meteorological variables, e.g. Palmer Drought Severity Index.	changes in vector abundance if vector breeds in dried up river beds, for example.
Drought	agricultural	drier than normal conditions leading to decreased crop production	depends on socioeconomic factors, i.e. other sources of food available and the means to acquire them.
Drought	social	reduction in food supply or income, reduction in water supply and quality	food shortage, illness, malnutrition (increases risk of infection) increased risk of disease associated with lack of water for hygiene.
Drought	food shortage/ famine/drought disaster	food shortage leading to deaths >10 killed, and/or 200 affected, or government call for external assistance.	deaths (starvation) malnutrition (increases risk of infection) health impacts associated with population displacement

Source: reproduced from reference (2).

This chapter summarizes what is known about the historical effects of climate extremes on human health. The following section describes studies of infectious diseases and climate extremes related to El Niño Southern Oscillation. The next considers the impacts of short-term extremes of temperature. The final section contains a discussion of climate-related disasters.

El Niño and infectious diseases

There is a well-studied relationship between rainfall and diseases spread by insect vectors which breed in water, and are therefore dependent on surface water availability. The main species of interest are mosquitoes, which spread malaria and viral diseases such as dengue and yellow fever. There is considerable evidence linking mosquito abundance to rainfall events. Mosquitoes need access to stagnant water in order to breed—conditions that may be favoured by both wet and dry conditions. For example, heavy rain can create as well as wash away breeding sites, while in normally wet regions drought conditions can increase breeding sites by causing stagnation of water in rivers. The timing of rainfall in the year and the co-variation of other climate factors also are likely to be important.

Vector-borne disease transmission is sensitive to temperature fluctuations also. Increases in temperature reduce the time taken for vector populations to breed. Increases in temperature also decrease the incubation period of the pathogen (e.g. malaria parasite, dengue or yellow fever virus) meaning that vectors become infectious more quickly (5). On the other hand (depending on thresholds that are species-specific) hot, dry conditions can reduce the lifetime of mosquitoes. Temperature also may affect the behaviour of the vector and human populations, affecting the probability of transmission. Warmer temperatures tend to increase biting behaviour of the vector and produce smaller adults which may require multiple blood meals in order to reproduce.

Malaria

Malaria is the world's most important vector-borne disease. Over 2.5 billion people are at risk, and there are estimated to be 0.5 billion cases and more than 1 million deaths from malaria per year (6). Malaria incidence is influenced by the effectiveness of public health infrastructure, insecticide and drug resistance, human population growth, immunity, travel, land-use change and climate factors.

Very high temperatures are lethal to the mosquito and the parasite. In areas where temperatures are close to the physiological tolerance limit of the parasite, a small temperature increase would be lethal to the parasite and malaria transmission would therefore decrease. However, at low temperatures a small increase in temperature can greatly increase the risk of malaria transmission (7).

Malaria's sensitivity to climate is illustrated in desert and highland fringe areas where rainfall and temperature, respectively, are critical parameters for disease transmission (8). In these regions higher temperatures and/or rainfall associated with El Niño may increase transmission of malaria. In areas of unstable malaria in developing countries, populations lack protective immunity and are prone to epidemics when weather conditions facilitate transmission. Across the globe, many such areas experience drought or excessive rainfall during ENSO events.

Drought in the previous year has been identified as a factor contributing to increased malaria mortality. There are several possible reasons for this relationship. Drought-related malnutrition may increase an individual's susceptibility to infection (9). Also, drought may reduce malaria transmission resulting in a reduction in herd immunity in the human population. Therefore, in the subsequent year the size of the vulnerable population is increased (10).

Alternatively, a change in ecology of the natural predators may affect mosquito vector dynamics; mosquito populations recover more quickly than their predator populations following a dry year. Famine conditions may have contributed to excess mortality during historical epidemics of malaria, for example following the 1877 El Niño in India. Many deaths occurred after the end of the drought; the proximate cause was malaria when drought-breaking rains increased vector abundance, exacerbated by population movement and the concentration of people in feeding camps (11).

Many parts of South America show ENSO-related climate anomalies. Serious epidemics in the northern countries of South America have occurred mainly in the year after El Niño (year +1). In 1983 following a strong El Niño event, Ecuador, Peru and Bolivia experienced malaria epidemics (12, 13, 14). In Venezuela and Colombia, malaria increased in the post-Niño year (+1) (10, 15, 16, 17). A statistically significant relationship was found between El Niño and malaria epidemics in Colombia, Guyana, Peru, and Venezuela (18). The causal mechanisms are not completely understood. El Niño is associated with a reduction of the normal high rainfall regime in much of Colombia, as well as an increase in mean temperature, increase in dew point, and decrease in river discharges (17). These relationships between malaria and ENSO nevertheless can be used to predict high and low-risk years for malaria, giving sufficient time to mobilise resources to reduce the impact of epidemics (15).

Africa has desert fringe malaria around the Sahara (e.g. the Sudan) and the Kalahari (Namibia, Botswana). Of these areas, southern Africa and a region east of the Sahara show ENSO-related rainfall anomalies. Several recent studies have examined evidence of relationships between climate extremes and malaria in Africa (19, 20, 21).

The 1997/98 El Niño was associated with heavy rainfall and flooding in Kenya, after two years of drought. From January to May 1998, a major epidemic of falciparum malaria occurred. Brown et al. (19) reported an attack rate of approximately 40% in the town of Wajir, Kenya. Three districts in Kenya reported a six-fold increase in malaria cases in the first two months of 1998 compared to the same period in 1997 (22). The malaria epidemic was compounded by widespread food shortages.

Other researchers emphasize the significance of non-climate factors in explaining recent malaria epidemiology in Africa (23). A resurgence of malaria in the highlands of Kenya over the past 20 years has been attributed to resistance to antimalarial drugs (24). Another study did not find a relationship between climate trends and the timing of malaria epidemics in Kenya. Based on a 30-year time series of climate and disease data, it concluded: ". . . intrinsic population dynamics offer the most parsimonious explanation for the observed interepidemic periods" (25). One study has reported no significant meteorological trends in four high-altitude sites in East Africa where increases in malaria have been reported (26). This study used spatially averaged climate data that may be unreliable for this purpose. An association between rainfall, temperatures and the number of inpatient malaria cases three to four months later has been reported recently (27).

Dengue

Dengue is the most important arboviral disease of humans, occurring in tropical and subtropical regions worldwide. In recent decades, dengue has become an increasing urban health problem in tropical countries. The disease is thought to have spread mainly as a result of ineffective vector and disease surveillance; inadequate public health infrastructure; population growth; unplanned and uncontrolled urbanization; and increased travel (28, 29). The main vector of dengue is the domesticated mosquito, *Aedes aegypti*, that breeds in urban environments in artificial containers that hold water. Dengue also can be transmitted by *Aedes albopictus*, which can tolerate colder temperatures.

Dengue is seasonal and usually associated with warmer, more humid weather. There is evidence that increased rainfall in many locations can affect the vector density and transmission potential. ENSO may act indirectly by causing changes in water storage practices brought about by disruption of regular supplies (5). Rainfall may affect the breeding of mosquitoes but this may be less important in urban areas: *Aedes aegypti* breed in small containers, such as plant pots, which often contain water in the absence of rain.

Between 1970 and 1995, the annual number of epidemics of dengue in the South Pacific was positively correlated with the Southern Oscillation Index (SOI) (30). This is plausible since, in this part of the world, high positive values of the SOI (denoting La Niña conditions) are associated with much warmer and wetter conditions than the average—ideal for breeding of mosquitoes. In a subsequent study, Hales et al. examined the relationship between ENSO and monthly reports of dengue cases in 14 island nations in the Pacific (31). There were positive correlations between SOI and dengue in ten countries. In five of these (American Samoa, Nauru, Tokelau, Wallis and Western Samoa) there were positive correlations between SOI and local temperature and/or rainfall. During La Niña, these five islands are likely to experience wetter and warmer than normal conditions. Local weather patterns may trigger an increase in transmission in larger, more populated islands where the disease is endemic, but infected people then carry the disease to smaller neighbouring islands. This implies that the effect of climate on vector-borne diseases is not necessarily confined to the region affected by altered climate, suggesting that forecasts may need to take account of regional social and environmental factors too.

A study of dengue in Viet Nam, found that the number of cases increased in El Niño years (32). In Thailand, which does not have a strong ENSO signal, there was no correlation (25). Many countries in Asia experienced an unusually high level of dengue and dengue haemorrhagic fever in 1998, some of which may be attributable to El Niño-related weather (5). Gagnon et al. reported positive associations between El Niño and dengue epidemics in French Guyana, Indonesia, Colombia and Suriname, regions that experience warmer temperatures and less rainfall during El Niño years (33).

These studies do not identify unequivocally the environmental risk factors for increases in dengue cases. Further regional or global scale studies are needed to determine whether El Niño is associated with a change in dengue activity and if so, what climate parameters (temperature, rainfall, humidity, sea level or wind velocity) primarily are responsible.

Rodent-borne diseases

Rodents act as reservoirs for a number of diseases whether as intermediate infected hosts or as hosts for arthropod vectors such as ticks. Certain rodent-borne diseases are associated with flooding including leptospirosis, tularaemia and viral haemorrhagic diseases. Other diseases associated with rodents and ticks include plague, Lyme disease, tick borne encephalitis (TBE) and hantavirus pulmonary syndrome (HPS).

Rodent populations have been shown to increase in temperate regions following mild wet winters (34). One study found that human plague cases in New Mexico occurred more frequently following winter-spring periods with above-average precipitation (35). These conditions may increase food sources for rodents and promote breeding of flea populations. Ticks also are climate sensitive.

Infection by hantaviruses mainly occurs from inhalation of airborne particles from rodent excreta. The emergence of the disease hantavirus pulmonary syndrome in the early 1990s in the southern United States has been linked to changes in local rodent density (36). Drought conditions had reduced populations of the rodents' natural predators; subsequent high rainfall increased food availability in the form of insects and nuts. These combined effects lead to a tenfold increase in the population of deer mice from 1992 (36) to 1993. In 1998, an increase in cases of hantavirus was linked to increased rodent populations which, in turn, were attributed to two wet, relatively warm winters in the southern United States associated with 1997/98 El Niño (37, 38). A comprehensive study by Engelthaler et al. in the Four Corners region, USA, concluded that above-average precipitation during the winter and spring of 1992–1993 may have increased rodent populations and thereby increased contact between rodents and humans and viral transmission (39).

Diarrhoeal illness

Many enteric diseases show a seasonal pattern, suggesting sensitivity to climate. In the tropics diarrhoeal diseases typically peak during the rainy season. Floods and droughts are each associated with an increased risk of diarrhoeal diseases, although much of the evidence for this is anecdotal. The suggestion is plausible, however, since heavy rainfall can wash contaminants into water supplies, while drought conditions can reduce the availability of fresh water leading to an increase in hygiene-related diseases.

Major causes of diarrhoea linked to contaminated water supplies are: cholera, cryptosporidium, *E.coli*, giardia, shigella, typhoid, and viruses such as hepatitis A. Outbreaks of cryptosporidiosis, giardia, leptospirosis and other infections have been shown to be associated with heavy rainfall events in countries with a regulated public water supply (40, 41, 42, 43, 44, 45).

An association between drinking water turbidity and gastrointestinal illness has been reported (46). This was one of the first studies to apply time series methods to the analysis of water-related disease. A study of waterborne disease outbreaks in the United States has shown that about half were significantly associated with extreme rainfall (41). Outbreak locations from an Environmental Protection Agency database were assigned to watersheds. The rainfall in the month of the outbreak and in previous months was estimated from climate records: for outbreaks associated with surface water the association was strongest for rainfall events in the same month as the outbreak.

Transmission of enteric diseases may be increased by high temperatures, via a direct effect on the growth of disease organisms in the environment (47, 48, 49). In 1997 a markedly greater number of patients with diarrhoea and dehydration were admitted to a rehydration unit in Lima, Peru, when temperatures were higher than normal during an El Niño event (50). A time series analysis of daily data from the hospital confirmed an effect of temperature on diarrhoea admissions, with an estimated 8% increase in admissions per 1 °C increase in temperature (51).

Analysis of average diarrhoea reports in the Pacific Islands (1978–1986) suggested a positive association with average temperature and an inverse association with estimated water availability (52). Time series analysis of diarrhoea reports in the islands of Fiji (1978–1992) confirmed a statistically significant effect of monthly temperature changes (an estimated 3% increase in diarrhoea reports per 1 °C increase in temperature). Extremes of rainfall also were associated with increases in diarrhoea (52).

In summary, there is good evidence of associations between several important communicable diseases and climate on several temporal and geographical scales. This is true of vector-borne diseases, many enteric illnesses and certain water-related diseases. These associations are not found everywhere—hardly surprising given the complexity of the causal pathways involved. Relationships between year-to-year variations in climate and communicable diseases are most evident where these climate variations are marked, and in vulnerable populations in poor countries. Major scientific reviews agree that El Niño can provide a partial analogue for the effects of global climate change on communicable diseases (53). However, the Intergovernmental Panel on Climate Change (IPCC) cautions:

"Policymakers should appreciate that although our scientific capacity to foresee and model these various health outcomes of climate change continues to evolve, it is not possible to make precise and localized projections for many health outcomes . . ." (3).

Temperature extremes: heatwaves and cold spells

In recent years there has been a great increase in interest in time series studies of temperature and mortality. These are seen as the most satisfactory method for

---

BOX 5.1 Impact of climate extremes on malaria in Irian Jaya

Beginning in late August 1997, a significant increase of unexplained deaths was reported from the central highland district of Jayawijaya. The alarming number of fatalities rapidly escalated into September, dropping off precipitously by late October. More than 550 deaths due to "drought-related" disease had been officially reported from the district during this 10-week period. The outbreaks occurred in extremely remote areas of steep mountainous terrain inhabited by shifting agriculturist populations.

Microscopic evidence and site survey data implicated malaria as the principal cause of the excess morbidity and mortality at elevations between approximately 1000 and 2200 m. The dramatic increase in malaria and associated deaths was related indirectly to the prolonged and severe drought created by the prevailing 1997–98 El Niño affecting the Australasian region.

Clinical cases of malaria were described as severe, due in large part to the low level of naturally acquired immunity in these highland populations and the predominance of *Plasmodium falciparum* infection. Disease may have been further exacerbated by the population's compromised nutritional status because of drought-related severe shortages of staple foods. Based on a retrospective investigation, an *a posteriori* epidemiological explanation of the probable interrelated causes of the epidemic is presented:

"Beginning in late July 1997, drought conditions resulted in numerous transient pools of standing water along zones of steep gradient streams normally associated with fast-flowing water. This permitted sufficient and rapid increases in vector populations (*Anopheles punctulatus* complex) that could sustain recently introduced or intensified local low-level malaria transmission. Moreover, water and food shortages contributed to increased demographic movement and exposure to high risk malaria endemic lowlands, thus increasing the prevalence of human infections and infectious reservoirs in those populations returning to the highlands."

Source: Based on reference (54)

CLIMATE CHANGE AND HUMAN HEALTH

BOX 5.2 Cholera

Traditionally cholera is viewed as a strictly faecal–oral infection but increased attention is being paid to the environmental determinants of this disease. The discovery of
a marine reservoir of the cholera pathogen and its long term persistence with various
marine organisms (in the mucilaginous sheath of blue-green algae and copepods) helps
to explain the endemicity in certain regions, such as the estuaries of the Ganges and
Bramaputra in Bangladesh (55). Recent work has suggested links between the seasonality of cholera epidemics and seasonality of plankton (algal blooms) and the marine
food chain. A study of *Vibrio cholerae* 01 in Bangladesh (1987–90) found that abundance increases with the abundance of copepods (which feed on phytoplankton) in
coastal waters (55). Analysis of cholera data from Bangladesh showed that the
temporal variability of cholera exhibits an interannual component at the dominant
frequency of El Niño (56, 57).

Several cholera outbreaks occurred in 1997 following heavy rains. Countries in East
Africa were severely affected: major cholera outbreaks occurred in the United Republic of Tanzania, Kenya, Guinea-Bissau, Chad and Somalia (2, 58). Outbreaks also were
reported in Peru, Nicaragua and Honduras (59, 60). However, the total number of
cholera cases reported to WHO in 1997, globally and by region, was similar to that
in 1996. Countries that experienced increased cholera incidents in 1997 are at risk
of increases in cholera in subsequent years. In 1997, the regional WHO cholera surveillance team was aware of the forecasts of an El Niño-related drought In south-east
Africa. The team was able to institute measures to help reduce the severity of a cholera
outbreak in Mozambique by increased monitoring and heightened preparedness of
health care institutions (60).

quantifying the short-term associations between ambient temperatures and daily
mortality. Any long-term patterns in the series (e.g. seasonal cycles) are removed.
The effect of a hot day is apparent only for a few days in the mortality series; in
contrast, a cold day has an effect that lasts up to two weeks. In many temperate
countries mortality rates in winter are 10–25% higher than death rates in
summer but the causes of this winter excess are not well understood (61).

It is likely that different mechanisms are involved in heat and cold related
mortality; cold related mortality in temperate countries is related in part to the
occurrence of seasonal respiratory infections. High temperatures cause some
well-described clinical syndromes such as heatstroke (62). Very few deaths are
reported as attributed directly to heat. Exposure to high temperatures increases
blood viscosity and it is plausible that heat stress may trigger a vascular event
such as heart attack or stroke (63). Studies have shown that elderly people have
impaired temperature regulation (62, 64, 65, 66). Physiological studies in the
elderly indicate that low temperatures are associated with increased blood pressure and fibrinogen levels (67, 68).

The impact of heatwave events on mortality

Heatwaves can kill. In July 1995 a heatwave in Chicago, USA, caused 514 heat-
related deaths (12 per 100 000 population) and 3300 excess emergency admissions (69). The morgues were full and bodies had to be stored in refrigerated
trucks. From 12 to 20 July, daily temperatures ranged from 34–40 °C, with the

highest temperatures on 13 July. The maximum number of deaths occurred on 15 July (70).

During heatwaves, excess mortality is greatest in the elderly and those with pre-existing illness (71). Much of this excess mortality is due to cardiovascular, cerebrovascular and respiratory disease. The mortality impact of a heatwave is uncertain in terms of the amount of life lost: a proportion of the deaths occur in susceptible persons who were likely to have died in the near future. Nevertheless, there is a high level of certainty that an increase in the frequency and intensity of heatwaves would increase the numbers of additional deaths due to hot weather.

There is no standard international definition of a heatwave. Operational definitions are needed for meteorological services. As meteorological agencies are becoming more commercialized they are keen to develop practical applications of their forecasts and tailor them to user needs. The Netherlands meteorological bureau uses the following definition to trigger advance warnings in the media and directly to health services: at least 5 days with maximum temperature above 25 °C of which at least 3 days with maximum temperature above 30 °C. The evidence on which this is based is not clear. In the United States, the National Weather Service suggest that a heat advisory (early warning) be issued when the daytime heat index reaches 40.6 °C and a night time minimum temperature of 26.7 °C persists for at least 48 hours (72). Local definitions are used: in Dallas the medical examiners office define a heatwave as three consecutive days of temperatures over 37.8 °C.

It is surprisingly difficult to define a heatwave as responses to very high temperatures vary between populations and within the same population over time. A 1987 heatwave in Athens resulted in 926 deaths classified as heat-related, although the attributable excess mortality was estimated to be more than 2000 (73). A subsequent heatwave in 1988 was associated with a much smaller excess mortality. This has been observed also in Chicago following the 1995 heatwave (74).

Few analyses have looked at the impacts of heatwaves in developing countries and the evidence is largely anecdotal. A heatwave in India in June 1998 was estimated to have caused 2600 deaths over 10 weeks of high temperatures (75). In Ores, the temperature rose to 49.5 °C and was reported to have caused 1300 deaths. The high temperatures were exacerbated by recurrent power failures that affected cooling systems and hospital services in Delhi.

Important behavioural factors may be specific to certain countries: in Japan, young children are often affected when left in motor vehicles. Not all heat related deaths are due to weather conditions. For example, in the United States in 1994, 221 heat related deaths were recorded, but only 101 (46%) were due to ambient weather conditions. The rest were due to overexertion during exercise, for example. During the period 1979–1994, heat-related mortality due to weather conditions was 2.7–3.7 per million population in the four highest reporting states (Arizona, Arkansas, Kansas, and Missouri) (72). Most of these deaths occurred in the over-55 age group. Overall the impact of mortality is underestimated because death rates from other diseases increase during heatwaves. This is true in all populations where it has been investigated.

Rooney et al. estimated the excess mortality associated with the 1995 heatwave in the United Kingdom (76). An estimated 619 extra deaths (8.9% increase) were observed relative to the expected number of deaths, based on the 31-day moving average for that period. Excess deaths were apparent in all age groups but most marked in females and for deaths from respiratory and cerebrovascular disease. A heatwave in Belgium in 1994 was associated with excess mortality. Part of the excess was due to mortality displacement, since there was a deficit in deaths in the elderly following the heatwave (no deficit was apparent for age group 0–64 years) (77).

Vulnerability to temperature-related mortality

Indicators of vulnerability to heat and cold that have been investigated include:

- age and disease profile
- socioeconomic status
- housing conditions
- prevalence of air conditioning
- behaviour (e.g. clothing).

These factors also have counterparts in individuals as risk factors for heat related mortality or morbidity, such as presence of air conditioning at time of death.

Both individual and population level studies provide strong and consistent evidence that age is a risk factor for heat-related mortality. Studies vary on the age at which the vulnerability is increased. There are physiological reasons why the elderly are more vulnerable.

An important study was undertaken following the Chicago heatwave in 1995. Semenza et al. interviewed the relatives of those who died during the heatwave and controls who lived near the case, matched for age and neighbourhood (78). Individual risk factors for dying in the heatwave were identified: chronic illness; confined to bed; unable to care for themselves; isolated; without air conditioning. A comparison of mortality rates in three Illinois heatwaves (1966) by age group, sex and ethnic group (white vs. other) found that women and white people were at more risk (79).

Winter mortality

In many temperate countries there is a clear seasonal variation in mortality (80, 81), death rates during winter being 10–25% higher than those in summer. The major causes of winter death are cardiovascular, cerebrovascular, circulatory and respiratory diseases (82, 83).

Annual outbreaks of winter diseases such as influenza, which have a large effect on winter mortality rates, are not strongly associated with monthly winter temperatures (84). Social and behavioural adaptations to cold play an important role in preventing winter deaths in high latitude countries. Sensitivity to cold weather (measured as the percentage increase in mortality per 1 °C change in temperature) is greater in warmer regions. Mortality increases to a greater extent with a given fall in temperature in regions with warmer winters, in populations with less home heating and where people wear lighter clothes (85).

The elderly (aged 75 and over) are particularly vulnerable to winter death, having a winter excess of around 30%. This vulnerability is not yet well understood but may arise through a combination of physiological susceptibility, behavioural factors and socioeconomic disadvantage. Excess winter mortality is an important problem in the United Kingdom where there has been much debate about the role of poor housing, fuel poverty and other socioeconomic issues for the elderly population (86). Several studies have linked routine mortality data at ward or enumeration district level with small-area indicators of housing and deprivation. A study of ischaemic heart disease morbidity in Stockport found higher winter excess in the higher social class groups although a clear gradient was not observed (87). A small-area study found that inadequate home heating and socioeconomic deprivation were the strongest independent predictors of ward-level variation in excess winter death in England and Wales (88). In general, however, studies have found only weak or absent relationship between excess winter mortality and deprivation (86).

The potential impact of climate change on temperature related mortality

Global climate change is likely to be accompanied by an increase in the frequency and intensity of heatwaves, as well as warmer summers and milder winters (2). Extreme summer heat's impact on human health may be exacerbated by increases in humidity. There has been significant warming in most regions in the last 25 years (see chapter 5) some of which the IPCC has attributed to human activities. However, it is not clear that the frequency of heatwaves has been increasing, although few studies have analysed daily temperature data to confirm this (1). There is much regional variation in the trends observed. Gaffen and Ross looked at data from 1961–1990 for 113 weather stations in the United States and found that the annual frequency of days exceeding a heat stress threshold increased at most stations (89).

Predictive modelling studies use climate scenarios to estimate future temperature related mortality. Those studies which use the empirical statistical model (based on coefficients derived from linear regression of the temperature mortality relationship) find that reductions in winter deaths are greater than increases in summer deaths in temperate countries (84, 90). However, other methods indicate a more significant increase in summer deaths. Kalkstein and Green estimated future excess mortality under climate change in United States' cities (91). Excess summer mortality attributable to climate change, and assuming acclimatization, was estimated to be between 500–1000 for New York and 100–250 for Detroit by 2050, for example.

Populations can be expected to adapt to changes in climate via a range of physiological, behavioural and technological changes. These will tend to reduce the impacts of future increases in heatwaves. The initial physiological acclimatization to hot environments can occur over a few days but behavioural and technological changes, such as changes to the built environment, may take many years.

While it is well established that summer heatwaves are associated with short term increases in mortality, the extent of winter-associated mortality directly attributable to stressful weather is difficult to determine and currently being debated. Limited evidence indicates that, in at least some temperate countries, reduced winter deaths would outnumber increased summer deaths. The net impact on mortality rates will vary between populations. There are no clear implications of climate change for non-fatal outcomes as there is a lack of relevant studies.

Natural disasters

The health effects of disasters are difficult to quantify because secondary effects and delayed consequences are poorly reported and communicated. Information on natural disasters generally is gathered by the organisations and bodies directly involved in disaster relief and reconstruction. As a result, information usually is collected for specific operational purposes not as a database; figures are estimated, not measured directly (92). This is especially true of flood events and windstorms where the actual deaths and injuries directly caused by the event are small compared to the problems that arise as a result, including deaths from communicable diseases and the economic losses sustained (93, 94, 95) (see Box 5.3 on Hurricane Mitch).

El Niño has an effect on the total number of persons affected by natural disasters (96, 97). Worldwide, disasters triggered by droughts are twice as frequent during the year after the onset of El Niño than other years (97). This risk is concentrated in southern Africa and south-east Asia. The El Niño effect on disasters is strong enough to be apparent at the global level (96). In an average El Niño year, around 35 per 1000 persons are affected by a natural disaster. This is over four times greater than the rate in non El Niño years, based on analysis of data from 1963 to 1992. This difference in risk is much stronger for famine disasters; El Niño's global disaster footprint is largely determined by the consequences of drought.

In 1997/98 Kenya was particularly hard hit by flooding and excess rainfall. Ecuador and northern Peru experienced severe flooding and mudslides along the coastal regions which severely damaged the local infrastructure (98). In Peru, 9.5% of health facilities were damaged, including 2% of hospitals and 10% of other health centres (98). At the other extreme, Guyana, Indonesia and Papua New Guinea were severely affected by drought. Although not all natural disasters in 1997/98 should be attributed to the El Niño event, global estimates of the impact vary from 21 000 (99) to 24 000 (100) deaths.

Trends in weather disasters

Globally, there is an increasing trend in natural disaster impacts. An analysis by the reinsurance company Munich Re found a three-fold increase in the number of natural catastrophes in the last ten years, compared to the 1960s (94). This is primarily from global trends affecting population vulnerability rather than changes in the frequency of climatological triggers.

Developing countries are poorly equipped to deal with weather extremes. The number of people killed, injured or made homeless by natural disasters is increasing alarmingly. This is due partly to population growth and the concentration of population in high-risk areas like coastal zones and cities. Large shanty-towns with flimsy habitations often are located on land subject to frequent flooding. In many areas the only land available to poor communities may be that with few natural defences against weather extremes. Direct hits of extreme events on towns and cities tend to cause large losses. In recent decades there has been a large migration to cities and more than half the world's population now lives in urban areas. Such migration and increasing vulnerability means that even without increasing numbers of extreme events, losses attributable to each event will tend to increase (101).

There are several sources of information but the largest, most used and most reliable is a database created in 1988 with support from the World Health Organization and the Belgium Government (EM-DAT). The objective of EM-DAT is:

". . . to serve the purposes of humanitarian action at national and international levels. It is an initiative aimed to rationalise decision-making for disaster preparedness, as well as providing an objective base for vulnerability assessment and priority setting" (92).

The Centre for Research on the Epidemiology of Disasters (CRED) records events where at least 10 people were reported killed; 100 people were reported affected; there was a call for international assistance; or declaration of a state of emergency. There are increasing trends of economic and insured losses from disaster events, and economic annual losses have increased ten-fold since the 1950s (*102*). However, much of the upward trend in economic losses probably is due to societal shifts and increasing vulnerability to weather and climate extremes (*103*).

Data for the 1980s and 1990s are shown in Table 5.3. This shows the numbers of events, people killed and people affected by weather-related natural disasters in each decade, by region of the world. Some regions are more severely affected than others, although some show a decrease in the number of people killed (Africa and eastern Mediterranean) and the number of people affected (Africa, Americas and south-east Asia).

Reasons for the observed increases include:

- increasing concentration of people and property in urban areas
- settlement in exposed or high risk areas (e.g. flood plains, coastal zones)
- changes in environmental conditions (e.g. deforestation can increase flood risk).

There has been an apparent recent increase in the number of disasters but little change in the number of people killed (*94*). In 2000 there were over 400 disasters, with 250 million people affected (*94*). This paradox may be explained by technological advances in the construction of buildings and infrastructure along with advancements in early warning systems, especially in more developed regions. Although there are pronounced year-to-year fluctuations in the numbers of deaths due to disasters, a trend towards increased numbers of deaths and numbers of people affected has been observed in recent decades (*94*).

The health impacts of disasters

Extreme weather events directly cause death and injury and have substantial indirect health impacts. These indirect impacts occur as a result of damage to the

TABLE 5.3 Number of events, people killed and affected, by region of the world for the 1980s and 1990s.

Region	1980s			1990s		
	Events	Killed	Affected	Events	Killed	Affected
Africa	243	416851	137758905	247	10414	104269095
Eastern Europe	66	2019	129345	150	5110	12356266
Eastern Mediterranean	94	161632	17808555	139	14391	36095503
Latin America & Caribbean	265	11768	54110634	298	59347	30711952
South East Asia	242	53853	850496448	286	458002	427413756
Western Pacific	375	35523	273089761	381	48337	1199768618
Developed	563	10211	2791688	577	5618	40832653
Total	1848	691857	1336185336	2078	601219	1851447843

N.B. Regions used in this table correspond to the map of regions used in chapter 7.

TABLE 5.4 Theoretical risk of acquiring communicable diseases, by type of disaster.

Type	Person to person	Water borne	Food borne	Vector borne
Earthquake	M	M	M	L
Volcano	M	M	M	L
Hurricane	M	H	M	H
Tornado	L	L	L	L
Heatwave	L	L	L	L
Coldwave	L	L	L	L
Flood	M	H	M	H
Famine	H	H	M	M
Fire	L	L	L	L

H = High.
M = Medium.
L = Low.
Source: Reproduced from reference 106.

local infrastructure, population displacement and ecological change. Direct and indirect impacts can lead to impairment of the public health infrastructure, psychological and social effects, and reduced access to health care services (*104*). The health impacts of natural disasters include (*105, 106*):

- physical injury;
- decreases in nutritional status, especially in children;
- increases in respiratory and diarrhoeal diseases due to crowding of survivors, often with limited shelter and access to potable water;
- impacts on mental health which may be long lasting in some cases;
- increased risk of water-related and infectious diseases due to disruption of water supply or sewage systems, population displacement and overcrowding;
- release and dissemination of dangerous chemicals from storage sites and waste disposal sites into flood waters.

Floods

Floods are associated with particular dangers to human populations (*107*). Immediate effects are largely death and injuries from drowning and being swept against hard objects. Local infrastructure can be affected severely during a natural disaster. El Niño related damage may include: flood damage to buildings and equipment, including materials and supplies; flood damage to roads and transport; problems with drainage and sewerage; and damage to water supply systems.

During and following both catastrophic and non-catastrophic flooding, there is a risk to health if the floodwaters become contaminated with human or animal waste. A study in populations displaced by catastrophic floods in Bangladesh in 1988 found that diarrhoea was the most common illness, followed by respiratory infection. Watery diarrhoea was the most common cause of death for all age groups under 45 (*108*). In both rural Bangladesh and Khartoum, Sudan, the proportion of severely malnourished children increased after flooding (*109, 110*). In developed countries, both physical and disease risks from flooding are greatly reduced by a well maintained flood control and sanitation infrastructure and public health measures, such as monitoring and surveillance activities to detect and control outbreaks of infectious disease. However, the recent experience of flooding in Central Europe, in which over 100 people died, showed that floods can have a major impact on health and welfare in industrialised countries too (*111*).

Floods also cause psychological morbidity. Following flooding in Bristol, UK, primary care attendance rose by 53% and referrals and admissions to hospitals more than doubled (*112*). Similar psychological effects were found following floods in Brisbane, Australia, in 1974 (*113*). An increase in psychological symptoms and post-traumatic stress disorder, including 50 flood-linked suicides, were reported in the two months following the major floods in Poland in 1997 (*99*).

A number of studies have established a link between dampness in the home, including occasional flooding, with a variety of respiratory symptoms. For example, a Canadian study found that flooding was linked significantly to childhood experience of cough, wheeze, asthma, bronchitis, chest illness, upper respiratory symptoms, eye irritation and non-respiratory symptoms (*114*).

Windstorms and tropical cyclones

Impoverished and high-density populations in low-lying and environmentally degraded areas are particularly vulnerable to tropical cyclones, the majority of deaths caused by drowning in the storm surge (*106, 115*).

Bangladesh has experienced some of the most serious impacts of tropical cyclones this century, due to a combination of meteorological and topographical conditions and the inherent vulnerability of a low-income, poorly resourced population. Improved early warning systems have decreased the impacts in recent years. However, the experience of Hurricane Mitch demonstrated the destructive power of an extreme event on such a region (*116*).

Droughts

A drought can be defined as "a period of abnormally dry weather which persists long enough to produce a serious hydrologic imbalance" (*118*), or as a "period of deficiency of moisture in the soil such that there is inadequate water required for plants, animals and human beings" (*92*). There are four general types of drought, all which impact on humans, but in different ways (*118*):

1. meteorological: measured precipitation is unusually low for a particular region;
2. agricultural: amount of moisture in the soil is no longer sufficient for crops under cultivation;
3. hydrological: surface water and groundwater supplies are below normal;
4. socioeconomic: lack of water affects the economic capacity of people to survive, i.e. affects non-agricultural production.

The health impacts on populations occur primarily on food production. Famine often occurs when a pre-existing situation of malnutrition worsens: the health consequences of drought include diseases resulting from malnutrition (*105*). In addition to adverse environmental conditions political, environmental or economic crises can trigger a collapse in the food marketing systems. The major food emergency in Sudan during 1998 illustrates the interrelationship between climatic triggers of famine and conflict.

In times of shortage, water is used for cooking rather than hygiene. This increases the risk of diarrhoeal diseases (due to faecal contamination) and water-washed diseases (trachoma, scabies). Outbreaks of malaria can occur due to changes in vector breeding sites (*119*) and malnutrition increases susceptibility to infection.

CLIMATE CHANGE AND HUMAN HEALTH

BOX 5.3 Hurricane Mitch

Hurricane Mitch was the worst disaster to strike Central America in the twentieth century (95). It began when a tropical depression, subsequently named Mitch, formed in the southern Caribbean Sea on 21 October 1998. Between 22–26 October, Mitch increased in intensity, developed into a tropical storm and then a Category 5 hurricane (117). Winds of up to 295 km per hour struck the coastlines of Nicaragua, Honduras, El Salvador, Guatemala and Belize, followed by heavy continuous rainfall for over 5 days (95).

Mitch caused around 9550 deaths; destroyed or affected around 137851 homes; and affected a population of around 3174700 people (95, 116). Yet the effects of the tropical storm/hurricane were worse than these high numbers indicate. This was due to the damage to infrastructure and services that worsened the secondary effects. These secondary effects and additional impacts included (93, 95):

- increase in vectors leading to increased transmission of vector-borne diseases, especially malaria and dengue;
- increases in communicable diseases such as gastrointestinal and respiratory diseases; losses in the production sector: Honduras lost over 70% of banana, coffee and pineapple crops;
- set-backs to development plans: in Honduras reported to be 50 years;
- set-backs in progress in public health;
- set-backs in environmental health caused by flooding of wells and latrines, destruction of water and sanitation systems and leakage of septic tanks and sewerage systems.

Factors that increased the vulnerability of the population of these countries to the effects of this natural disaster include (116).

- increased population pressure;
- migration of the population to the more vulnerable areas, such as the low lying coastal areas and along river banks;
- urbanization of the population leading to increased numbers living in poorly constructed houses with little access to health, water and sanitation services;
- marginalisation of the population.

Economic losses were estimated at over US$ 7 billion for the region as a whole (95).

Forest fires

The direct effects of fires on human health are burns and smoke inhalation. Loss of vegetation on slopes may lead to soil erosion and increased risk of landslides, often exacerbated when an urban population expands into surrounding hilly and wooded areas.

Air pollution is linked to increased mortality and morbidity in susceptible persons, and increased risk of hospital and emergency admissions. Assessments are being undertaken of the short-term impacts on mortality and morbidity associated with the 1997 El Niño episode. However, such assessments often are limited by lack of baseline data. WHO has published health guidelines for episodic vegetation fires (120, 121).

Conclusions

The increasing trend in natural disasters partly is due to more complete reporting, as well as increasing vulnerability of populations. Poverty, population growth and migration are major contributory factors affecting this vulnerability. Particularly in poor countries, the impacts of major vector-borne diseases and disasters can limit or even reverse improvements in social development; even under favourable conditions recovery from major disasters can take decades.

Quantitative public health forecasts, based on statistical associations between climate variability and health outcomes, will be highly uncertain because future social and economic trends will influence strongly the effects of climate change. It will be possible to carry out qualitative assessments for policy purposes based on less-than-perfect scientific evidence.

There is the potential to use seasonal forecasts to reduce the burden of disease. At present, seasonal forecasts are of most use in the mitigation of drought, food shortages and famine disasters, but the relationships described above could provide the basis for early warning systems for epidemics. Currently available predictive models of climate variability and communicable disease are insufficiently reliable to provide early warning of epidemics. Policy use probably should await validation of these models by analysis of prospective epidemic forecasts.

Seasonal forecasting is only part of an early warning system that must incorporate monitoring and surveillance, as well as adequate response activities. Forecasts of climate extremes could improve preparedness and reduce adverse effects. Focusing attention on extreme events also may help countries to develop better means of dealing with the longer-term impacts of global climate change.

Conversely, the pressures on the biosphere that drive climate change may cause critical thresholds to be breached, leading to shifts in natural systems that are unforeseen and rapid. Studying historical extremes of climate cannot forewarn on the consequences of such events. Rapid changes in climate during extreme events may be more stressful than slowly developing changes due to the greenhouse effect. Thus climatic variables that have the greatest influence in the short-term may not be those with the biggest impact in the longer term.

Adaptive social responses to rapidly occurring periodic extremes of climate may be less effective in the face of progressive climate shifts. For example, increased food imports might prevent hunger and disease during occasional drought, but poor countries are unlikely to be able to afford such measures indefinitely in response to gradual year-by-year drying of continental areas.

Analogue studies of extreme climate events and human health provide important clues about the interactions between climate, ecosystems and human societies that may be triggered by long-term climate trends. Whilst the short term localized effects of simple climate extremes are most readily quantifiable, studies of complex climate extremes provide important qualitative insights into these relationships and the factors affecting population vulnerability.

References

1. Easterling, D.R. et al. Climate extremes: observations, modelling and impacts. *Science* 289: 2068 (2000).
2. Kovats R. *El Niño* and health. Geneva, Switzerland, World Health Organization 1999.
3. Intergovernmental Panel on Climate Change (IPCC) *Climate change 2001*: IPCC third assessment report. Geneva, Switzerland, Intergovernmental Panel on Climate Change 2001a.
4. Sachs, J. & Malaney, P. The economic and social burden of malaria. *Nature* 415: 680–686 (2002).

5. MacDonald, G. *The epidemiology and control of malaria.* Oxford, UK, Oxford University Press 1957.

6. World Health Organization (WHO). El Niño and its health impacts. *Weekly Epidemiological Record* 20: 148–152 (1998b).

7. Bradley, D.J. Human tropical diseases in a changing environment. *Environmental Change and Human Health.* Ciba Foundation Symposium 175: 146–162 (1993).

8. Bouma, M.J. & van der Kaay, H.J. Epidemic malaria in India's Thar Desert. *Lancet* 373: 132–133 (1995).

9. Gill, C.A. The relationship of malaria and rainfall. *Indian Journal of Medical Research* 7(3): 618–632 (1920).

10. Bouma, M.J. & Dye, C. Cycles of malaria associated with El Niño in Venezuela. *Journal of the American Medical Association* 278: 1772–1774 (1997).

11. Diaz, H.F. et al. Climate and human health linkages on multiple timescales. *Climate and climatic impacts through the last 1000 years.* Jones, P.D. et al. Cambridge, UK, Cambridge University Press 2000.

12. Cedeno, J.E. Rainfall and flooding in the Guayas river basin and its effects on the incidence of malaria 1982–1985. *Disasters* 10(2): 107–111 (1986).

13. Nicholls, N. ENSO, drought and flooding rain in south-east Asia. In: *South-east Asia's environment future: the search for sustainability.* Brookfield, H. & Byron, Y. Tokyo, Japan, United Nations University Press/Oxford University Press: pp. 154–175 1993.

14. Russac, P.A. Epidemiological surveillance: malaria epidemic following the Niño phenomenon. *Disasters* 10(2): 112–117 (1986).

15. Bouma, M.J. et al. Predicting high-risk years for malaria in Colombia using parameters of El Niño Southern Oscillation. *Tropical Medicine and International Health* 2(12): 1122–1127 (1997b).

16. Poveda, G. et al. Climate and ENSO variability associated to malaria and dengue fever in Colombia. In: *El Niño and the Southern Oscillation, multiscale variability and global and regional impacts.* Diaz, H.F. & Markgraf, F. Cambridge, UK, Cambridge University Press: pp. 183–204 2000.

17. Poveda, G. et al. Coupling between annual and ENSO timescales in the malaria climate association on Colombia. *Environmental Health Perspectives* 109(5): 307–324 (2001).

18. Gagnon, A. et al. The El Niño Southern Oscillation and malaria epidemics in South America. *International Journal of Biometeorology* 46: 81–89 (2002).

19. Brown, V. et al. Epidemic of malaria in north-eastern Kenya. *Lancet* 352: 1356–1357 (1998).

20. Kilian, A.H. et al. Rainfall patterns, El Niño and malaria in Uganda. *Transactions of the Royal Society of Tropical Medicine and Hygiene* 93: 22–23 (1999).

21. Lindblade, K.A. et al. Highland malaria in Uganda: prospective analysis of an epidemic associated with El Niño. *Transactions of the Royal Society of Tropical Medicine and Hygiene* 93: 480–487 (1999).

22. Allan, R. et al. MERLIN and malaria epidemic in north-east Kenya. *Lancet* 351: 1966–1967 (1998).

23. Mouchet, J. et al. Evolution of malaria in Africa for the past 40 years: impact of climatic and human factors. *Journal of the American Mosquito Control Association* 14: 121–130 (1998).

24. Malakooti, M.A. et al. Re-emergence of epidemic malaria in the highlands of western Kenya. *Emerging Infectious Diseases* 4(4) (1998).

25. Hay, S.I. et al. Etiology of interepidemic periods of mosquito-borne disease. *Proceedings of the National Academy of Sciences* 97(16): 9335–9339 (2000).

26. Hay, S.I. et al. Climate change and the resurgence of malaria in the East African highlands. *Nature* 415: 905–909 (2002).

27. Githeko, A.K. & Ndegwa, W. Predicting malaria epidemics in the Kenyan highlands using climate data: a tool for decision makers. *Global Change and Human Health* 2: 54–63 (2001).

28. Gubler, D.J. Dengue and dengue hemorrhagic fever: its history and resurgence as a global public health problem. In: *Dengue and dengue hemorrhagic fever*. Gubler, D.J. & Kuno, G. New York, USA, CAB International: 1–22 1997.

29. Rigau-Perez, J.G. et al. Dengue and dengue haemorrhagic fever. *Lancet* 352: 971–977 (1998).

30. Hales, S. et al. Dengue fever epidemics in the South Pacific region: driven by El Niño Southern Oscillation? *Lancet* 348: 1664–1665 (1996).

31. Hales, S. et al. El Niño and the dynamics of vector-borne disease transmission. *Environmental Health Perspectives* 107: 99–102 (1999).

32. Lien, T.V. & Ninh, N.H. In: *Currents of change: El Niño's impact on climate and society.* Glantz, M.H. Cambridge, UK, Cambridge University Press 1996.

33. Gagnon, A.S. et al. Dengue epidemics and the El Niño Southern Oscillation. *Climate Research* 19(1): 35–43 (2001).

34. Mills, J.N. et al. Long-term studies of hantavirus reservoir populations in the southwestern United States: rationale, potential and methods. *Emerging Infectious Diseases* 5: 95–101 (1999).

35. Parmenter, R.R. et al. Incidence of plague associated with increased winter-spring precipitation in New Mexico. *American Journal of Tropical Medicine and Hygiene* 61: 814–821 (1999).

36. Wenzel, R.P. A new hantavirus infection in North America. *New England Journal of Medicine* 330: 1004–1005 (1994).

37. Hjelle, B. & Glass, G.E. Outbreak of hantavirus infection in the Four Corners region of the United States in the wake of the 1997–1998 El Niño-Southern Oscillation. *Journal of Infectious Diseases* 181(5): 1569–1573 (2000).

38. Rodriguez-Moran, P. et al. Hantavirus infection in the Four Corners region of USA in 1998. *Lancet* 352: 1353–1353 (1998).

39. Engelthaler, D.M. et al. Climatic and environmental patterns associated with hantavirus pulmonary syndrome, Four Corners region, United States. *Emerging Infectious Diseases* 5: 87–94 (1999).

40. Atherton, F. et al. An outbreak of waterborne cryptosporidiosis associated with a public water supply in the UK. *Epidemiology and Infection* 115: 123–131 (1995).

41. Curriero, F. et al. The association between extreme precipitation and waterborne disease outbreaks in the United States, 1948–1994. *American Journal of Public Health* 91(8): 1194–1199 (2001).

42. Kriz, B. et al. Monitorování Epidemiologické Situace V ZaplavenýCh Oblastech V Èeské Republice V Roce 1997. [Monitoring the epidemiological situation in flooded areas of the Czech Republic in 1997.] In: Konference DDD '98; Kongresové Centrum Lázeòská Kolonáda Podìbrady, 11.–13. Kvìtna 1998 [Proceedings of the Conference DDD '98, 11–12th May, 1998, Prodebrady, Czech Republic.]. Prodebrady, Czech Republic: 19–34 1998.

43. Lisle, J.T. & Rose, J.B. Cryptosporidium contamination of water in the USA and UK: a mini-review. *Aqua* 44(3): 103–117 (1995).

44. Rose, J.B. et al. Climate and waterborne outbreaks in the US. *Journal of the American Water Works Association* 2000.

45. Rose, J.B. et al. Climate variability and change in the United States: potential impacts on water- and food-borne diseases caused by microbiologic agents. *Environmental Health Perspectives* 109 Supplement 2: 211–221 (2001).

46. Schwartz, J. & Levin, R. Drinking water turbidity and health. *Epidemiology* 10: 86–89 (1999).

47. Bentham, G. & Langford, I.H. Climate change and the incidence of food poisoning in England and Wales. *International Journal of Biometeorology* 39: 81–86 (1995).

48. Bentham, G. & Langford, I.H. Environmental temperatures and the incidence of food poisoning in England and Wales. *International Journal of Biometeorology* 45(1): 22–26 (2001).

49. Madico, G. et al. Epidemiology and treatment of cyclospora cayetenanis infection in Peruvian children. *Clinical Infectious Diseases* 24: 977–981 (1997).

50. Salazar-Lindo, E. et al. El Niño and diarrhoea and dehydration in Lima, Peru. *Lancet* 350: 1597–1598 (1997).

51. Checkley, W. et al. Effects of El Niño and ambient temperature on hospital admissions for diarrhoeal diseases in Peruvian children. *Lancet* 355(2000).

52. Singh, R.B.K. et al. The influence of climate variation and change on diarrhoeal disease in the Pacific Islands. *Environmental Health Perspectives* 109: 155–159 (2001).

53. Jaenisch, T. & Patz, J. Assessment of associations between climate and infectious diseases. *Global Change and Human Health* 3: 67–72 (2002).

54. Anonymous. El Niño and associated outbreaks of severe malaria in highland populations in Irian Jaya, Indonesia: a review and epidemiological perspective. *Southeast Asian Journal of Tropical Medicine & Public Health* 30: 608–619 (1999).

55. Colwell, R.R. Global climate and infectious disease: the cholera paradigm. *Science* 274: 2025–2031 (1996).

56. Pascual, M. et al. Cholera dynamics and El Niño Southern Oscillation. *Science* 289: 1766–1767 (2000).

57. Rodo, X. et al. ENSO and cholera: a nonstationary link related to climate change? *Proceedings of the national Academy of Sciences* 99: 12901–12906.

58. Kovats, S. El Niño and human health. *Bulletin of the World Health Organization* 78(9): 1127–1135 (2000).

59. Franco, A.A. et al. Cholera in Lima, Peru, correlates with prior isolation of Vibrio cholerae from the environment. *American Journal of Epidemiology* 146:1067–1075 (1997).

60. World Health Organization. Cholera in 1997. *Weekly Epidemiological Record* 73: 201–208 (1998).

61. Curwen, M. & Devis, T. Winter mortality, temperature and influenza: has the relationship changed in recent years? *Population Trends* 54: 17–20 (1988).

62. Kilbourne, E.M. Illness due to thermal extremes. In: *Public health and preventative medicine.* Last, J.M. & Norwalk, W.R.B. Connecticut, US, Appleton Lang: pp. 491–501 1992.

63. Keatinge, W.R. et al. Increased platelet and red cell counts, blood viscosity, and plasma cholesterol levels during heat stress, and mortality from coronary and cerebral thrombosis. *American Journal of Medicine* 81: 795–800 (1986).

64. Drinkwater, B. & Horvath, S. Heat tolerance and ageing. *Medicine and Science in Sports and Exercise* 11: 49–55 (1979).

65. Mackenbach, J.P. et al. Heat-related mortality among nursing home patients. *Lancet* 349: 1297–1298 (1997).

66. Vassallo, M. et al. Factors associated with high risk of marginal hyperthermia in elderly patients living in an institution. *Postgraduate Medical Journal* 71: 213–216 (1995).

67. Woodhouse, P.R. et al. Seasonal variation of blood pressure and its relationship to ambient temperature in an elderly population. *Journal of Hypertension* 11(11): 1267–1274 (1993).

68. Woodhouse, P.R. et al. Seasonal variations of plasma fibrinogen and factor VII activity in the elderly: winter infections and death from cardiovascular disease. *Lancet* 343: 435–439 (1994).

69. Whitman, S. et al. Mortality in Chicago attributed to the July 1995 heatwave. *American Journal of Public Health* 87(9): 1515–1518 (1997).

70. Dematte, J.E. et al. Near-fatal heat stroke during the 1995 heatwave in Chicago. *Annals of Internal Medicine* 129: 173–181 (1998).

71. Kilbourne, E.M. Heat waves. In: *The public health consequences of disasters.* Gregg, M.B. Atlanta, US, US Department of Health and Human Services, Centers for Disease Control: 51–61 1989.

72. Centers for Disease Control and Prevention (CDC). Heat-related deaths–Dallas, Wichita, and Cooke Counties, Texas, and United States, 1996. *Journal of the American Medical Association* 278: 462–463 (1997).

73. Katsouyanni, K. et al. The 1987 Athens heatwave [Letter]. *Lancet* ii: 573–573 (1988).

74. Palecki, M.A. et al. The nature and impacts of the July 1999 heatwave in the midwestern United States: learning from the lessons of 1995. *Bulletin of the American Meteorological Society*: 1353–1367 (2001).

75. Kumar, S. India's heatwave and rains result in massive death toll. *Lancet* 351: 1869–1869 (1998).

76. Rooney, C. et al. Excess mortality in England and Wales, and in Greater London, during the 1995 heatwave. *Journal of Epidemiology and Community Health* 52: 482–486 (1998).

77. Sartor, F. et al. Temperature, ambient ozone levels and mortality during summer 1994 heatwave in Belgium. *Environmental Research* 70(2): 105–113 (1995).

78. Semenza, J.C. et al. Heat-related deaths during the July 1995 heatwave in Chicago. *New England Journal of Medicine* 335: 84–90 (1996).

79. Bridger, C.A. & Helfand, L.A. Mortality from heat during July 1966 in Illinois. *International Journal of Biometeorology* 12: 51–70 (1968).

80. Laake, K. & Sverre, J.M. Winter excess mortality: a comparison between Norway and England plus Wales. *Age Ageing* 25: 343–348 (1996).

81. Sakamoto, M.M. *Seasonality in human mortality*. Tokyo, Japan, University of Tokyo Press (1977).

82. Donaldson, G.C. et al. Winter mortality and cold stress in Yekaterinberg, Russia: interview survey. *British Medical Journal* 316: 514–518 (1998).

83. West, R.R. & Lowe, C.R. Mortality from ischaemic heart disease: inter-town variation and its association with climate in England and Wales. *International Journal of Epidemiology* 5(2): 195–201 (1976).

84. Langford, I.H. & Bentham, G. The potential effects of climate change on winter mortality in England and Wales. *International Journal of Biometeorology* 38: 141–147 (1995).

85. Donaldson, G.C. et al. Outdoor clothing and its relationship to geography, climate, behaviour and cold related mortality. *International Journal of Biometeorology* 45: 45–51 (2001).

86. Mitchell, R. Short days—shorter lives: studying winter mortality to get solutions. *International Journal of Epidemiology* 30: 1116–1118 (2001).

87. Watkins, S.J. et al. Winter excess morbidity: is it a summer phenomenon? *Journal of Public Health Medicine* 23(3): 237–241 (2001).

88. Wilkinson, P. et al. Case-control study of hospital admission with asthma in children aged 5–14 years: relation with road traffic in north-west London. *Thorax* 54(12): 1070–1074 (1999).

89. Gaffen, D.J. & Ross, R.J. Increased summertime heat stress in the US. *Nature* 396: 529–530 (1998).

90. Guest, C. et al. Climate and mortality in Australia: retrospective study, 1970–1990, and predicted impacts in five major cities. *Climate Research* 13: 1–15 (1999).

91. Kalkstein, L.S. & Greene, J.S. An evaluation of climate/mortality relationships in large US cities and the possible impacts of climate change. *Environmental Health Perspectives* 105(1): 84–93 (1997).

92. Office of U.S. Foreign Disaster Assistance (OFDA)/Centre for Research on the Epidemiology of Disasters(CRED). EM-DAT: The International Disaster Database. Brussels, Belgium, Université Catholique de Louvain (2001). http://www.cred.be/emdat/intro.html

93. Glantz, M. & Jamison, D.T. Societal response to Hurricane Mitch and intra-versus intergenerational equity issues: whose norms should apply? *Risk Analysis* 20(6): 869–882 (2000).

94. Munich Re Group Topics: *Natural Catastrophes 2000* 2001.

95. Pan American Health Organization (PAHO). *Disasters and health in 1998*: a report of the Pan American Health Organization's Emergency Preparedness and Disaster Relief Coordination Program, PAHO 1999b.

96. Bouma, M.J. et al. Global assessment of El Niño's disaster burden. *Lancet* 350: 1435–1438 (1997a).

97. Dilley, M. & Heyman, B. ENSO and disaster: droughts, floods, and El Niño/Southern Oscillation warm events. *Disasters* 19(3): 181–193 (1995).

98. Pan American Health Organization (PAHO). El Niño and its impact on health 1998.

99. International Federation of Red Cross and Red Crescent Societies (IFRC). *World disasters report*. Geneva, Switzerland, International Federation of Red Cross and Red Crescent Societies 1999.

100. National Oceanic and Atmospheric Administration (NOAA). *An experiment in the application of climate forecasts: NOAA-OGP activities related to the 1997–98 El Niño event*. Boulder, US, Office of Global Programs, US Dept of Commerce 1999.

101. Guha-Sapir, D. Rapid assessment of health needs in mass emergencies: review of current concepts and methods. *World Health Statistics Quarterly* 44: 171–181 (1991).

102. Intergovernmental Panel on Climate Change (IPCC). *Summary for policymakers to climate change 2001*: Synthesis report of the IPCC Third Assessment Report. Geneva, Switzerland, IPCC 2001.

103. Changnon, S.A. et al. Human factors explain the increased losses from weather and climate extremes. *Bulletin of the American Meteorological Society* 81(3): 437–442 (2000).

104. Greenough, G. et al. The potential impacts of climate variability and change on health impacts of extreme weather events in the United States. *Environmental Health Perspectives* 109(Supplement 2): 191–198 (2001).

105. McMichael, A.J. et al. *Climate change and human health:* an assessment prepared by a task group on behalf of the World Health Organization, the World Meteorological Organization and the United Nations Environment Programme. Geneva, Switzerland, World Health Organization 1996.

106. Noji, E.N. *The public health consequences of disasters*. New York, US, Oxford University Press 1997.

107. Menne, B. *Floods and public health consequences, prevention and control measures*. Rome, Italy, World Health Organization European Centre for Environment and Health (WHO-ECEH) 1999.

108. Siddique, A.K. et al. 1988 floods in Bangladesh: pattern of illness and causes of death. *Journal of Diarrhoeal Disease Research* 9(4): 310–314 1991.

109. Choudhury, A.Y. & Bhuiya, A. Effects of biosocial variable on changes in nutritional status of rural Bangladeshi children, pre- and post-monsoon flooding. *Journal of Biosocial Science* 25: 351–357 (1993).

110. Woodruff, B.A. et al. Disease surveillance and control after a flood: Khartoum, Sudan, 1988. *Disasters* 14(2): 151–163 (1990).

111. Red Cross. *Death toll rises in central Europe's epic floods* (2002). http://www.redcross.org/news/in/flood/020819europe.html

112. Bennet, G. Bristol floods 1968: controlled survey of effects on health of local community disaster. *British Medical Journal* 3: 454–458 (1970).

113. Abrahams, M.J. et al. The Brisbane floods, January 1974: their impact on health. *Medical Journal of Australia* 2: 936–939 (1976).

114. Dales, R.E. et al. Respiratory health effects of home dampness and molds among Canadian children. *American Journal of Epidemiology* 134: 196–203 (1991).

115. Alexander, D. *Natural disasters*. London, UK, UCL Press 1993.

116. Pan American Health Organization (PAHO). The devastating path of Hurricane Mitch in central America. *Disasters: preparedness and mitigation in the Americas* (Supplement 1): S1–S4 (1999a).

117. National Oceanic and Atmospheric Administration (NOAA). *Hurricane Mitch special coverage*, NOAA's National Environmental Satellite, Data and Information Service (NESDIS) (2002). http://www.osei.noaa.gov/mitch.html.

118. National Oceanic and Atmospheric Administration (NOAA). *Droughts*. National Oceanic and Atmospheric Administration 2002.

119. Bouma, M.J. & van der Kaay, H.J. The El Niño Southern Oscillation and the historic malaria epidemics on the Indian subcontinent and Sri Lanka: an early warning system. *Tropical Medicine and International Health* 1(1): 86–96 1996.

120. Sastry, N. Forest fires, air pollution and mortality in south-east Asia. *Demography* 39(1): 1–23 (2002).

121. Institute of Environmental Epidemiology (IEE). *Health guidelines for vegetation fire events*. Schwela, D.H. et al. eds. Singapore, Singapore: Institute of Environmental Epidemiology 1999.

CHAPTER 6

Climate change and infectious diseases

J. A. Patz,[1] A. K. Githeko,[2] J. P. McCarty,[3] S. Hussein,[1] U. Confalonieri,[4] N. de Wet[5]

Introduction

The previous chapter considered how short-term variations in climatic conditions and extreme weather events can exert direct effects on human death rates, physical injury, mental health and other health outcomes. Changes in mean climatic conditions and climate variability also can affect human health via indirect pathways, particularly via changes in biological and ecological processes that influence infectious disease transmission and food yields. This chapter examines the influences of climatic factors on infectious diseases.

For centuries humans have known that climatic conditions affect epidemic infections—since well before the basic notion of infectious agents was understood late in the nineteenth century. The Roman aristocracy took refuge in their hill resorts each summer to avoid malaria. South Asians learnt early that in high summer, strongly curried foods were less prone to induce diarrhoeal diseases. In the southern United States one of the most severe summertime outbreaks of yellow fever (viral disease transmitted by the *Aedes aegypti* mosquito) occurred in 1878, during one of the strongest El Niño episodes on record. The economic and human cost was enormous, with an estimated death toll of around 20 000 people. In developed countries today it is well known that recurrent influenza epidemics occur in mid-winter.

Infectious disease transmission should be viewed within an ecological framework. Infectious agents obtain the necessary nutrients and energy by parasitization of higher organisms. Most such infections are benign, and some are even beneficial to both host and microbe. Only a minority of infections that adversely affect the host's biology are termed "infectious disease".

During the long processes of human cultural evolution; population dispersal around the world; and subsequent inter-population contact and conflict; several distinct transitions in human ecology and inter-population interactions have changed profoundly the patterns of infectious disease in human populations. Since the early emergence of agriculture and livestock herding around 10 000 years ago, three great transitions in human/microbe relationships are readily recognizable (1):

[1] Johns Hopkins University, Baltimore, MD, USA.
[2] Kenya Medical Research Institute, Kisumu, Kenya.
[3] University of Nebraska at Omaha, Omaha, NE, USA.
[4] Fundação Oswaldo Cruz, Rio de Janeiro, Brazil.
[5] The International Global Change Institute, University of Waikato, New Zealand.

1. Early human settlements enabled enzootic infective species to enter *H. sapiens*.
2. Early Eurasian civilizations came into military and commercial contact around 2000 years ago, swapping dominant infections.
3. European expansionism over past five centuries caused transoceanic spread of often lethal infectious diseases.

This may be the fourth great transitional period. The spread and increased lability of various infectious diseases, new and old, reflects the impacts of demographic, environmental, technological and other rapid changes in human ecology. Climate change, one of the global environmental changes now under way, is anticipated to have a wide range of impacts upon the occurrence of infectious disease in human populations.

Disease classification

Broadly, infectious diseases may be classified into two categories based on the mode of transmission: those spread directly from person to person (through direct contact or droplet exposure) and those spread indirectly through an intervening vector organism (mosquito or tick) or a non-biological physical vehicle (soil or water). Infectious diseases also may be classified by their natural reservoir as anthroponoses (human reservoir) or zoonoses (animal reservoir).

Climate sensitivities of infectious diseases

Both the infectious agent (protozoa, bacteria, viruses, etc) and the associated vector organism (mosquitoes, ticks, sandflies, etc.) are very small and devoid of thermostatic mechanisms. Their temperature and fluid levels are therefore determined directly by the local climate. Hence, there is a limited range of climatic conditions—the climate envelope—within which each infective or vector species can survive and reproduce. It is particularly notable that the incubation time of a vector-borne infective agent within its vector organism is typically very sensitive to changes in temperature, usually displaying an exponential relationship. Other climatic sensitivities for the agent, vector and host include level of precipitation, sea level elevation, wind and duration of sunlight.

Documented and predictive climate/infectious disease linkages

The seasonal patterns and climatic sensitivities of many infectious diseases are well known; the important contemporary concern is the extent to which changes in disease patterns will occur under the conditions of global climate change. Over the past decade or so this question has stimulated research into three concentrations. First, can the recent past reveal more about how climatic variations or trends affect the occurrence of infectious diseases? Second, is there any evidence that infectious diseases have changed their prevalence in ways that are reasonably attributable to climate change? Third, can existing knowledge and theory be used to construct predictive models capable of estimating how future scenarios of different climatic conditions will affect the transmissibility of particular infectious diseases?

Modifying influences

Climate is one of several important factors influencing the incidence of infectious diseases. Other important considerations include sociodemographic influences such as human migration and transportation; and drug resistance and nutrition; as well as environmental influences such as deforestation; agricultural development; water projects; and urbanization. In this era of global development and land-use changes, it is highly unlikely that climatic changes exert an isolated effect on disease; rather the effect is likely dependent on the extent to which humans cope with or counter the trends of other disease modifying influences. While recognizing the important independent role of these non-climatic factors, the focus of this section is to examine the extent to which they may compound the effects of climatic conditions on disease outcomes.

Disease classifications relevant to climate/health relationships

Several different schemes allow specialists to classify infectious diseases. For clinicians who are concerned with treatment of infected patients, the clinical manifestation of the disease is of primary importance. Alternatively, microbiologists tend to classify infectious diseases by the defining characteristics of the microorganisms, such as viral or bacterial. For epidemiologists the two characteristics of foremost importance are the method of transmission of the pathogen and its natural reservoir, since they are concerned primarily with controlling the spread of disease and preventing future outbreaks (2).

Climate variability's effect on infectious diseases is determined largely by the unique transmission cycle of each pathogen. Transmission cycles that require a vector or non-human host are more susceptible to external environmental influences than those diseases which include only the pathogen and human. Important environmental factors include temperature, precipitation and humidity (discussed in more detail in the following section). Several possible transmission components include pathogen (viral, bacterial, etc.), vector (mosquito, snail, etc.), non-biological physical vehicle (water, soil, etc.), non-human reservoir (mice, deer, etc.) and human host. Epidemiologists classify infectious diseases broadly as anthroponoses or zoonoses, depending on the natural reservoir of the pathogen; and direct or indirect, depending on the mode of transmission of the pathogen. Figure 6.1 illustrates these four main types of transmission cycles for infectious diseases. The following is a description of each category of disease, discussed in order of probable increasing susceptibility to climatic factors (3).

Directly transmitted diseases

Anthroponoses

Directly transmitted anthroponoses include diseases in which the pathogen normally is transmitted directly between two human hosts through physical contact or droplet exposure. The transmission cycle of these diseases comprises two elements: pathogen and human host. Generally, these diseases are least likely to be influenced by climatic factors since the agent spends little to no time outside the human host. These diseases are susceptible to changes in human behaviour, such as crowding and inadequate sanitation that may result from altered land-use caused by climatic changes. Examples of directly transmitted anthroponoses include measles, TB, and sexually transmitted infections such as HIV, herpes and syphilis (3).

Zoonoses

Directly transmitted zoonoses are similar to directly transmitted anthroponoses in that the pathogen is transmitted though physical contact or droplet exposure

FIGURE 6.1 Four main types of transmission cycle for infectious diseases.
Source: reproduced from reference 3.

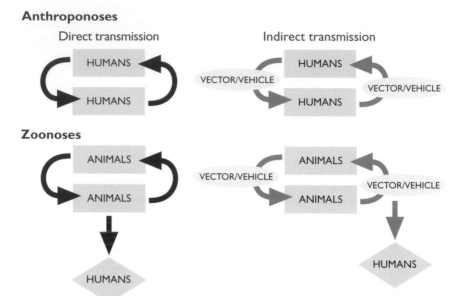

between reservoirs. However, these agents are spread naturally among animal reservoirs and the infection of humans is considered to be a result of an accidental human encounter. The persistence of these pathogens in nature is largely dependent on the interaction of the animal reservoir and external environment which can impact the rate of transmission, host immunity, rate of reproduction, and species death, rendering these diseases more susceptible to effects of climate variability. Hantavirus is a directly transmitted zoonosis that is naturally maintained in rodent reservoirs and can be transmitted to humans at times of increased local abundance of the reservoir (4). Rabies is another directly transmitted zoonosis that naturally infects small mammals, although with very little opportunity for widespread transmission, being highly pathogenic to its vertebrate host (3). Several of today's anthropogenic diseases, e.g. TB and HIV, originally emerged from animals.

Indirectly transmitted diseases (anthroponoses & zoonoses)

Indirectly transmitted anthroponoses are a class of diseases defined by pathogen transmission between two human hosts by either a physical vehicle (soil) or a biological vector (tick). These diseases require three components for a complete transmission cycle: the pathogen, the physical vehicle or biological vector, and the human host. Most vectors require a blood meal from the vertebrate host in order to sustain life and reproduce. Indirectly transmitted anthroponoses include malaria and dengue fever, whereby the respective malaria parasite and the dengue virus are transmitted between human hosts by mosquito vectors (vector-borne disease). Indirectly transmitted water-borne anthroponoses are susceptible to climatic factors because the pathogens exist in the external environment during part of their life cycles. Flooding may result in the contamination of water supplies or the reproduction rate of the pathogen may be influenced by ambient air temperatures (3). Cholera is an indirectly transmitted water-borne anthroponose that is transmitted by a water vehicle: the bacteria (*Vibrio cholerae*) reside in marine ecosystems by attaching to zooplankton. Survival of these small crustaceans in turn depends on the abundance of their food supply, phytoplankton. Phytoplankton populations tend to increase (bloom) when ocean temperatures are warm. As a result of these ecological relationships, cholera outbreaks occur when ocean surface temperatures rise (5).

Indirectly transmitted zoonoses are similar to indirectly transmitted anthroponoses except that the natural cycle of transmission occurs between non-human vertebrates: humans are infected due to accidental encounters with an infected vehicle or vector. This class of disease involves four components in the transmission cycle: the pathogen, biological vector or physical vehicle, animal reservoir, and human host. These diseases are highly susceptible to a combination of ecological and climatic factors because of the numerous components in the transmission cycle, and the interaction of each of these with the external environment (3).

Complex cycles of disease transmission also exist for several diseases which cannot be classified simply by method of transmission or natural reservoir. Such a disease is Rift Valley fever where the virus is primarily a zoonotic disease, spread among vertebrate hosts by the mosquito species *Aedes*. Primarily under flood conditions, *Culex* mosquitoes may feed upon infected ungulate hosts. This vector is referred to as a bridge species because it feeds on humans also, resulting in spread of the virus outside its normal zoonotic cycle (3).

Climate sensitivity of infectious disease

Seasonality of infectious disease

Chapter 5 discussed patterns of winter mortality and infectious disease using the example of cyclic influenza outbreaks occurring in the late fall, winter and early spring in North America. This disease pattern may result from increased likelihood of transmission due to indirect social or behavioural adaptations to the cold weather such as crowding indoors. Another possibility is that it may be attributed directly to pathogen sensitivities to climatic factors such as humidity. In addition to influenza, several other infectious diseases exhibit cyclic seasonal patterns, which may be explained by climate.

In diverse regions around the world, enteric diseases show evidence of significant seasonal fluctuations. In Scotland, campylobacter infections are characterized by short peaks in the spring (6). In Bangladesh, cholera outbreaks occur during the monsoon season (5). In Peru, cyclospora infections peak in the summer and subside in the winter (7). Similarly, some vector-borne diseases (e.g. malaria and dengue fever) also show significant seasonal patterns whereby transmission is highest in the months of heavy rainfall and humidity. Epidemics of other infections (e.g. meningococcal meningitis) tend to erupt during the hot and dry season and subside soon after the beginning of the rainy season in sub-Saharan Africa (8).

Seasonal fluctuations of infectious disease occurrence imply an association with climatic factors. However, to prove a causal link to climate, non-climatic factors must be considered. Furthermore, in order to assess long-term climate influences on disease trends, data must span numerous seasons and utilize proper statistics to account for seasonal fluctuations.

Vector-borne diseases

Important properties in the transmission of vector-borne diseases include:

- survival and reproduction rate of the vector
- time of year and level of vector activity, specifically the biting rate
- rate of development and reproduction of the pathogen within the vector (9).

Vectors, pathogens, and hosts each survive and reproduce within certain optimal climatic conditions and changes in these conditions can modify greatly these properties of disease transmission. The most influential climatic factors for vector-borne diseases include temperature and precipitation but sea level elevation, wind, and daylight duration are additional important considerations. Table 6.1 gives an overview of the impact of climatic change on each biological component of both vector and rodent-borne diseases. The following paragraphs discuss several of these effects in greater detail.

Temperature sensitivity

Extreme temperatures often are lethal to the survival of disease-causing pathogens but incremental changes in temperature may exert varying effects. Where a vector lives in an environment where the mean temperature approaches the limit of physiological tolerance for the pathogen, a small increase in temperature may be lethal to the pathogen. Alternatively, where a vector lives in an environment of low mean temperature, a small increase in temperature may result in increased development, incubation and replication of the pathogen (10, 11). Temperature may modify the growth of disease carrying vectors by altering their biting rates, as well as affect vector population dynamics and alter the rate at which they come into contact with humans. Finally, a shift in temperature regime can alter the length of the transmission season (12).

Disease carrying vectors may adapt to changes in temperature by changing geographical distribution. An emergence of malaria in the cooler climates of the African highlands may be a result of the mosquito vector shifting habitats to cope with increased ambient air temperatures (13). Another possibility is that vectors undergo an evolutionary response to adapt to increasing temperatures. There is recent evidence to suggest that the pitcher-plant mosquito (*Wyeomia smithii*) can adapt genetically to survive the longer growing seasons associated with climate change. Bradshaw and Holzapfel demonstrated this by documenting a change in the photoperiodic response between two different time periods in two populations of pitcher-plant mosquitoes. The change in response was correlated to a marked genetic shift within the mosquito species. A greater degree of micro-evolutionary response was associated with mosquito populations inhabiting higher latitudes; the hypothesis is that because these populations have greater selection pressure they have also a greater ability to evolve genetically. Although this study was limited to one specific species of mosquito, it suggests that other mosquitoes, perhaps disease carrying vectors, may undergo an analogous micro-evolution which would allow adaptation to altered seasonal patterns associated with global climate change (14).

Precipitation sensitivity

Variability in precipitation may have direct consequences on infectious disease outbreaks. Increased precipitation may increase the presence of disease vectors by expanding the size of existent larval habitat and creating new breeding

TABLE 6.1 Effects of weather and climate on vector and rodent-borne diseases[a].

Vector-borne pathogens spend part of their life-cycle in cold-blooded arthropods that are subject to many environmental factors. Changes in weather and climate that can affect transmission of vector-borne diseases include temperature, rainfall, wind, extreme flooding or drought, and sea level rise. Rodent-borne pathogens can be affected indirectly by ecological determinants of food sources affecting rodent population size, floods can displace and lead them to seek food and refuge.

Temperature effects on selected vectors and vector-borne pathogens

Vector
- survival can decrease or increase depending on species;
- some vectors have higher survival at higher latitudes and altitudes with higher temperatures;
- changes in the susceptibility of vectors to some pathogens e.g. higher temperatures reduce size of some vectors but reduce activity of others;
- changes in the rate of vector population growth;
- changes in feeding rate and host contact (may alter survival rate);
- changes in seasonality of populations.

Pathogen
- decreased extrinsic incubation period of pathogen in vector at higher temperatures
- changes in transmission season
- changes in distribution
- decreased viral replication.

Effects of changes in precipitation on selected vector-borne pathogens

Vector
- increased rain may increase larval habitat and vector population size by creating new habitat
- excess rain or snowpack can eliminate habitat by flooding, decreasing vector population
- low rainfall can create habitat by causing rivers to dry into pools (dry season malaria)
- decreased rain can increase container-breeding mosquitoes by forcing increased water storage
- epic rainfall events can synchronize vector host-seeking and virus transmission
- increased humidity increases vector survival; decreased humidity decreases vector survival.

Pathogen
Few direct effects but some data on humidity effects on malarial parasite development in the anopheline mosquito host.

Vertebrate host
- increased rain can increase vegetation, food availability, and population size
- increased rain can cause flooding: decreases population size but increases human contact.

Increased sea level effects on selected vector-borne pathogens
Alters estuary flow and changes existing salt marshes and associated mosquito species, decreasing or eliminating selected mosquito breeding-sites (e.g. reduced habitat for *Culiseta melanura*)

---

[a] The relationship between ambient weather conditions and vector ecology is complicated by the natural tendency for insect vectors to seek out the most suitable microclimates for their survival (e.g. resting under vegetation or pit latrines during dry or hot conditions or in culverts during cold conditions).
Source: reproduced from reference 12.

grounds. In addition, increased precipitation may support a growth in food supplies which in turn support a greater population of vertebrate reservoirs. Unseasonable heavy rainfalls may cause flooding and decrease vector populations by eliminating larval habitats and creating unsuitable environments for vertebrate reservoirs. Alternatively, flooding may force insect or rodent vectors to seek refuge in houses and increase the likelihood of vector-human contact. Epidemics of leptospirosis, a rodent-borne disease, have been documented following severe flooding in Brazil (15). In the wet tropics unseasonable drought can cause rivers to slow, creating more stagnant pools that are ideal vector breeding habitats.

Humidity sensitivity

Humidity can greatly influence transmission of vector-borne diseases, particularly for insect vectors. Mosquitoes and ticks can desiccate easily and survival decreases under dry conditions. Saturation deficit (similar to relative humidity) has been found to be one of the most critical determinants in climate/disease models, for example, dengue fever (16, 17) and Lyme disease models (18).

Sea level sensitivity

The projected rise in sea level associated with climate change is likely to decrease or eliminate breeding habitats for salt-marsh mosquitoes. Bird and mammalian hosts that occupy this ecological niche may be threatened by extinction, which would also aid the elimination of viruses endemic to this habitat (19). Alternatively, inland intrusion of salt water may turn former fresh water habitats into salt-marsh areas which could support vector and host species displaced from former salt-marsh habitats (19).

Water-borne diseases

Human exposure to water-borne infections can occur as a result of contact with contaminated drinking water, recreational water, coastal water, or food. Exposure may be a consequence of human processes (improper disposal of sewage wastes) or weather events. Rainfall patterns can influence the transport and dissemination of infectious agents while temperature can affect their growth and survival (20). Table 6.2 outlines some of the direct and indirect weather effects on enteric viruses, bacteria and protozoa.

TABLE 6.2 Water and food-borne agents: connection to climate.

Pathogen groups	Pathogenic agent	Food-borne agents	Water-borne agents	Indirect weather effect	Direct weather effect
Viruses	Enteric viruses (e.g. hepatitis A virus, Coxsackie B virus)	Shellfish	Groundwater	Storms can increase transport from faecal and waste water sources	Survival increases at reduced temperatures and sunlight (ultraviolet)[a]
Bacteria Cyanobacteria Dinoflagellates	Vibrio (e.g. V. vulnificus, V. Parahaemolyticus, V. cholerae non-01; Anabaena spp., Gymnodinium Pseydibutzschia spp.)	Shellfish	Recreational, Wound infections	Enhanced zooplankton blooms	Salinity and temperature associated with growth in marine environment
Protozoa	Enteric protozoa (e.g. Cyclospora, Crytosporidium)	Fruit and vegetables	Recreational and drinking water	Storms can increase transport from faecal and waste water sources.	Temperature associated with maturation and infectivity of Cyclospora

[a] Also applies to bacteria and protozoa.
Source: Reproduced from reference 20.

Temperature sensitivity

Increasing temperatures may lengthen the seasonality or alter the geographical distribution of water-borne diseases. In the marine environment, warm temperatures create favourable conditions for red tides (blooms of toxic algae) which can increase the incidence of shellfish poisoning (*21*). Increasing sea surface temperatures can indirectly influence the viability of enteric pathogens such as *Vibrio cholerae* by increasing their reservoir's food supply (*5*). Ambient air temperatures also have been linked to hospital admissions of Peruvian children with diarrhoeal disease (*22*).

Precipitation sensitivity

Heavy rains can contaminate watersheds by transporting human and animal faecal products and other wastes in the groundwater. Evidence of water contamination following heavy rains has been documented for cryptosporidium, giardia, and *E.coli* (*4, 23*). This type of event may be increased in conditions of high soil saturation due to more efficient microbial transport (*20*). At the other extreme, water shortages in developing countries have been associated with increases in diarrhoeal disease outbreaks that are likely attributed to improper hygiene (*24*).

Documented and predicted climate/infectious disease links

Research investigating possible links between temporal and spatial variation of climate and the transmission of infectious diseases can be categorized into one of three conceptual areas:

1. evidence for associations between short-term climate variability and infectious disease occurrence in the recent past.
2. evidence for long-term trends of climate change and infectious disease prevalence.
3. evidence from climate and infectious disease linkages used to create predictive models for estimating the future burden of infectious disease under projected climatic conditions.

Historical evidence of climate/infectious disease links

The study of several infectious diseases has resulted in evidence for associations between climatic variations or trends and disease occurrence. These include encephalitis, malaria and various water-borne diseases.

Encephalitis

Evidence suggests that epidemics of certain arboviruses, such as Saint Louis encephalitis virus (SLEV), may be associated with climatic factors. Shaman and colleagues conducted a study on SLEV in south Florida. In this region the transmission cycle can be divided into:

- maintenance: January–March
- amplification: April–June
- early transmission: July–September
- late transmission: October–December.

Mosquito vectors interact with avian hosts during the period of amplification.

The objective of this study was to assess the relationship between precipitation level and SLEV transmission, using a hydrology model to simulate water table depth (WTD) and SLEV incidence in sentinel chickens to estimate human transmission risk.

Three episodes of SLEV transmission were observed in the duration of this study. Each episode occurred during a wet period directly followed by drought (defined by high and low WTD, respectively). These results suggest the following sequence of events predisposing springtime SLEV transmission: spring drought forces the mosquito vector, *Cx. nigripalpus*, to converge with immature and adult wild birds in restrictive, densely vegetated, hammock habitats. This forced interaction of mosquito vectors and avian hosts creates an ideal setting for rapid transmission and amplification of SLEV. Once the drought ends and water sources are restored, the infected vectors and hosts disperse and transmit SLEV to a much broader geographical area. This study suggests that similar drought induced amplification might occur in other arboviruses (*25*).

A similar effect of climate has been suggested for West Nile virus (WNV), introduced into the Americas in 1999. Amplification of the virus is thought to occur under the climatic conditions of warm winters followed by hot dry summers. Similar to SLEV, WNV is a vector-borne zoonotic disease, normally transmitted between birds by the *Culex pipiens* mosquito. The vector tends to breed in foul standing water. In drought conditions, standing water pools become even more concentrated with organic material and mosquito predators, such as frogs and dragonflies, lessen in number. Birds may circulate around small water-holes and thus increase interactions with mosquitoes. In 1999 such climatic conditions existed in the mid-Atlantic States. Together with urban and suburban environments suitable for avian species they are possible explanations for the epidemic (*26*).

Malaria

Scientific evidence suggests that malaria varies seasonally in highly endemic areas. Malaria is probably the vector-borne disease most sensitive to long-term climate change (*27*). Malaria thus provides several illustrative examples (based on historical studies) of the link between infectious disease and climate change, many of which have been described in the previous chapter.

Githeko et al. (*28*) compared monthly climate and malaria data in highland Kakamega and found a close association between malaria transmission and monthly maximum temperature anomalies over three years (1997–2000).

Patz and colleagues studied the effect of soil moisture to determine the effects of weather on malaria transmission. Compared to raw weather data, hydrological modelling has several potential advantages for determining mosquito-breeding sites. High soil moisture conditions and vector breeding habitats can remain long after precipitation events, depending on factors such as watershed, run-off and evapotranspiration. For *An. gambiae*, the soil moisture model predicted up to 45% and 56% of the variability of human biting rate and entomological inoculation rate, respectively (*29*).

The link between malaria and extreme climatic events has long been the subject of study in the Indian subcontinent. Early in the twentieth century, the Punjab region experienced periodic epidemics of malaria. Irrigated by five rivers, this geographical plains region borders the Thar Desert. Excessive monsoon rainfall and resultant high humidity were clearly identified as major factors in the occurrence of epidemics through enhancement of both the breeding and life-

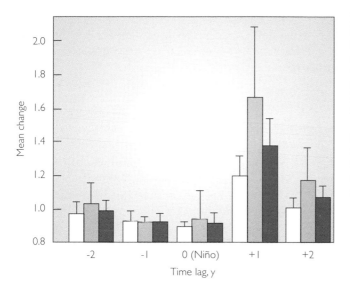

FIGURE 6.2 Relationship between reported malaria cases and El Niño in Venezuela. Average relative change in malaria incidence two years before (−1 and −2 years), during (year 0) and two years after El Niño event (+1 and +2) years). Data are for malaria deaths at the coast 1910–1935 (open bars), malaria cases for the whole country 1975–1995 (grey bars), and the average of both (black bars), with lines indicating the SEs. *Source: reproduced from reference 34.*

span of mosquitoes (*30*). More recently, historical analyses have shown that the risk of a malaria epidemic increased approximately five-fold during the year following an El Niño in this region (*31, 32*). Further, the risk of an epidemic is greater in a year in which excess rain occurs in critical months. A strong correlation is found between both annual rainfall and the number of rainy days and the incidence of malaria in most districts of Rajasthan and in some districts in Gujarat (*33*).

The relationship between interannual climatic variation associated with the ENSO cycle and malaria has been examined in various other countries. For example, Venezuela experiences reduced rainfall during an El Niño year (*34*). Figure 6.2 illustrates that, consistently throughout the twentieth century, malaria rates increased on average by over one-third in the year immediately following an El Niño year. The likely reasons for this include the combination of mosquito-favouring, high rainfall in the post-Niño year with temporarily reduced immunity levels in the local population following the previous low-incidence year.

Water-borne disease

Most observed associations between climate and water-borne diseases are based on indirect evidence of seasonal variations. However, several studies provide quantitative evidence of water-borne diseases' links to climatic factors such as precipitation and ambient air temperature.

The previous chapter introduced one study of childhood diarrhoeal disease in Peru, a good example of historical evidence of a climate-health linkage. Checkley and colleagues used time series regression techniques to analyse the health effects of the 1997–98 EL Niño event on hospital admissions for diarrhoea. This study revealed several important findings. The El Niño event increased hospital admissions up to two-fold in winter, compared to expected rates determined from the previous four years. For each 1 °C increase in temperature, hospital admission increased by 8%. An additional 6225 cases of diarrhoeal disease were attributed to El Niño (*22*).

In another study, Curriero and colleagues examined the association between extreme precipitation and water-borne disease outbreaks in the United States between 1948 and 1994. Their findings show a statistically significant association between heavy precipitation and water-borne disease outbreaks, over 50% of outbreaks being preceded by very wet months within the upper tenth percentile of a 50-year monthly rainfall record (35).

Early indicators for long-term trends in global warming

Many physical and biological indicators of long-term climate change effects have been documented. These include: thawing of permafrost; later freezing and earlier break-up of ice on rivers and lakes; pole-ward and altitudinal shifts in the ranges of a variety of plants and animals; earlier flowering of trees, emergence of insects and egg-laying of birds (36). However, human health outcomes are dependent on many upstream physical and biological systems. Disease analyses are complicated by the potential for numerous human population response options to reduce risk. Even if disease does occur, variability in detection and/or reporting remain major obstacles to determining valid trends in human disease incidence. Despite the difficulty that precludes a quantitative analysis of long-term trends in climate change and incidence of infectious diseases, several studies have reported such associations.

Rodo and colleagues found a robust relationship between progressively stronger El Niño events and cholera prevalence in Bangladesh, spanning a 70-year period. The investigators used innovative statistical methods to conduct a time series analysis of historical cholera data dating back to 1893, to examine the effect of non-stationary interannual variability possibly associated with climate change. In the last two decades, the El Niño Southern Oscillation (ENSO) has differed from previous decades. Since the 1980s there has been a marked intensification of the ENSO, beyond that expected from the known shift in the Pacific Basin temperature regime that began in the mid 1970s. The authors found the association of cholera incidence in the earlier half of the century (1893–1940) to be weak and uncorrelated with ENSO, while late in the century (1980–2001) the relationship is strong and consistent with ENSO. Past climate change, therefore, already may have affected cholera trends in the region via intensified ENSO events (37).

For vector-borne disease, linkages with climate may be questioned by the role of other factors such as socioeconomic, demographic and environmental influences. Kovats and colleagues offer a strategy for critically assessing the evidence for an association between vector-borne diseases and observed climate change. Important criteria for accepting a causal relationship between climate change and disease include:

- evidence for biological sensitivity to climate, requiring both field and laboratory research on important vectors and pathogens;
- meteorological evidence of climate change, requiring sufficient measurements for specific study regions;
- evidence for epidemiological or entomological change with climate change, accounting for potential confounding factors.

Kovats and colleagues stress the importance of frequent and long-term sampling to monitor the full range of specific vector species (9).

Evidence suggests a long-term association between climate change and tick-borne disease, such as encephalitis. In Sweden, Lindgren and colleagues conducted a study to determine whether the increasing incidence of tick-borne encephalitis (TBE) could be linked to changes in the climate during the period 1960–1998. Regression analysis on incident TBE cases was conducted accounting for climatic factors for the two previous years, to account for the long lifespan of the ticks. Results indicate that the increase in TBE was associated with several climatic variables including: two mild consecutive winter seasons; temperatures favouring spring development; extended autumn activity in the year preceding the case; and temperatures favouring tick activity in the early spring of the incident year. One conclusion of this study is that the increased incidence of TBE can be explained by climate changing towards milder winters and early spring arrival. In 1994 Sweden reported the highest rates of TBE: a three-fold increase from the annual average. The year was preceded by five consecutive mild winters and seven early spring arrivals (38).

The investigators suggest a possible role of non-climatic factors in the increase incidence of TBE, including increased summer habitation in the area, availability of TBE vaccine and mammalian host populations respectively (38). Another opinion finds no causal link in the relationship between increased TBE and climate change in Sweden: elsewhere in Europe TBE shows variable patterns with season, suggesting the role of an alternative factor. Other explanations may be sociopolitical circumstances or changing agricultural patterns (39).

Predictive modelling

Models are important tools for analysing complex infectious disease transmission pathways. They allow investigators to link future geographically gridded projections of climate change (especially temperature and precipitation) generated by global or regional climate models to a model of the relationship between those climate variables and the occurrence of the disease of interest. Models will likely be important in directing future studies as well as predicting disease risk. The complex modelling required for the study of infectious diseases is a challenge but several successful attempts have been made. Modelling approaches can be classified into several categories including statistical, process and landscape based models (40).

Statistical based models

Empirical models incorporating a range of meteorological variables have been developed to describe the climatic constraints (the bioclimate envelope) for various vector-borne diseases. The CLIMEX model developed by Sutherst and colleagues (41, 42) maps the geographical distribution of vector species in relation to climate variables. The assessment is based on an ecoclimatic index governed largely by the temperature and moisture requirements of the vector. CLIMEX analyses conducted in Australia indicate that the indigenous vector of malaria would be able to expand its range 330 km south under one typical scenario of climate change. However, such methods cannot include all factors that affect species distributions. Local geographical barriers and interaction/competition between species are important factors which determine whether species colonise the full extent of suitable habitat (43). Assessments may also include additional dynamic population (process-based) models (e.g. DYMEX).

Similarly, Martin and Lefebvre developed a Malaria-Potential-Occurrence-Zone model. This model was combined with 5 general circulation models (GCMs) of global climate to estimate the changes in malaria risk based on moisture and the minimum and maximum temperatures required for parasite development.

This model corresponded fairly well with the distribution of malaria in the nineteenth century and the 1990s, allowing for areas where malaria had been eradicated. An important conclusion of this exercise was that all simulation runs showed an increase in seasonal (unstable) malaria transmission (*44*).

Rogers and Randolph presented data contradicting the prevailing predictions of global malaria expansion. They used a two-step statistical approach that used present day distributions of *falciparum* malaria to identify important climatic factors and applied this information, along with projected climatic conditions, to predict future *falciparum* distributions. The model showed future *falciparum* habitat distributions similar to the present day. One of their conclusions is that the organism's biological reaction to temperature extremes will likely be balanced by other factors, such as precipitation (*45*).

Another approach, used by Reeves and colleagues, incorporates experimental and observational data to examine the effect of temperature on survival of mosquito species *Culex tarsalis*, the primary vector for St. Louis encephalitis (SLE) and western equine encephalitis (WEE). Observational data were obtained for two regions of California that differ consistently by an average of 5 °C. In the northern region vector populations peak in midsummer, southern populations peak in the spring. The experimental data was obtained by use of a programmable environmental chamber that allowed investigators to study the effect of temperature on a variety of factors including vector competence and survival. Investigators predict that an increase of 5 °C across California would result in a loss of WEE prevalence and an expansion of SLE in the warmer region (*19*).

Process-based (mathematical) models

A process-based approach is important in climate change studies as some anticipated climate conditions have never occurred before and cannot be empirically based.

Martens and colleagues have developed a modelling method (*46, 47, 48, 49*). Their MIASMA model links GCM-based climate change scenarios with the formula for the basic reproduction rate (R_o) to calculate the 'transmission potential' of a region where malaria mosquitoes are present. The basic reproduction rate is defined as the number of new cases of a disease that will arise from one current case introduced into a non-immune host population during a single transmission cycle (*50*). This harks back to classical epidemiological models of infectious disease. Model variables within R_o that are sensitive to temperature include mosquito density, feeding frequency, survival and extrinsic incubation period (i.e. time required for parasite to develop the in the mosquito). The MIASMA approach models temperature effects on the last three of these processes, making an assumption (conservative in many habitats) that mosquito density remains unaffected by temperature increases. The model shows that the effects of small temperature increases on the extrinsic incubation period have large affects on transmission potential. This is consistent with the observation that the minimum temperature for parasite development is the limiting factor for malaria transmission in many areas.

In an effort to examine the apparent trend of increasing endemic malaria in the African highlands, Lindsay and Martens used the MIASMA model based on climate variables to identify epidemic prone regions in the African highlands and make predictions for future epidemic regions with projected changes in the climate. The model was used to calculate the effect of climate on factors such as mosquito development, feeding frequency, longevity and parasite incubation time. Baseline transmission potential was compared to several futuristic scenarios including:

- 2 °C increase in ambient air temperature
- 2 °C increase in ambient air temperature and 20% increase in precipitation
- 2 °C increase in ambient air temperature and 20% decrease in precipitation.

Increasing temperature resulted in an increase in transmission potential at high elevations (>900 m). Also, transmission potential was found to decrease with decreasing precipitation. A combination of decreasing precipitation and increasing temperature is unfavourable for disease transmission (51).

Hartman et al. (52) used 16 GCM future scenarios to study their effect on climate suitability for malaria in Zimbabwe. Using the MARA/ARMA model (see Box 1) of climatic limits on *Anopheles* mosquitoes and *Plasmodium* parasites, the authors mapped the change in geographical distribution of endemic malaria. For all scenarios the highlands become more suitable for malaria transmission while the lowlands show varying degrees of change.

Process-based models also have been applied to dengue fever. Focks and colleagues developed a pair of probabilistic models to examine the transmission of dengue fever in urban environments. The container-inhabiting mosquito simulation model (CIMSiM) is an entomological model which uses information from weather to model outcomes such as mosquito abundance, age, development, weight, fecundity, gonotrophic status and adult survival. The dengue simulation model (DENSiM) uses the entomological output from the CIMSiM to determine the biting mosquito population. Survival and emergence factors are important predictors of population size within the DENSiM and gonotrophic development and weight are major influences on biting rates. These combined models provide a comprehensive account of the factors influencing dengue transmission. The on-going validation of these models will help to create a useful tool for future dengue control (16).

Landscape based models

Climate influences infectious disease transmission by not only affecting directly the rate of biological processes of pathogens and vectors but also influencing their habitats. Clearly there is potential in combining the climate-based models described above with the rapidly developing field of study which uses spatial analytical methods to study the effects of meteorological and other environmental factors (e.g. different vegetation types), measured by ground-based or remote sensors.

Satellite data can be used to identify the environmental limitations defining a vector/host/pathogen range. They can be analysed with weather data to determine the factors most important in describing current disease distribution, and the effect of projected changes in climate on this distribution (53). Other indices, such as the Normalised Difference Vegetation Index (NDVI), which correlates with the photosynthetic activity of plants, rainfall and saturation deficit also may be useful (54).

Remote sensing has been used in several studies of malaria. Mosquito distribution and malaria incubation time are dependent on climatic factors such as moisture and temperature. This allows the potential for satellite imagery of climate to predict mosquito distribution and transmission patterns. These extrinsic data, along with intrinsic variables such as immunity and nutritional status of the population, can help to create models appropriate at the local level for present climatic scenarios. In turn these can be used as future warning tools for predicted climate scenarios (55). An illustrative example of this type of modelling is the MARA project which used satellite data to stratify malaria risk in the African highlands (Box 1).

Thomson et al. conducted satellite based spatial analysis in the Gambia to measure the impact of bed net use on malaria prevalence rates in children. Satellite data (NDVI) are used to measure changes in vegetative growth and senescence as proxy ecological variables representing changes in rainfall and humidity to predict length and intensity of malaria transmission. After accounting for spatial correlation of the villages used in the study, investigators found a significant association between malaria prevalence and both NDVI and use of bed nets (treated or untreated), by use of multiple regression analysis. This study suggests that NDVI data can be a useful predictor of malaria prevalence in children, having accounted for behavioural factors (56).

A three-phase study of malaria provides another example of satellite based predictive modelling. The first phase of this study tested the use of remote sensing data and ground surveillance data as predictive tools for the temporal and spatial distribution of larval populations of mosquito vector *Anopheles freeborni*, in California. The satellite measurements predicted 90% of peak densities of larvae as a function of rice field distance from livestock pastures. The second phase investigated the ecology of the primary malaria vector species *Anopheles albimanus* in the Mexican state of Chiapas: specifically, the association of mosquito larval populations with aquatic plant species, discerned though digitally processed satellite data. Results showed that transitional swamps and unmanaged pastures were the most important landscape factors associated with vector abundance (57). Phase three combined digitized map, remotely sensed and field data to define the association between vegetation and individual villages in Belize with high or low mosquito densities. These three studies show the usefulness of remotely sensed data for malaria surveillance (58).

Other vector-borne diseases have been studied using remote data. Rogers et al. mapped the changes of three important disease vectors (ticks, tsetse flies and mosquitoes) in southern Africa under three climate change scenarios (59). The results indicate significant potential changes in areas suitable for each vector species, with a net increase for malaria mosquitoes (*Anopheles gambiae*) (53). Thomson et al. used satellite data to map the geographical distribution of the sandfly vector, *Phlebotomus orientalis*, for visceral leishmaniasis in the Sudan (60).

In a Maryland study of *Ixodes scapularis* (the tick vector that transmits Lyme disease in the eastern United States) Glass and colleagues used GIS to extract information on a total of 41 environmental variables pertinent to the abundance of tick populations on white-tailed deer, the zoonotic host. They found tick abundance to be positively correlated with well-drained sandy soils and negatively correlated with urban land-use patterns, wetlands, privately owned land, saturated soils and limited drainage (61).

Recent work by Kitron's group shows similar results in the distribution and abundance of *Ixodes scapularis* in Wisconsin, northern Illinois, and portions of the upper peninsula of Michigan. Tick presence was positively associated with sandy or loam-sand soils overlying sedimentary rock (62). Deciduous or dry to mesic forests also were positive predictors. Absence was correlated with grasslands; conifer forests; wet to wet/mesic forests; acidic soils of low fertility and a clay soil texture; and Pre-Cambrian bedrock.

BOX 6.1 The Highland Malaria Project

WHO estimates that approximately 300–500 million cases of malaria occur worldwide each year. A global increase in malaria may be associated with deforestation, water development projects, and agricultural practices in poor countries. High altitude regions have been protected from malaria endemicity because parasite sporogony and vector development are inefficient in low temperatures. However, there appears to be an emergence of malaria in the African highlands. This may be attributable to a true change in disease pattern caused by increasing temperatures associated with climate change. As global temperatures continue to rise, it is important to have a system that allows public health practitioners to forecast where and when malaria epidemics may occur.

The Highland Malaria Project (HIMAL) is part of the umbrella international collaboration "Mapping Malaria Risk in Africa" (MARA). The project consists of two phases. The main objective of phase one was to create a stratified model of malaria risk for several select regions in the African highlands using GIS based modelling, and to compare these special models to the known historical distribution of malaria epidemics. This phase of the project prepared for a future phase two with the goal of predicting where and when malaria epidemics may occur.

FIGURE 6.3 Fuzzy climate suitability for transmission of malaria across Africa, according to the MARA (Mapping Malaria Risk in Africa) climate suitability model.
Areas of climate unsuitability include both deserts (limited by low rainfall), and highland areas (limited by cold temperatures), for example on the border between Uganda and Rwanda.
Source: Produced by Marlies Craig/MARA project (63).

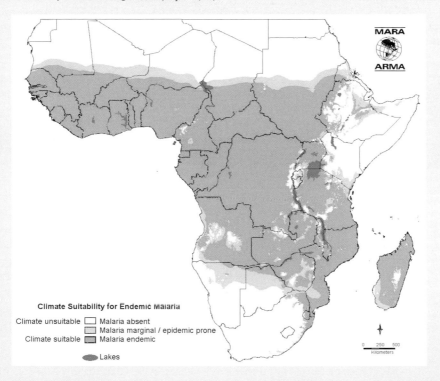

The HIMAL model used fuzzy suitability fractions to predict the likelihood of conditions suitable for malaria transmission for any given location. Suitability fractions between zero and one were calculated for each location by overlaying the average monthly temperature by the mean rainfall over several years(64). More conceptual definitions of the regions defined by the suitability scale include:

Continued

- perennial: conditions always suitable for transmission;
- seasonal: conditions suitable for short season every year;
- epidemic: long-term variations in climate render conditions suitable for transmission on irregular basis;
- malaria free: conditions always unsuitable for transmission.

The strength of this project was the technique used to validate the model using data from several East African countries, including Kenya, Uganda, Rwanda/Burundi, and the United Republic of Tanzania. The purpose of the validation was to compare the model estimates to real historical malaria transmission intensity data. This data came from a variety of sources including surveys, seasonal data, published and unpublished studies, and site visits. Information on climate, as well as parasite rates, spleen rates (enlarged spleen) and epidemic locations were obtained for a total of 1713 sample points.

One important consideration is that the parameters used to create the climate suitability scale (average continental monthly temperature and mean rainfall over several years) were static and gave an average expected risk. Temporal variation in transmission and annual fluctuation or long-term systematic changes in climate therefore may have been masked.

In the validation study the model predictions based on fuzzy climatic estimates closely approximated the known distribution of malaria, based on case data and historical maps across most of the continent (including the highlands). However, discrepancies along some large river valleys and other locations may be attributed to several possible limitations of the model.

One possible limitation was the use of rainfall estimates to predict vector habitats. Rainfall alone may not account for vector breeding grounds along lakes, riverbanks and flood plains. Lake Bunyonyi in Uganda provides a suitable breeding site for vectors, and may act as a spawning point for malaria epidemics in the surrounding areas. In addition, the model was unable to account for other possible confounders such as health services, malaria control programmes and drug resistance.

Using the historical malaria data this project was able to make several important observations. Highland epidemics generally tend to occur within defined altitudinal ranges and these epidemic areas generally are limited to small, discrete areas. Also the historical evidence showed that the most significant widespread epidemics occurred during or after abnormal weather events. Incorporating additional local factors into geographical models will enhance predictability of future malaria epidemics.

ENSO and predictive models for early warning systems

New insights gained from observations of extreme climate variability accompanying the El Niño phenomenon can be employed for ENSO-based disease predictions. Early warning systems (EWS) are feasible given that ENSO events can be anticipated months in advance. Several examples of these ENSO-based climate studies are presented in the previous chapter. The examples below describe studies that found strong ENSO/disease associations and either are already used for early warning (e.g. Glass et al) or have good potential for a weather-based disease early warning system.

By analysing hantavirus pulmonary syndrome (HPS) in the American southwest, Glass et al. (65) found human cases of disease to be preceded by ENSO-related heavy rainfall, with a subsequent increase in the rodent population. Landsat Thematic Mapper satellite imagery collected the year before the outbreak was used to estimate HPS risk by logistic regression analysis. Satellite and elevation data showed a strong association between environmental conditions and HPS risk the following year. Repeated analysis using satellite imagery from a non-ENSO year (1995) showed substantial decrease in medium to high-risk areas.

The United States' Indian Health Service uses this ENSO-based predictive model in a regional hantavirus early warning system.

Linthicum et al. (66) found a strong relationship between Rift Valley fever and ENSO-driven rainfall in East Africa. Rift Valley fever virus outbreaks in East Africa from 1950 to May 1998 followed periods of unusually high rainfall. By using Pacific and Indian Ocean sea surface temperature anomalies, coupled with satellite normalized difference vegetation index (NDVI) data, it was demonstrated that Rift Valley fever outbreaks could be predicted up to five months in advance of outbreaks in East Africa. Near-real-time monitoring with satellite images could pinpoint actual affected areas, with good prospects of becoming a useful disease EWS for targeting livestock vaccination to protect both animals and humans.

The World Health Organization regional office for south-east Asia has voiced a need for an operational early warning system that would provide sufficient lead time (one to three months) to permit mobilization of control operations. Focks and colleagues (unpublished data) have developed a prototype dengue EWS. Predictive variables include: sea surface temperature anomalies (five month running mean of spatially averaged SST anomalies over the tropical Pacific: 4°S–4°N, 150°W–90°W as measured by the Japanese Meteorological Association) and past monthly cases of dengue in several cities. With these variables, the probability of an epidemic year can be forecasted one to two months before peak transmission season. The two and one month forecasts had respective error rates of three and two per thirty-five years. This EWS is sufficiently accurate and will be put into use in Yogyakarta, Indonesia in 2002.

Pascual and colleagues (37, 67) conducted analysis of ENSO-cholera relationships in Bangladesh. Results from non-linear time series analysis support ENSO's important role in predicting disease epidemics. Considering a possible threshold effect (37), it may be feasible to develop a cholera EWS for the region.

Modifying influences

Climate is one of several factors that can influence the spread of infectious disease. Human activities and behaviours also are critical determinants of disease transmission. Sociodemographic factors include (but are not limited to): increasing trends in travel, trade, and migration; erratic disease control efforts; emerging drug or pesticide resistance; and inadequate nutrition. Environmental influences include: changes in land-use, such as the clearing of forested land, agricultural development and water projects, urban sprawl; as well as ecological influences. These sociodemographic and environmental influences may act in synergy or antagonistically with climatic factors, exacerbating or lessening the impact on infectious disease transmission of either factor acting independently.

Sociodemographic influences

Human travel trade and migration

Dramatic improvements in human ability and efficiency to travel have affected infectious disease transmission. Within the incubation time for diseases such as malaria, dengue and West Nile virus, an infected person can travel the distance from a rain forest in Africa to a United States suburb (68). The movement of food across borders also can result in the importation of infectious agents, as in the case of cyclospora transported into the United States on berries from South America (69). Humans who migrate from disease free areas to endemic regions typically lack immunity, making them susceptible to infection and transmission of the disease. This is of relevance for new settlers living at the forest's edge on cleared farmland who are particularly susceptible to zoonotic parasites. In addition, migrants may act as reservoirs: carrying pathogens home to their native regions.

Disease control efforts

While biological and technical knowledge are needed to control the spread of infectious diseases, additional requirements include political will, financial resources and national stability. Successful moves towards eradication and control of communicable diseases, such as smallpox (eradication) and polio (control), largely are attributable to unwavering global commitment. In other scenarios, erratic disease control efforts have had detrimental consequences for human infection. In 1947 the Pan-American Health Organization (PAHO) initiated a programme to eradicate *Aedes aegypti*, the mosquito vector for both dengue and yellow fever. By 1972 this mosquito had been eradicated from 73% of the area originally infected: 19 countries of the Americas. However, nearly one decade later, lack of continuing public support and withdrawal of funding resulted in re-establishment of the mosquito in nearly the entire original habitat. Subsequently, major epidemics of dengue have erupted in several South and Central American countries as well as parts of Africa and Asia (*70*).

Drug resistance

Drug resistance results from an infectious agent genetically mutating to avoid harm from drugs. Resistance to important anti-malarial drugs has been rising in many parts of the world since the 1970s. Chloroquine's use as a treatment for malaria infections has led to the broad geographical spread of resistant *P. falciparum* in Africa (*71*). Similarly, the large scale use of antimicrobials in livestock and humans has lead to resistance among bacterial enteropathogens, an important problem in developing countries that lack medical supervision to dispense appropriate medication (*72*).

Nutrition

Malnutrition is an important determinant of infectious disease morbidity and mortality. Micronutrients play an important role in the ability to mount an effective immune response against infection. Malnutrition is a particularly important risk factor for diarrhoeal diseases in children in developing countries (*73*).

Environmental influences

Land-use influences

Land-use changes typically are undertaken with the intention to improve livelihoods, economic well-being and quality of life. This is achieved in many cases. However, as well as adverse environmental effects, adverse human health effects may be an unintended outcome (*74, 75, 76*). With increased land-use changes such as deforestation and increased agriculture and crop irrigation, climate variability's adverse effect on infectious diseases may be heightened. Several of these environmental factors are given in Table 6.3, a few are described in detail below.

Deforestation: The status of the world's forests is threatened by: conversion for crop production or pastures; road or dam building; timber extraction; and the

TABLE 6.3 Examples of environmental changes and possible effects on infectious diseases.

Environmental changes	Example diseases	Pathway of effect
Dams, canals, irrigation	Schistosomiasis	↑ Snail host habitat, human contact
	Malaria	↑ Breeding sites for mosquitoes
	Helminthiasies	↑ Larval contact due to moist soil
	River blindness	↓ Blackfly breeding, ↓ disease
Agricultural intensification	Malaria	Crop insecticides and ↑ vector resistance
	Venezuelan haemorraghic fever	↑ rodent abundance, contact
Urbanization, urban crowding	Cholera	↓ sanitation, hygiene; ↑ water contamination
	Dengue	Water-collecting trash, ↑ *Aedes aegypti* mosquito breeding sites
	Cutaneous leishmaniasis	↑ proximity, sandfly vectors
Deforestation and new habitation	Malaria	↑ Breeding sites and vectors, immigration of susceptible people
	Oropouche	↑ contact, breeding of vectors
	Visceral leishmaniasis	↑ contact with sandfly vectors
Reforestation	Lyme disease	↑ tick hosts, outdoor exposure
Ocean warming	Red tide	↑ Toxic algal blooms
Elevated precipitation	Rift valley fever	↑ Pools for mosquito breeding
	Hantavirus pulmonary syndrome	↑ Rodent food, habitat, abundance

Source: reproduced from reference 3.

encroachment of urban areas. Historically these activities have been associated with changes in infectious diseases in the local population. The diseases most frequently affected are those that exist naturally in wild ecosystems and circulate among animals, especially those with vertebrate reservoirs and invertebrate vectors. In general such changes result from factors affecting the populations of animal reservoirs, vectors, and pathogens, or from factors associated with human exposure. Deforestation and forest fragmentation in Latin America has resulted in an increase in the incidence of visceral leishmaniasis associated with an increase in the number of fox reservoirs and sandfly vectors that have adapted to the peri-domestic environment (77). Finally, removal of vegetation (and thereby transpiration of plants) can alter local weather patterns significantly.

Agricultural development and other water projects: Agricultural development can lead to an increase in diarrhoeal disease. In intensely stocked farmland, heavy rains can cause contamination of water resources by *Cryptosporidium parvum* oocysts. Infiltration of high quality water treatment and supply systems can occur: a 1993 occurrence in Milwaukee, USA, resulted in 400 000 cases of cryptosporidiosis (78). Intense cattle farming and livestock operations in combination with factors related to watershed management have been implicated in such outbreaks (79). A similar mechanism is involved in giardiasis where a variety of animals may serve as reservoirs of Giardia lamblia and contaminate surface water with their excreta. Predicted flooding accompanying climate change could increase the water contamination trends associated with agricultural development.

Agricultural development in many parts of the world has resulted in an increased requirement for crop irrigation, reducing water availability for other uses and increasing breeding sites of disease vectors. An increase in soil moisture associated with irrigation development in the Southern Nile Delta following the construction of the Aswan High Dam has caused a rapid rise in the mosquito *Culex pipiens* and consequential increase in the mosquito-borne disease,

Bancroftian filariasis (*80, 81*). Onchocerciasis and trypanosomiasis are further examples of vector-borne parasitic diseases that are triggered by changing land-use and water management patterns. Projected droughts will necessitate expanded irrigation to counter water stress, which could create more breeding sites for vector-borne diseases.

Urbanization: Urbanization is associated with a range of health problems, including vector-borne diseases such as dengue and malaria (*82*), diarrhoeal (*83*) and respiratory diseases (*84*). Overcrowding and pollution resulting from inadequate infrastructure can trigger these conditions. At present, there are an estimated four billion cases of diarrhoeal disease each year, causing over two million deaths. Studies have shown that water sanitation and hygiene interventions can greatly reduce water-related diseases (*85, 86*).

Ecological influences

In addition to land-use changes, there is a host of indirect links between infectious disease and environmental conditions that are mediated through changes in ecosystems resulting from human activities (*40*). Zoonoses and vector-transmitted anthroponoses, dependent on the ecology of non-human animals, will be especially sensitive to the effects of these ecological changes. An estimated 75% of emerging infectious diseases of humans have evolved from exposure to zoonotic pathogens (*87*), therefore any changes in the ecological conditions influencing wildlife diseases have the potential to impact directly on human health (*3, 88*). Lyme disease in the north-east United States is an example of this (Box 2).

Climate change will likely modify the relationships between pathogens and hosts directly by: altering the timing of pathogen development and life histories; changing seasonal patterns of pathogen survival; changing hosts' susceptibility to pathogens (*12, 93*). However, ecosystem processes can influence human infectious diseases indirectly (*3, 40, 94*).

Forecasts of infectious diseases' responses to climate change are complicated by the difficulties associated with predicting how ecosystems will respond to changes in climate (*40, 94*). Geographical distributions and abundances of species that compose ecological communities depend, in part, on patterns of temperature and precipitation (*36*). The link between metabolic rates and temperature means that average temperatures also will have an influence on the growth rates and generation times of many organisms (*36, 95, 96*). Patterns of precipitation also will influence species abundance and distribution by determining plant/habitat range and/or vector breeding sites (*97*). Any changes in the aquatic community will affect organisms involved in human infectious disease: mosquitoes that transmit diseases such as malaria and dengue fever develop in aquatic habitats. Changes in precipitation amounts in turn will alter the availability of suitable habitats for larval mosquitoes.

One of ecology's current challenges is to predict how species will respond. An initial approach has been to couple data on climate tolerances for species and their current geographical distributions with projections from climate models in an attempt to determine where suitable conditions for the species might develop. In one study of 80 tree species, species varied in how closely their current distributions correlated with climate. There were also strong differences between species in the predicted impact of future climate change (*98*). Analogous patterns appear when examining expected changes in animal communities (*99*).

This variability in climate sensitivity is only one aspect of the expected differences in how species will respond to climate change. Species also will vary in

BOX 6.2 Ecological influences and Lyme disease

Variability in the prevalence of Lyme disease in the north-east United States provides an informative example of how changes in ecological conditions can produce changes in an infectious disease of humans.

As land-use patterns have shifted away from agriculture and second-growth forests have become more prevalent, the populations of ticks have changed (36). Human infection rates have risen over time with the increase in contact between humans and ticks (89). Spatial variation in risk to humans is influenced by the ecology of the mammalian hosts of the tick. White-tailed deer are the primary host for adult ticks and their population size is an important determinant of tick abundance in a particular region (90, 91). However, while population size is an important factor determining the probability of a human being bitten, not all ticks will be infected with the Lyme disease bacterium. The proportion of infected ticks in an area appears to be linked to the relative abundance of highly competent small mammal reservoirs for the B. burgdorferi bacteria, specifically white-footed mice (89, 91).

The probability of a human contracting Lyme disease in a given area is a function of the density of infected ticks, which in turn depends on the population size of deer and mice. Additional ecological relationships will influence the abundance of these mammalian hosts. A good example is the key role played by oak trees (92). The links between these trees and human disease start with the periodic production of acorns by masting oak trees. Acorns are a high quality food for wildlife, and deer tend to concentrate in their vicinity during the fall and winter. These concentrations of deer result in larger numbers of larval ticks. At the same time, white-footed mice also increase in abundance in areas with masting oaks. The interplay between these species dramatically increases both the concentration of ticks and the proportion of ticks infected with Lyme-causing spirochaetes (89, 91).

Recent modelling studies suggest that complete understanding of the ecology of Lyme disease will need to include other ecological influences (91). Results suggest the probability of a tick becoming infected depends not simply on the density of white-footed mice, but on the density of mice relative to other hosts in the community. Under this scenario, the density effect of white-footed mice (efficient reservoirs for Lyme disease) can be diluted by increasing density of alternative hosts that are less efficient at transmitting the disease. These results lead to the conclusion that increasing host diversity (species richness) may decrease the risk of disease through a dilution-effect (91).

Habitat fragmentation (e.g. due to urban sprawl and road building) may exacerbate the loss of species biodiversity, thereby increasing the risk of Lyme disease. Wildlife biodiversity decreases when human development fractures forests into smaller and smaller pieces. Deforestation also can lead to closer contact between wildlife and humans and their domestic animals. Each of these ecosystem changes has implications for the distribution of micro-organisms and the health of human, domestic animal and wildlife populations.

their abilities to disperse to new geographical areas and compete with the species already in place. Some species have limited ability to disperse, others will be able to shift ranges rapidly (100). Once in a new area, species vary greatly in their ability to become established and increase in population size (101).

There is conflicting evidence about the speed at which species might evolve in response to climate change. The necessary genetic variability exists in many species but the pace of environmental change may outstrip their ability to evolve

(*102*). Given evolutionary change's importance in the emergence of new and newly virulent diseases, better understanding of its role in infectious disease ecology is a critical area of research concern (*102*).

While significant uncertainties exist about the exact nature of the ecological changes expected as climate changes, there is little doubt that ecosystems will change. Indeed, widespread changes are already seen in both the geographical range of species and the timing of annual events (phenology). Analyses of ecological data from recent decades have revealed widespread shifts in species' range, abundance and phenology, associated with the comparatively moderate climatic changes of the past century (*103*). It also appears that some species have undergone adaptive evolution in response to recent climate change (*104*): the pitcher-plant mosquito for example. Changes in ecological relationships will become more obvious as expected changes in climate continue over coming decades (*36*).

There is ample evidence that species will differ in their sensitivity to climate and respond to climate change at different rates (*36, 103*). It is highly likely that this differential ability to respond will break up important ecological relationships: in western Europe there is a well studied ecological relationship between oak trees, winter moths that feed on the leaves, and insectivorous birds that feed on moth caterpillars. Analyses show that during recent warmer springs, these organisms have not responded equally to changes in climate, threatening to disrupt the ecological linkages in the community (*105, 106*).

Severe epidemics of cholera strike regularly in many parts of the developing world (*107, 108*). The timing of these epidemics is partly explained by environmental and ecological conditions that are influenced by climate. In particular a significant reservoir of the cholera-causing organism, *Vibrio cholerae*, appears to reside in marine ecosystems where it attaches to zooplankton (*5, 107*). Populations of these small crustaceans in turn depend on the abundance of their food supply (phytoplankton). Phytoplankton populations tend to increase (bloom) when ocean temperatures are warm. The result is that cholera outbreaks are associated with warmer ocean surface temperature via a series of ecological relationships: phytoplankton blooms lead to higher numbers of zooplankton and their associated *Vibrio cholerae*, which in turn are more likely to infect exposed humans (*5, 107, 108*).

Climate also will impact on the diseases of plants and non-human animals. This impact provides a pathway for complex interactions where ecosystems change in response to climate and alter the conditions for infectious disease organisms and vectors. Resulting changes in the patterns of infectious disease may in turn lead to further ecological disruption through the changing distributions and populations of target plants and animals of those infectious agents (*93*). With growing awareness of the importance of disease in ecosystems and the link to human health, it may be possible to detect such complex pathways in current systems. For example, the interactions between emerging wildlife diseases such as mycoplasmal conjunctivitis in wild house finches in the eastern United States and the spread of West Nile virus in the same geographical area (*109, 110*).

Conclusions and recommended future steps

The purpose of this chapter is to highlight the evidence linking climatic factors such as temperature, precipitation, and sea level rise, to the lifecycles of infectious diseases, including both direct and indirect associations via ecological processes. Many studies demonstrate seasonal fluctuations in infectious diseases but few have documented long-term trends in climate-disease associations. A variety of models has been developed to simulate the climatic changes and predict future disease outbreaks although few have controlled successfully for important sociodemographic and environmental influences. Gaps in knowledge indicate that future initiatives are required in the following areas:

Increase in active global disease surveillance. The lack of precise knowledge of current disease incidence rates makes it difficult to comment about whether incidence is changing as a result of climatic conditions. Incidence data are needed to provide a baseline for epidemiological studies. These data also will be useful for validating predictive models. As these data are difficult to gather, particularly in remote regions, a centralized computer database should be created to facilitate sharing of these data among researchers (40).

Continuation of epidemiological research into associations between climatic factors and infectious diseases. In order to draw a causal relationship between climate change and patterns of infectious disease, research needs to prove consistent trends across diverse populations and geographical regions. This will best be accomplished by implementing rigorous study designs that adequately control for social and environmental confounders (40). International collaboration between researchers is important (36) as well as interdisciplinary collaboration between specialists such as epidemiologists, climatologists and ecologists, in order to expand the breadth of information. A comprehensive study of mosquito-borne diseases, for example, requires a combination of entomologists, epidemiologists and climatologists to work together to examine the associations of changing vector habitats, disease patterns and climatic factors. Epidemiological data can be shared with policy-makers to make preventive policies (36).

Further development of comprehensive models. Models can be useful in forecasting likely health outcomes in relation to projected climatic conditions. Integrating the effects of social and environmental influences is difficult but necessary (40).

Improvements in public health infrastructure. These include public health training, emergency response, and prevention and control programmes. Improved understanding is needed of the adaptive capacity of individuals affected by health outcomes of climate change, as well as the capacity for populations to prepare a response to projected health outcomes of climate change (36).

References

1. McNeil, W.H. Plagues and peoples, New York, USA, Doubleday, 1976.
2. Nelson, K.E. Early history of infectious disease: epidemiology and control of infectious diseases. In: *Infectious Disease Epidemiology*, Nelson, K.E. et al. eds. Gaithersburg, MD, USA, Aspen Publishers Inc. pp. 3–16, 2000.
3. Wilson, M.L. Ecology and infectious disease. In: *Ecosystem change and public health: a global perspective*, Aron J.L. & Patz, J.A. eds. Baltimore, USA, John Hopkins University Press, pp. 283–324, 2001.
4. Parmenter, R.R. et al. Incidence of plague associated with increased winter-spring precipitation in New Mexico. *American Journal of Tropical Medicine and Hygiene.* 61(5): 814–821 (1999).
5. Colwell, R.R. Global climate and infectious disease: the cholera paradigm. *Science* 274(5295): 2025–2031 (1996).
6. Colwell, R.R. & Patz, J.A. *Climate, infectious disease and health.* Washington, DC, USA, American Academy of Microbiology, 1998.

7.   Madico, G. et al. Epidemiology and treatment of Cyclospora cayetanensis infection in Peruvian children. *Clinical Infectious Diseases* 24(5): 977–981 (1997).

8.   Moore, P.S. Meningococcal meningitis in sub-Saharan Africa: a model for the epidemic process. *Clinical Infectious Diseases* 14(2): 515–525 (1992).

9.   Kovats, R.S. et al. Early effects of climate change: do they include changes in vector-borne disease? *Philosophical Transactions of the Royal Society of London B Biological Sciences* 356(1411): 1057–1068 (2001).

10. Lindsay, S.W. & Birley, M.H. Climate change and malaria transmission. *Annals of Tropical Medicine and Parasitology* 90(6): 573–588 (1996).

11. Bradley, D.J. Human tropical diseases in a changing environment. Ciba Foundation Symposium, 175: 146–62; discussion 162–170 (1993).

12. Gubler, D.J. et al. Climate variability and change in the United States: potential impacts on vector- and rodent-borne diseases. *Environmental Health Perspectives* 109 Suppl 2: 223–233 (2001).

13. Cox, J. et al. *Mapping malaria risk in the highlands of Africa.* MARA/HIMAL technical report. p. 96, 1999.

14. Bradshaw, W.E. & Holzapfel, C.M. Genetic shift in photoperiodic response correlated with global warming. *Proceedings of the National Academy of Sciences USA* 98(25): 14509–14511 (2001).

15. Ko, A.I. et al. Urban epidemic of severe leptospirosis in Brazil. Salvador Leptospirosis Study Group. *Lancet* 354(9181): 820–825 (1999).

16. Focks, D.A. et al. A simulation model of the epidemiology of urban dengue fever: literature analysis, model development, preliminary validation, and samples of simulation results. *American Journal of Tropical Medicine and Hygiene* 53(5): 489–506 (1995).

17. Hales, S. et al. Potential effect of population and climate changes on global distribution of dengue fever: an empirical model. *Lancet* 360: 830–834 (2002).

18. Mount, G.A. et al. Simulation of management strategies for the blacklegged tick (*Acari: Ixodidae*) and the Lyme disease spirochete, *Borrelia burgdorferi. Journal of Medical Entomology* 34(6): 672–683 (1997).

19. Reeves, W.C. et al. Potential effect of global warming on mosquito-borne arboviruses. *Journal of Medical Entomology* 31(3): 323–332 (1994).

20. Rose, J.B. et al. Climate variability and change in the United States: potential impacts on water- and foodborne diseases caused by microbiologic agents. *Environmental Health Perspectives* 109 Suppl 2: 211–221 (2001).

21. Epstein, P.R. Algal blooms in the spread and persistence of cholera. *Biosystems.* 31(2–3): 209–221 (1993).

22. Checkley, W. et al. Effect of El Niño and ambient temperature on hospital admissions for diarrhoeal diseases in Peruvian children. *Lancet* 355(9202): 442–450 (2000).

23. Atherholt, T.B. et al. Effects of rainfall on Giardia and Cryptosporidium. *Journal of the American Water Works Association* 90(9): 66–80 (1998).

24. Monograph on water resources and human health in Europe. Rome, Italy, WHO-European Centre for Environment and Health/European Environment Agency, 1999.

25. Shaman, J. et al. Using a dynamic hydrology model to predict mosquito abundances in flood and swamp water. *Emerging Infectious Diseases* 8(1): 6–13 (2002).

26. Epstein, P.R. West Nile virus and the climate. *Journal of Urban Health* 78(2): 367–371 (2001).

27. World Health Organization (WHO), *Climate change and human health.* McMichael, A.J. et al. eds. Geneva, Switzerland, World Health Organization, 1996.

28. Githeko, A. & Ndegwa, W. Predicting malaria epidemics in the Kenyan highlands using climate data: a tool for decision-makers. *Global Change & Human Health* 2: 54–63 (2001).

29. Patz, J.A. et al. Predicting key malaria transmission factors, biting and entomological inoculation rates, using modelled soil moisture in Kenya. *Tropical Medicine and International Health* 3(10): 818–827 (1998).

30. Christophers, S.R. Epidemic malaria of the Punjab, with a note on a method of predicting epidemic years. *Paludism* 2: 17–26 (1911).

31. Bouma, M. & van der Kaay, H.J. The El Niño Southern Oscillation and the historic malaria epidemics on the Indian subcontinent and Sri Lanka: an early warning system for future epidemics? *Tropical Medicine and International Health* 1(1): 86–96 (1996).

32. Bouma, M.J. & van der Kaay, H.J. Epidemic malaria in India and the El Niño southern oscillation [letter] [see comments]. *Lancet* 344(8937):1638–1639 (1994).

33. Akhtar, R. & McMichael, A.J. Rainfall and malaria outbreaks in western Rajasthan. *Lancet* 348(9039):1457–1458 (1996).

34. Bouma, M.J. & Dye, C. Cycles of malaria associated with El Niño in Venezuela. *Journal of the American Medical Association* 278(21): 1772–1774 (1997).

35. Curriero, F.C. et al. The association between extreme precipitation and water-borne disease outbreaks in the United States, 1948–1994. *American Journal of Public Health* 91(8): 1194–1199 (2001).

36. McCarthy, J.J. et al. eds. *Climate change 2001: impacts, adaptation, and vulnerability.* Contribution of Working Group II to the Third Assessment Report of the Intergovernmental Panel on Climate Change. New York, USA, Cambridge University Press, 2001.

37. Rodo, X. et al. *ENSO and cholera: a non-stationary link related to climate change?* in press, 2002.

38. Lindgren, E. & Gustafson, R. Tick-borne encephalitis in Sweden and climate change. *Lancet* 358(9275): 16–18 (2001).

39. Randolph, S.E. The shifting landscape of tick-borne zoonoses: tick-borne encephalitis and Lyme borreliosis in Europe. *Philosophical Transactions of the Royal Society of London B Biological Sciences* 356(1411): 1045–1056 (2001).

40. Burke, D. et al. eds. *Under the weather, climate, ecosystems, and infectious disease.* Washington, DC, USA, National Academy Press, 2001.

41. Sutherst, R.W. The vulnerability of animal and human health to parasites under global change. *International Journal of Parasitology* 31(9): 933–948 (2001).

42. Sutherst, R.W. Implications of global change and climate variability for vector-borne diseases: generic approaches to impact assessments. *International Journal of Parasitology* 28(6): 935–945 (1998).

43. Davis, A.J. et al. Making mistakes when predicting shifts in species range in response to global warming. *Nature* 391(6669): 783–786 (1998).

44. Martin, P.H. & Lefebvre, M.G. Malaria and climate: sensitivity of malaria potential transmission to climate. *Ambio* 24(4): 200–207 (1995).

45. Rogers, D.J. & Randolph, S.E. The global spread of malaria in a future, warmer world. *Science* 289(5485): 1763–1766 (2000).

46. Martens, P. et al. Climate change and future populations at risk of malaria. *Global Environmental Change*, 9: S89–S107 (1999).

47. Martens, W. et al. Sensitivity of malaria, schistosomiasis and dengue to global warming. *Climatic Change* 35: 145–156 (1997).

48. Martens, W. et al. Global atmospheric change and human health: more than merely adding up the risks. *World Resource Review* 7(3): 404–416 (1995).

49. Martens, W.J. et al. Potential impact of global climate change on malaria risk. *Environmental Health Perspectives* 103(5): 458–464 (1995).

50. Anderson, R.M. & May, R.M. *Infectious diseases of humans: dynamics and control.* New York, USA, Oxford University Press, 1991.

51. Lindsay, S.W. & Martens, W.J. Malaria in the African highlands: past, present and future. *Bulletin of the World Health Organization* 76(1): 33–45 (1998).

52. Hartman, J. et al. Climate suitability for stable malaria transmission in Zimbabwe under different climate change scenarios. *Global Change & Human Health* 3(1): 42–53 (2002).

53. Rogers, D.J. & Packer, M.J. Vector-borne diseases, models, and global change. *Lancet* 342(8882): 1282–1284 (1993).

54. Hay, S.I. et al. Remotely sensed surrogates of meteorological data for the study of the distribution and abundance of arthropod vectors of disease. *Annals of Tropical Medicine and Parasitology* 90(1): 1–19 (1996).

55. Rogers, D.J. et al. Satellite imagery in the study and forecast of malaria. *Nature*. 415(6872): 710–715 (2002).

56. Thomson, M.C. et al. Predicting malaria infection in Gambian children from satellite data and bed net use surveys: the importance of spatial correlation in the interpretation of results. *American Journal of Tropical Medicine and Hygiene* 61(1): 2–8 (1999).

57. Beck, L.R. et al. Remote sensing as a landscape epidemiologic tool to identify villages at high risk for malaria transmission. *American Journal of Tropical Medicine and Hygiene* 51(3): 271–280 (1994).

58. Roberts, D.R. & Rodriguez, M.H. The environment remote sensing, and malaria control. *Annals New York Academy of Sciences* 740: 396–401 (1994).

59. Hulme, M. *Climate change and Southern Africa: exploration of some potential impacts. Implications for the SADC region.* Norwich, UK, University of East Anglia, 1996.

60. Thomson, M.C. et al. Towards a kala azar risk map for Sudan: mapping the potential distribution of *Phlebotomus orientalis* using digital data of environmental variables. *Tropical Medicine and International Health* 4(2): 105–113 (1999).

61. Glass, G.E. et al. Environmental risk factors for Lyme disease identified with geographic information systems. *American Journal of Public Health* 85(7): 944–948 (1995).

62. Guerra, M. et al. Predicting the risk of Lyme disease: habitat suitability for *Ixodes scapularis* in the north central United States. *Emerging Infectious Diseases* 8(3): 289–297 (2002).

63. Craig, M. H. et al. A climate-based distribution model of malaria transmission in Africa. *Parasitology Today* 15: 105–111 (1999).

64. Hutchinson, M.F. et al. *A Topographic and Climate Database for Africa.* Australian National University, Australia Centre for Resource and Environmental Studies, 1995. (CRES_AFR_01).

65. Glass, G.E. et al. Using remotely sensed data to identify areas at risk for hantavirus pulmonary syndrome. *Emerging Infectious Diseases* 6(3): 238–247 (2000).

66. Linthicum, K.J. et al. Climate and satellite indicators to forecast Rift Valley fever epidemics in Kenya. *Science* 285(5426): 397–400.

67. Pascual, M. et al. Cholera dynamics and El Niño-Southern Oscillation. *Science* 289(5485): 1766–1769 (2000).

68. Murphy, F.A. & Nathanson, N. The emergence of new virus diseases: an overview. *Seminars in Virology* 5: 87–102 (1994).

69. Herwaldt, B.L. & Ackers, M.L. An outbreak in 1996 of cyclosporiasis associated with imported raspberries. The Cyclospora Working Group [see comments]. *New England Journal of Medicine*. 336(22): 1548–1556 (1997).

70. Gubler, D.J. Dengue and dengue hemorrhagic fever. *Clinical Microbiology Review* 11(3): 480–496 (1998).

71. Oaks, S.C. et al. eds. *Malaria, obstacles and opportunities.* Washington, DC, USA, National Academy Press, 1991.

72. Black, R.E. Diarrheal Diseases, In: *Infectious Disease Epidemiology*, Nelson, K.E. et al. eds. Gaithersburg, MD, USA, Aspen Publishers Inc. pp. 497–517, 2000.

73. Semba, R.D. Nutrition and infectious diseases. In: *Infectious Disease Epidemiology*, Nelson, K.E. et al. eds. Gaithersburg, MD, USA, Aspen Publishers Inc. 2000.

74. Haines, A. et al. Global health watch: monitoring impacts of environmental change. *Lancet* 342(8885):1464–1469 (1993).

75. Morse, S.S. Factors in the emergence of infectious diseases. *Emerging Infectious Diseases* 1(1): 7–15 (1995).

76. Cohen, M.L. Resurgent and emergent disease in a changing world. *British Medical Bulletin* 54(3): 523–532 (1998).

77. Patz, J.A. et al. Effects of environmental change on emerging parasitic diseases. *International Journal of Parasitology* 30(12–13): 1395–1405 (2000).

78. MacKenzie, W.R. et al. A massive outbreak in Milwaukee of Cryptosporidium infection transmitted through the public water supply. *New England Journal of Medicine* 331(3): 161–167 (1994).

79. Graczyk, T.K. et al. Environmental and geographical factors contributing to watershed contamination with Cryptosporidium parvum oocysts. *Environmental Research* 82(3): 263–271 (2000).

80. Harb, M. et al. The resurgence of lymphatic filariasis in the Nile delta. *Bulletin of the World Health Organization* 71(1):49–54 (1993).

81. Thompson, D.F. et al. Bancroftian filariasis distribution and diurnal temperature differences in the southern Nile delta. *Emerging Infectious Diseases* 2(3):234–235 (1996).

82. Tauil, P.L. Urbanization and dengue ecology. *Cadernos de Saude Publica* 17(Suppl): 99–102 (2001).

83. De Souza, A.C. et al. Underlying and proximate determinants of diarrhoea-specific infant mortality rates among municipalities in the state of Ceara, north-east Brazil: an ecological study. *Journal of Biosocial Sciences* 33: 227–244 (2001).

84. D'Amato, G. et al. The role of outdoor air pollution and climatic changes on the rising trends in respiratory allergy. *Respiratory Medicine* 95(7): 606–611 (2001).

85. Esrey, S.A. et al. Effects of improved water supply and sanitation on ascariasis, diarrhoea, dracunculiasis, hookworm infection, schistosomiasis, and trachoma. *Bulletin of the World Health Organization* 69(5): 609–621 (1991).

86. Esrey, S.A. Water, waste, and well-being: a multicountry study. American Journal of Epidemiology 143(6): 608–623 (1996).

87. Taylor, L.H. et al. Risk factors for human disease emergence. *Philosophical Transactions of the Royal Society of London B Biological Sciences* 356(1411): 983–989 (2001).

88. Daszak, P. et al. Emerging infectious diseases of wildlife—threats to biodiversity and human health. *Science* 287(5452): 443–449 (2000).

89. Ostfeld, R.S. The ecology of Lyme-disease risk. *American Scientist* 85: 338–346 (1997).

90. Van Buskirk, J. & Ostfeld, R.S. Controlling Lyme disease by modifying the density and species composition of tick hosts. *Ecological Applications* 5:1133–1140 (1995).

91. Schmidt, K.A. & Ostfeld, R.S. Biodiversity and the dilution effect in disease ecology. *Ecology* 82: 609–619 (2001).

92. Ostfeld, R.S. et al. Of mice and mast: ecological connections in eastern deciduous forests. *BioScience* 46(5): 323–330 (1996).

93. Harvell, C.D. et al. Climate warming and disease risks for terrestrial and marine biota. *Science* 296(5576): 2158–2162 (2002).

94. Patz, J.A. Chapter 15: Potential consequences of climate variability and change for human health in the United States. In: *Climate change impacts on the United States: the potential consequences of climate variability and change foundation report*, U.G.C.R.P. National Assessment Synthesis Team, Cambridge, UK, Cambridge University Press, 2001.

95. Butterfield, J.E.L. & Coulson, J.C. *Terrestrial invertebrates and climate change: physiological and life-cycle adaptations*, In: *Past and future rapid environmental changes: the spatial and evolutionary responses of terrestrial biota*. Huntley, B. et al. eds. Berlin, Germany, Springer pp. 401–412, 1997.

96. Cammell, M.E. & Knight, J.D. Effects of climatic change on the populations dynamics of crop pests. *Advances in Ecological Research* 22: 117–162 (1992).

97. Carpenter, S.R. et al. Global change and fresh water ecosystems. *Annual Review of Ecology and Systematics* 23: 119–139 (1992).

98. Iverson, L.R. & Prasad, A.M. Predicting abundance of 80 tree species following climate change in the eastern United States. *Ecological Monographs* 68: 465–485 (1998).

99. Peterson, A.T. et al. Future projections for Mexican faunas under global climate change scenarios. *Nature* 416(6881): 626–629 (2002).

100. Pitelka, L.F. & Group, P.M.W. Plant migration and climate change. *American Scientist* 85: 464–473 (1997).

101. Mack, R.N. et al. Biotic invasions: causes, epidemiology, global consequences, and control. *Ecological Applications*, 10: 689–710 (2000).

102. Etterson, J.R. & Shaw, R.G. Constraint to adaptive evolution in response to global warming. *Science* 294(5540): 151–154 (2001).

103. McCarty, J.P. Ecological consequences of recent climate change. *Conservation Biology* 15: 320–331 (2001).

104. Rodríguez-Trelles, F. & Rodríguez, M.A. Rapid micro-evolution and loss of chromosomal diversity in Drosophila in response to climate warming. *Evolutionary Ecology* 12: 829–838 (1998).

105. Visser, M.E. & Holleman, L.J.M. Warmer springs disrupt the synchrony of oak and winter moth phenology. *Proceedings of the Royal Society, London* B265: 289–294 (2001).

106. Visser, M.E. et al. Warmer springs lead to mis-timed reproduction in great tits (Parus major). *Proceedings of the Royal Society, London* B265: 1867–1870 (1998).

107. Huq, A. et al. Cholera and global ecosystems. In: *Ecosystem change and public health: a global perspective*, Aaron, J.L. & Patz, J.A. eds. Baltimore, USA, The Johns Hopkins University Press, pp. 327–352, 2001.

108. Lobitz, B. et al. Climate and infectious disease: use of remote sensing for detection of *Vibrio cholerae* by indirect measurement. *Proceedings of the National Academy of Sciences USA* 97(4): 1438–1443 (2000).

109. Anderson, J.F. et al. Isolation of West Nile virus from mosquitoes, crows, and a Cooper's hawk in Connecticut. *Science* 286(5448): 2331–2333 (1999).

110. Hockachka, W.M. & Dhondt, A.A. Density-dependent decline of host abundance resulting from a new infectious disease. *Proceedings of the National Academy of Science* 97: 5303–5306 (2000).

CHAPTER 7

How much disease could climate change cause?

D.H. Campbell-Lendrum,[1] C.F. Corvalán,[2] A. Prüss–Ustün[2]

Introduction

Given the clear evidence that many health outcomes are highly sensitive to climate variations, it is inevitable that long-term climate change will have *some* effect on global population health. Climate change is likely to affect not only health but also many aspects of ecological and social systems, and will be slow and difficult (perhaps impossible) to reverse. Many therefore would judge that there is already sufficient motivation to act, both to mitigate the causes of climate change, and to adapt to its effects. However, such actions would require economic and behavioural changes bringing costs or co-benefits to different sectors of society. Decision-makers, from individual citizens to national governments, have numerous competing claims on their attentions and resources. In order to give a rational basis for prioritizing policies, at the least it is necessary to obtain an approximate measurement of the likely magnitude of the health impacts of climate change.

Quantification of health impacts from specific risk factors, performed in a systematic and consistent way using common measures, could provide a powerful mechanism for comparing the impacts of various risk factors and diseases. It would allow us to begin to answer questions such as: on aggregate, are the positive effects of climate change likely to outweigh the negative impacts? How important is climate change compared to other risk factors for global health? How much of the disease burden could be avoided by mitigating climate change? Which specific impacts are likely to be most important and which regions are likely to be most affected?

Caution is required in carrying out and presenting such assessments. Richard Peto, in his foreword to the first global burden of disease study (1), echoed the economist John Kenneth Galbraith in suggesting that epidemiologists fall into two classes: those who cannot predict the future, and those who know they cannot predict the future. Given the importance of natural climate variability and the potential for societal and individual factors to mediate the potential effects of climate change, only approximate indications of likely impacts can be expected. However, it is important to make such estimates available to policymakers, along with a realistic representation of the associated uncertainty; or remain in the current unsatisfactory condition of introducing a potentially important and irreversible health hazard throughout the globe, without any quantitative risk assessment.

---

[1] London School of Hygiene and Tropical Medicine, London, England.
[2] World Health Organization, Geneva, Switzerland.

This chapter outlines the estimation of disease burden caused by climate change at global level, performed in the framework of a comprehensive World Health Organization (WHO) project. After the quantification of disease burden for over 100 diseases or disease groups at global level (2), WHO has defined a general methodology to quantify the disease burden caused by 26 risk factors (Comparative Risk Assessment) at selected time points to 2030 (3). Major environmental, occupational, behavioural and lifestyle risk factors are considered including: smoking, alcohol consumption, unsafe sex, diet, air pollution, water and sanitation, and climate change. Despite the scale of the challenge, this has presented a unique opportunity to compare the health consequences of climate change to other important risk factors determining human health, and to estimate future disease burdens.

General methods

Disease burdens and summary measures of population health

The burden of disease refers to the total amount of premature death and morbidity within the population. In order to make comparative measures it is necessary to use summary measures of population health. These, first, take into account the severity/disability and duration of the health deficit, and, second, use standard units of health deficit. The Disability-Adjusted Life Year (DALY), for example, has been used widely (4) and is the sum of:

- years of life lost due to premature death (YLL)
- years of life lived with disability (YLD).

The number of years of life lost (YLL) takes into account the age at death, compared to a maximum life expectancy. Years of life lived with disability (YLD) takes into account disease duration, age of onset, and a disability weight that characterizes the severity of disease.

Estimating burden of disease attributable to a risk factor

Estimation of attributable burdens, using a measure such as DALYs, thus enables Comparative Risk Assessment: i.e. comparison of the disease burdens attributable to diverse risk factors. For each such factor, we need to know the:

1. burden of specific diseases
2. estimated increase in risk of each disease per unit increase in exposure (the "relative risk")
3. current population distribution of exposure, or future distribution as estimated by modelling exposure scenarios.

Since the mid 1990s, WHO has published estimates of the global burden of specific diseases or groups of diseases in the annual World Health Report. The most recent updates of the measurements of these burdens (2) constitute the total disease burden that can be attributed to the various risk factors. For calculating the attributable fraction for diarrhoeal disease, for example, the exposure distribution in the population is combined with the relative risk for each scenario with the following formula (Impact fraction, adapted from Last) (5):

$$IF = \frac{\Sigma P_i RR_i - 1}{\Sigma P_i RR_i}$$

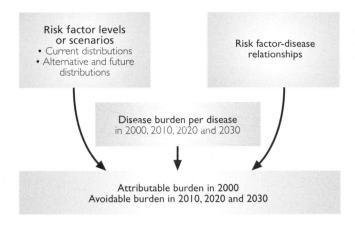

FIGURE 7.1 Key input data for estimating the global burden of disease caused by climate change.

Each exposure scenario is characterized by a relative risk (RR_i) compared to the individuals that are not exposed to the risk factor, or that correspond to a baseline "theoretical minimum" exposure scenario. The proportion of the population in each exposure scenario is P_i. The key input data for this estimate are summarized in Figure 7.1.

The attributable burden is estimated by multiplying the impact fraction by the disease burden for each considered disease outcome, given in the WHO World Health Report (2).

In addition to the attributable burden, the avoidable burden at future time points can be estimated by defining an alternative distribution of the risk factor in the study population and comparing projected relative risks under the alternative scenarios. In this case, the relative risks that are calculated for each scenario are applied to future "climate-change independent" trends produced by WHO, which attempt to take account of the most probable future changes due to climate-independent factors—e.g. improving socioeconomic and control conditions. The analysis therefore attempts to estimate the additional burden that climate change is likely to exert on top of the disease burden that otherwise would have occurred, if climate were to remain constant.

In this comprehensive project assessing the disease burden due to 26 risk factors, disease burden is estimated by sex, seven age groups and fourteen regions of the world. The full details of the analysis are presented in McMichael et al. (6). In this chapter, disease burdens are divided only into five geographical regions, plus a separate division for developed countries, which is the combination of the WHO regions: Europe, America A, and Western Pacific A (Figure 7.2). The attributable disease burden for climate change is estimated for 2000. In theory, avoidable burdens can be calculated for the years to 2030, however (at the time of writing) future projections of DALY burdens, in the absence of climate change, are not yet available for these. Instead, we present the climate-related relative risks of each outcome for 2030—i.e. the scenario-specific estimate of the likely proportional change in the burden of each of these diseases, compared to the situation if climate change were not to occur.

Type of evidence available for estimating disease burden due to climate change

The effects of climate change on human health are mediated by a variety of mechanistic pathways and eventual outcomes (chapters 3, 5, 6). There may be

FIGURE 7.2 Estimated impacts of climate change in 2000, by WHO region.

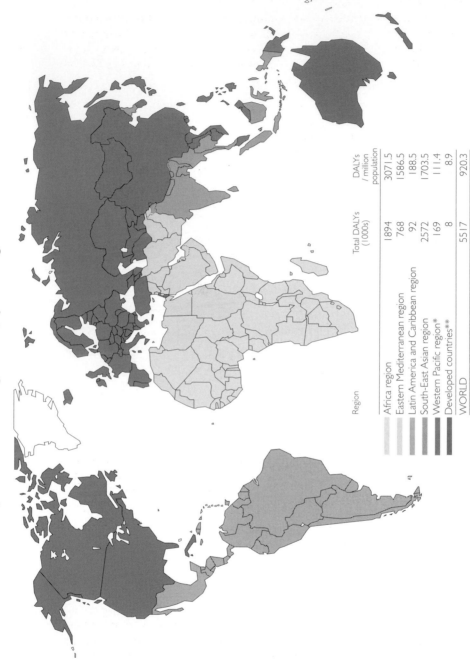

Region	Total DALYs (1000s)	DALYs / million population
Africa region	1894	3071.5
Eastern Mediterranean region	768	1586.5
Latin America and Caribbean region	92	188.5
South-East Asian region	2572	1703.5
Western Pacific region*	169	111.4
Developed countries**	8	8.9
WORLD	5517	920.3

* without developed countries; ** and Cuba

CLIMATE CHANGE AND HUMAN HEALTH

long delays between cause and certain outcomes, and reversibility may be slow and incomplete. Various methods have been developed for quantitative estimation of health impacts of future climate change (7). Ideally, future projections would be based on observations of the effects of the gradual anthropogenic climate change that has occurred so far. However, measurements of climate change and its effects, followed by formulation, testing and modification of hypotheses would take several decades due to the:

- lack of long-term standardized monitoring of climate sensitive diseases;
- methodological difficulties in controlling for effects of non-climatic differences and natural climate variability;
- relatively small (but significant) climatic changes that have occurred so far, that are poor proxies for the larger changes forecast for the coming decades.

While such direct monitoring of climate change effects is desirable, currently it does not provide the information necessary for quantitative estimation. The best estimation of future health effects of climate change therefore comes from predictive modelling based on the most comprehensive current understanding of the effects of climate (not weather) variation on health in the present and recent past, and applying these relationships to projections of future climate change.

Definition of risk factor and exposure scenario

Definition of the risk factor

For the purpose of this assessment, the risk factor climate change is defined as changes in global climate attributable to increasing concentrations of greenhouse gases (GHG).

Definition of exposure levels

As described in chapter 2, climate is a multivariate phenomenon and therefore cannot be measured on a single continuous scale. Also, climate changes will vary significantly with geography and time and cannot be captured fully in global averages of various climate parameters. The exposure scenarios used in this assessment are therefore comprehensive climate scenarios (i.e. predictions of the magnitude and geographical distribution of changes in temperature, precipitation and other climate properties) predicted to result from future patterns of GHG emissions.

Definition of baseline exposure scenario

In order to estimate discrete disease burden attributable to climate change, exposure scenarios need to be compared to a baseline exposure scenario that acts as a reference point. A logical baseline scenario would consist of a climate scenario not yet affected by any change due to GHG emissions. This is difficult to define accurately. The IPCC Third Assessment Report (8) shows clear evidence of changes in global average temperature of land and sea surface since the mid nineteenth century, and of extreme events throughout the last century (chapter 2, Figure 2.5 and Table 2.2), which it concludes mainly are due to human activities. However, given natural climate variability there is no clearly defined consensus on precisely what current climate conditions would have been, either now or in the future, in the absence of GHG emissions.

The baseline scenario therefore has been selected as the last year of the baseline period 1961 to 1990, i.e. 1990. This period is the reference point considered by the World Meteorological Organization and IPCC, and is supported by IPCC conclusions that the majority of climate change since this period has been caused by human activity. The selection of this baseline scenario implies that the generated results of attributable disease burden will be rather conservative, as any human-induced activity before that period is not addressed.

Scenarios considered for 2030

The exposure scenarios under investigation are selected according to the following projected emission levels:

1. unmitigated emission trends (i.e. approximately following the IPCC "IS92a" scenario (9))
2. emissions reduction resulting in stabilization at 750 ppm CO_2 equivalent by 2210 (s750)
3. more rapid emissions reduction, resulting in stabilization at 550 ppm CO_2 equivalent by 2170 (s550).

The predicted temperature changes and rise in sea level associated with these scenarios are outlined in Table 7.1 and Figure 7.3.

Methods for estimating exposure to climate change

Projections of the extent and distribution of climate change were generated by applying the various emission scenarios described above to the HadCM2 global climate model (GCM). This is one of the models approved by the IPCC, verified by back-casting (11), and provides results that lie approximately in the middle of the range of alternative models. The HadCM2 model generated estimates of the principal characteristics of the climate, including temperature, precipitation and absolute humidity for each month, at a resolution of 3.75° longitude and 2.5° latitude. The climate model outputs used here are estimated as averages over thirty-year periods.

Each scenario describes changes in global climate conditions, incorporating geographical variations. All of the population is considered as exposed to the scenario: i.e. *Pi* (above) is 100% in each case. However, the climate conditions experienced under different scenarios will vary between regions and between climate

TABLE 7.1 Successive measured and modelled global mean temperature and sea level rise associated with the various emissions scenarios. Future estimates are from the HadCM2 global climate model, produced by the UK Hadley Centre.

	1961–90	1990s	2020s	2050s
Temperature (°C change)				
HadCM2 Unmitigated Emissions	0	0.3	1.2	2.1
S750	0	0.3	0.9	1.4
S550	0	0.3	0.8	1.1
Sea level (cm change)				
HadCM2 Unmitigated Emissions	0	N/a	12	25
S750	0	N/a	11	20
S550	0	N/a	10	18

Source: reproduced from reference 10.

FIGURE 7.3 The global average temperature rise predicted from various emission scenarios: unmitigated emissions scenario (top line), emission scenario which stabilises CO$_2$ concentrations at 750 ppm by 2210 (middle) and at 550 ppm in 2170 (bottom). *Source: reproduced from reference (10).*

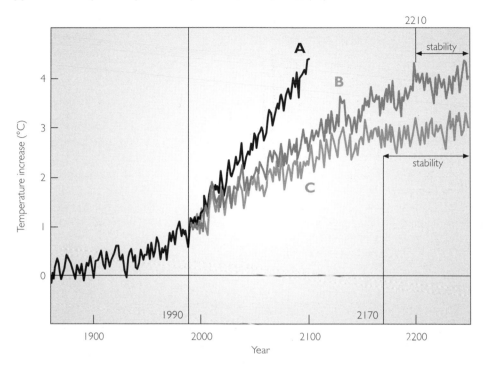

scenarios (e.g. under most future climate scenarios high latitudes will remain generally cooler than the tropics, but will experience greater rates of warming). The risk of suffering health impacts also will be affected by socioeconomic conditions and other factors affecting vulnerability. Such variations are considered in the calculations of relative risks for each disease, rather than in relation to exposure.

Outcomes to be assessed

While a wide variety of disease outcomes is suspected to be associated with climate change, only a few outcomes are addressed in this analysis (Table 7.2). These were selected on the basis of:

- sensitivity to climate variation
- predicted future importance
- availability of quantitative global models (or feasibility of constructing them).

The strength of evidence relating to each of these was reviewed through reference to all papers in the health section of the most recent IPCC report (*12*), from other wide ranging reviews of climate change and health (*13*) and a systematic review of the scientific literature using relevant internet search engines (Medline and Web of Science).

Additional likely effects of climate change that could not be quantified at this point include:

TABLE 7.2 Health outcomes considered in this analysis.

Type of outcome	Outcome	Incidence/Prevalence
Direct impacts of heat and cold	Cardiovascular disease deaths	Incidence
Food and water-borne disease	Diarrhoea episodes	Incidence
Vector-borne disease	Malaria cases; dengue cases	Incidence
Natural disasters[a]	Fatal unintentional injuries	Incidence
Risk of malnutrition	Non-availability of recommended daily calorie intake	Prevalence

[a] Separate estimation of impacts of coastal floods, and inland floods/landslides.

- changes in pollution and aeroallergen levels
- recovery rate of the ozone hole, affecting exposure to UV radiation (*14*)
- changes in distribution and transmission of other infectious diseases (particularly other vector-borne diseases and geohelminths)
- indirect effects on food production acting through plant pests and diseases
- drought
- famine
- population displacement due to natural disasters, crop failure, water shortages
- destruction of health infrastructure in natural disasters
- risk of conflict over natural resources.

Some of these may be included in future assessments as additional quantitative evidence becomes available.

Methods for estimating risk factor–disease relationships

The choice of modelling approach depends also on the availability of high-resolution data on health states and the possibility of estimating results that comply with the framework of the overall Comparative Risk Assessment.

As outlined above, estimates are based on observations of shorter-term climatic effects in the past, i.e. the effect of daily, seasonal or inter-annual variability on specific health outcomes, or on processes that may influence health states, e.g. parasite and vector population dynamics. In undertaking such an approach, it is necessary to appreciate that factors other than climate also are important determinants of disease, and to include in the quantitative estimates the likely effects of modifying factors such as socioeconomic status. Assumptions regarding these effect modifiers need to be clearly stated, together with an indication of the uncertainty range around the quantitative estimates.

There are two principal sets of assumptions relating to the definition of scenarios and health effects. Firstly, the secondary effects of climate change mitigation policies (e.g. the likely health benefits of reduced air pollution) are not considered here. Secondly, it is acknowledged that modifying factors such as physiological adaptation and wealth will influence health impacts due to climate change (*12*). Effects of improving socioeconomic conditions on the baseline (i.e. climate–change independent) rates of the diseases already are included in the WHO future scenarios (e.g. diarrhoea rates are projected to decrease over time as richer populations install improved water and sanitation services). However, changing socioeconomic conditions and physiological and other adaptations also will affect populations' vulnerability to the effects of climate change, and therefore the relative risk under each scenario. For example, improving water and sanitation also will affect the degree to which diarrhoea rates will be affected by temperature changes or more frequent flooding. The following sections describe how such affects are accounted for in both the relative risks and the uncertainty estimates for each health impact. No future actions taken specifically to adapt to the effects of a changing climate are considered.

For quantifying health impacts, all independent models linking climate change to quantitative global estimates on health or related impacts (e.g. numbers of people flooded or at risk of hunger) described in the IPCC Third Assessment Report (15) were considered. Where global models do not exist, local or regional projections were extrapolated. Models were further selected on the basis of validation against historical data and plausibility of both biological assumptions and extrapolation to other regions. In order to estimate relative risks for specific years, there is a linear interpolation of the relative risks between the various 30 year periods for which complete climate scenarios exist (e.g. between 2025, as the middle of the period described by the 2011–2040 climate scenario, and 2055, as the middle of the 2041–2070 scenario).

Specific health impacts

Direct physiological effects of heat and cold on cardiovascular mortality

Strength of evidence

The association between daily variation in meteorological conditions and mortality has been described in numerous studies from a wide range of populations in temperate climates (16, 17). These studies show that exposure to temperatures at either side of a "comfort range" is associated with an increased risk of (mainly cardio-pulmonary) mortality. Increases in other disease measures, such as General Practitioner consultations, have been associated with extreme temperatures (18, 19). However, it is not clear how these endpoints relate to quantitative measures of health burden.

Cardiovascular disease (CVD) has the best characterized temperature mortality relationship, followed by respiratory disease and total mortality in temperate countries. These relationships are supported by strong evidence for direct links between high and low temperatures and increased blood pressure, viscosity and heart rate for CVD (20, 21) and broncho-constriction for pulmonary disease (22).

The IPCC Third Assessment Report chapter on human health (11) also concludes that the frequency and intensity of heatwaves increases the number of deaths and serious illness. Yet the same report states that, in temperate countries, climate change would result in a reduction of wintertime deaths that would exceed the increase in summertime heatwave-related deaths.

Given the limited number of studies on which to base global predictions, quantitative estimates are presented only for the best supported of the direct physiological effects of climate change—changes in mortality attributable to extreme temperature for one or several days.

Exposure distribution and exposure-response relationships

The global population was divided into five climate zones according to definitions of the Australian Bureau of Meteorology (23). The polar zone is small and was excluded. Temperature distributions vary greatly within one climate zone. However, due to poor availability of meteorological data at daily time-scales, a single city was chosen to define a representative daily temperature distribution for each region. To give estimates of the mean temperature and variability under

TABLE 7.3 Synthesized relationships between temperature and cardiovascular mortality.

Climate zone	Heat[a]	Cold[a]	T[b]	Model	Source
Hot and dry	1.4	0.6	20	Seasonally adjusted, 1–6 day lags, all cause mortality	ISOTHURM (Delhi)
Warm humid	0.9	1.6	20	Seasonally adjusted, 1–6 day lags, all-cause all-age mortality	ISOTHURM (São Paulo)
Temperate	1.13	0.33	16.5	Seasonally adjusted, 1–2 day lag, cardiovascular disease	Kunst et al. 1993
Cold	1.13	0.33	16.5	Seasonally adjusted, 1–2 day lag, cardiovascular disease	Kunst et al. 1993

[a] Coefficient of % change in mortality per 1 °C of change in temperature.
[b] Temperature associated with lowest mortality rate.

each climate scenario, these distributions were "shifted" according to the projections of changes in the mean monthly temperature.

An exposure-response relationship was applied in each climate zone. Although many published studies describe the health effects of temperature, few have used daily values, controlled sufficiently for seasonal factors, or given adequate representation to populations in tropical developing countries. For cold and temperate regions, a relationship from a published study was used (24); for tropical countries and hot and dry countries a study (ISOTHURM) currently undertaken at the London School of Hygiene and Tropical Medicine (25) (Table 7.3).

The proportion of temperature-attributable deaths was calculated using the heat and cold mortality coefficients described in Table 7.3. Climate change attributable deaths were calculated as the change in proportion of temperature-attributable deaths (i.e. heat-attributable deaths plus cold-attributable deaths) for each climate scenario compared to the baseline climate.

The observation that temperatures associated with the lowest mortality vary between climate zones is supported by studies on various United States' cities (e.g. Braga et al. (26, 27)) and suggests that populations adapt at least partially to local conditions over time. However, the likely extent of adaptation has not been quantified for a globally representative range of populations. In our projections for the future, we assume that the temperature associated with lowest mortality rates (T[b] above) increases in line with the projected change in summer temperatures. No adjustment is made to the temperature-mortality slopes, i.e. it is assumed that populations biologically adapt to their new average temperatures, but remain equally vulnerable to departures from these conditions. Because this assumption about adaptation has not been formally tested, we include calculations assuming no adaptation as the other end of our uncertainty range. No adjustment is made for improving socioeconomic status: while rich populations appear to be partially protected by the use of air conditioners (e.g. studies in Chicago, USA (28)), research in populations with a wider range of socioeconomic conditions failed to detect a difference in susceptibility (work in São Paulo, Brazil (29)).

There also is evidence for a "harvesting effect", i.e. a period of unusually lower mortality following an extreme temperature period. This indicates that in some cases extreme temperatures advance the deaths of vulnerable people by a relatively short period, rather than killing people who would otherwise have lived to average life expectancy. However, this effect has not been quantified for temperature exposures and is not included in the model. As there is large uncer-

TABLE 7.4 Range of estimates of relative risks of cardio-vascular disease mortality attributable to climate change in 2030, under the alternative exposure scenarios.

Region	Relative risks		
	Unmitigated emissions	S570	S550
African region	(1.000–1.011)	(1.000–1.008)	(1.000–1.007)
Eastern Mediterranean region	(1.000–1.007)	(1.000–1.005)	(1.000–1.007)
Latin American and Caribbean region	(1.000–1.007)	(1.000–1.005)	(1.000–1.004)
South-East Asian region	(1.000–1.013)	(1.000–1.009)	(1.000–1.008)
Western Pacific region[a]	(1.000–1.000)	(1.000–1.000)	(1.000–1.000)
Developed countries[b]	(0.999–1.000)	(0.999–1.000)	(0.998–1.000)

[a] without developed countries.
[b] and Cuba.

tainty about the number of years that the casualties would have lived (i.e. the attributable years which are lost by exposure to the risk factor) the relative risk estimates will be used to calculate only attributable deaths, not DALYs.

Table 7.4 shows the range of estimates for the relative risk of cardiovascular mortality under the range of climate scenarios in 2030.

Quantification of temperature's effects on health due to climate change could be improved by the following research:

- additional analyses of the exposure-response relationship in tropical developing countries
- standardization of methods used to build exposure-response relationships
- adaptation
- investigation on additional outcomes, including inability to work in extreme temperatures.

Diarrhoeal disease

Diarrhoeal disease is one of the most important causes of disease burden, particularly in developing countries (2). As outlined in chapter 5, there is strong evidence that diarrhoea (particularly that caused by the bacteria and protozoan pathogens which predominate in developing regions) is highly sensitive to variations in both temperature and precipitation over daily, seasonal, and inter-annual time periods (30–33). It is therefore very likely that long-term climate change will lead to consistent changes in diarrhoea rates.

Despite the described quantitative relationships, this assessment addresses only the effects of increasing temperatures on the incidence of all-cause diarrhoea, as there are additional uncertainties in generating estimates for the effect of precipitation, or for specific pathogens:

- studies have addressed only a small part of the temperature spectrum represented globally—temperature-disease relationships are conditioned by the prevailing types of pathogens and modes of transmission and therefore may vary according to local circumstances;
- type of pathogen, whose occurrence varies with temperature, may affect the severity of disease;
- existing evidence on the link between climate and pathogen-specific diarrhoea cannot be used because important information is unknown, e.g. the partial contribution of each pathogen to all-cause diarrhoea;

- effects of changing rainfall patterns are not addressed because of the difficulties in extrapolating the observed non-linear relationships, and stochastic effects on outbreaks, to other regions.

Exposure distribution and exposure-response relationships

The change in mean annual temperature, per scenario, was estimated for each cell of a 1° latitude by 1° longitude population grid map. This was converted into a population-weighted average change in temperature for each country.

Although the influence of seasonality on diarrhoea is well recognized, only two studies describe a quantitative relationship between climate and overall diarrhoea incidence:

1. Checkley and co-workers (32) used time series analyses to correlate temperature, humidity and rainfall to daily hospital admissions in a paediatric diarrhoeal disease clinic in Lima, Peru. Correlations were controlled for seasonal variations and long-term trend. The analysis indicated an 8% (95% CI 7–9%) increase in admissions per 1 °C increase in temperature across the whole year. There was no significant independent association with rainfall or humidity. While the study design gives high confidence in the results, its scope is limited to the more severe (i.e. hospitalizing) diarrhoeal diseases and to children.
2. Singh et al. (34) used time series analyses to correlate temperature and rainfall to monthly reported diarrhoea incidence in Fiji. Reported overall incidence increased by 3% (95% CI 1.2–5.0%) per 1 °C temperature increase, and a significant increase in diarrhoea rates if rainfall was either higher or lower than average conditions. The use of monthly averages of climate conditions, and the lack of a clear definition of diarrhoea are likely to introduce a random effect and hence an underestimation of effects.

There appear to be no similar published studies showing clear and consistent evidence for changes in overall diarrhoea incidence with increased temperature in developed countries. The relative importance of pathogens which thrive at lower temperatures appears to be greater in populations of regions with higher standards of living, specifically access to clean water and sanitation (for which there is no clear and consistent evidence for peaks in all-cause diarrhoea in warmer months), compared to less well-off populations (where diarrhoea is usually more common in warmer, wetter months). This is demonstrated best by clear summer peaks of diarrhoea in black, but not white, infants in 1970s Johannesburg (35).

Here, countries are defined as "developing" if they have (or are predicted to have, for future assessment years) per capita incomes lower than the richer of the two study countries (Fiji) in 2000—approximately US $6000 per year in 1990. For such countries, a dose-response relationship of 5% increase in diarrhoea incidence per 1 °C temperature increase is applied to both sexes and all age groups. This is consistent with the relationships derived from the two studies described above. The 5% figure is chosen rather than the arithmetic mean of the constants from the two studies (5.5%): firstly to avoid giving a false impression of precision based on only 2 estimates, each with their own confidence intervals, secondly in order to be conservative. A wide uncertainty range (0–10%) is placed on this value in extrapolating these relationships both geographically and into the future. For developed countries, in the absence of further information, a (probably conservative) increase of 0% in diarrhoea incidence per 1 °C temperature increase (uncertainty interval −5 to 5%) is assumed.

TABLE 7.5 Range of estimates of relative risks of diarrhoea attributable to climate change in 2030, under the alternative exposure scenarios.

Region	Relative risks		
	Unmitigated emissions	S570	S550
African region	(0.99–1.16)	(0.99–1.13)	(0.99–1.11)
Eastern Mediterranean region	(0.98–1.16)	(0.98–1.11)	(0.98–1.11)
Latin American and Caribbean region	(0.92–1.08)	(0.94–1.06)	(0.95–1.05)
South-East Asian region	(0.99–1.17)	(0.99–1.13)	(0.99–1.12)
Western Pacific region[a]	(0.92–1.09)	(0.95–1.06)	(0.95–1.06)
Developed countries[b]	(0.94–1.06)	(0.94–1.06)	(0.93–1.08)

[a] without developed countries.
[b] and Cuba.

Regional relative risks are calculated by multiplying the projected increase in temperature by the relevant exposure-response relationship, using population-weighted averages. For projections of relative risks, developing countries reaching a per capita GDP above US $6000 per year are considered to have the same risk as developed countries, i.e. no effect of temperature on diarrhoea incidence. The resulting estimates of relative risks are given in Table 7.5.

Future research

Investigation of exposure-response relationships from a wider climatic and socioeconomic development could improve the accuracy of estimations. Studies also should explicitly measure economic development and improved levels of sanitation, which are very likely to influence populations' vulnerability to the effects of climate variation on diarrhoeal disease.

Malnutrition

Strength of evidence

Malnutrition is considered as the single most important risk factor to global health, accounting for an estimated 15% of total disease burden in DALYs (2). While multiple biological and social factors affect the influence of malnutrition, the fundamental determinant is the availability of staple foods. Climate change may affect this availability through the broadly negative effects of changes in temperature and precipitation and broadly positive effects of higher CO_2 levels on yields of food crops (36, 37)). The food trade system may be able to absorb these effects at the global level. However, climate change can be expected to have significant effects on food poverty in conjunction with variation in population pressure and economic capacity to cope (38).

Evidence for climate change effects on crop yields is strong. Crop models have been validated in 124 sites in 18 countries over a wide range of environments (39)). Major uncertainties relate to the extent this relationship will be maintained over long-term climate change, and in particular how the world food trade system will adapt to changes in production (40, 41). The IPCC has concluded with "medium confidence" that climate change would increase the number of hungry and malnourished people in the twenty-first century by 80 to 90 million.

While substantial literature describes effects of climate on individual crops, only one group has used these estimates to predict the numbers of people at risk of hunger (38). All results presented are based on work by this group. Although these are the most complete models currently available, they do not take into account more complex pathways by which climate change may affect health, such as the relative importance of fruit and vegetable availability, animal husbandry, and the effect on micronutrient malnutrition this may induce. The consequences of decreasing water sources and synergistic effects of malnutrition and poverty also cannot be modelled currently. Due to these omissions, the current estimate probably is conservative.

Exposure distribution and exposure-response relationship

Global maps of temperature and rainfall at 0.5° latitude by 0.5° longitude, and estimates of atmospheric CO_2 levels, were generated for each scenario and time point.

The IBSNAT-ICASA dynamic growth models (42) for grain cereals and soybean were used to estimate the effect of projected changes in temperature, rainfall and CO_2 on future crop yields. These crop yield estimates are introduced in the world food trade model "Basic Linked System' (43) to provide national food availability. This system consists of a linked series of 40 national and regional food models for food production, the effects of market forces and Government policies on prices and trade, and trends in agricultural and technological conditions (further details in Fischer (44)). Principal characteristics of the model include the following:

- assumes no major changes in political and economic context of world food trade
- population growth occurs according to the World Bank mid-range estimate (45)—10.7 billion by the 2080s
- GDP increases as projected by the Energy Modelling Forum (46)
- 50% trade liberalization is introduced gradually by 2020.

National food availability is converted into the proportion of the population in each region who do not have sufficient food to maintain a basal metabolic rate of 1.4, the UN Food and Agriculture Organization's definition of under-nourishment (47). The model generates outputs for continents principally made up of vulnerable developing countries (i.e. excluding North America, Europe and China). Although the broad geographical scale of the food model precludes detailed analysis, the model outputs correlate with incidence of stunting and wasting (48) at the continental level. For this analysis, it is therefore assumed that projected changes in food availability will cause proportional changes in malnutrition.

The relative risks of malnutrition are shown in Table 7.6. Uncertainty ranges around these estimates are difficult to quantify, as aside from applying alternative climate scenarios to a series of Hadley centre climate models, no sensitivity analyses have been carried out on other model assumptions. Hence, there are several possible sources of uncertainty, including the variation of critical parameters (particularly rainfall) between different climate models, and the influence of food trade and future socioeconomic conditions affecting the capacity to cope with climate-driven changes in food production. The mid-range estimates therefore are derived from a simple application of the model described above. In the absence of further information at this point, uncertainty intervals are defined as ranging from no risk to doubling of the mid-range risk.

Focus for research

For the purpose of estimating burden of disease, priorities for future research should include:

TABLE 7.6 Range of estimates of relative risks of malnutrition attributable to climate change in 2030, under the alternative exposure scenarios.

Region	Relative risks		
	Unmitigated emissions	S570	S550
African region	(1.00–1.05)	(1.00–1.09)	(1.00–1.00)
Eastern Mediterranean region	(1.00–1.12)	(1.00–1.20)	(1.00–1.06)
Latin American and Caribbean region	(1.00–1.00)	(1.00–1.22)	(1.00–1.10)
South-East Asian region	(1.00–1.27)	(1.00–1.32)	(1.00–1.22)
Western Pacific region[a]	(1.00–1.00)	(1.00–1.05)	(1.00–1.02)
Developed countries[b]	(1.00–1.00)	(1.00–1.00)	(1.00–1.00)

[a] without developed countries.
[b] and Cuba.

- sensitivity of estimates to the outputs of various different climate models
- estimation of uncertainty around exposure-response relationships
- validation of the climate-malnutrition model against past data
- improved resolution of model outputs, e.g. to national level
- correlation of model outputs with health outcomes at higher resolution
- investigation of synergistic effects of water availability and poverty on malnutrition.

Natural disasters caused by extreme weather and sea level rise

Natural disasters caused by extreme weather events are a significant cause of mortality and morbidity worldwide (*49, 50*). These impacts are influenced by short and long-term averages and variability of weather conditions (*51, 52*), and are likely to be affected by the observed and predicted trends towards increasingly variable weather (see chapter 2).

Weather events considered for estimating disease burden include the following:

- coastal flooding, driven by sea level rise
- inland flooding and mudslides caused by increased frequency of extreme precipitation.

Due to lack of quantitative information, climate change effects on the following impacts of natural disasters could not be quantified. However, the aggregate effect of such longer-term mechanisms may very well be greater than from the acute effects:

- effects of wind storms
- effects of melting snows and glaciers on floods and landslides
- longer term health impacts resulting from population displacement
- consequences of damage to health systems
- infectious disease outbreaks and mental problems due to emergency situations (such as living in camps).

Exposure distribution and exposure-response relationship

Coastal floods: Published models estimate the change in sea levels for each scenario (*53, 54*). The number of people affected has been estimated by applying these changes to topography and population distribution maps. The model has shown good results in comparison with detailed assessments at national level (summarized in Nicholls (54)).

Inland floods and mudslides: Despite clear causal links, inland floods and mudslides have not yet been quantitatively related to health impacts (55). At local level such natural disasters are determined by the frequency of extreme precipitation over a limited period (hour, day or week) and the average amount of precipitation. Health impacts are modulated by the topographical distribution of population as well as social aspects of vulnerability, including the quality of housing and early warning systems (56).

In the absence of detailed information, this analysis makes the *a priori* assumption that flood frequency is proportional to the frequency of monthly rainfall exceeding the highest monthly rainfall that would, under baseline (i.e. 1961–1990) climate conditions, occur in every 10 years (i.e. the upper 99.2% confidence interval of the distribution of monthly rainfall). The change in the frequency of such extreme events under the various climate scenarios was calculated for each cell of the global climate model grid. Using GIS software, this was overlaid on a map of global population distribution at 1° by 1° resolution. This allowed the calculation of the measure of exposure (i.e. the *per capita* change in risk of experiencing such an extreme weather event) within each region.

In contrast to the other health impacts considered in this assessment, health impacts caused by natural disasters do not refer to a specific disease, with an associated burden calculated by WHO. It is therefore not possible directly to apply the impact fraction calculations described above. Instead, it is necessary to estimate the impacts *attributable to these climate events* under baseline climate conditions; relative risk estimates for future scenarios are applied to these numbers. The numbers of such deaths and injuries are based on the EM-DAT database (57), which records events resulting in at least one of the following: (1) >10 people killed, (2) >200 injured or (3) a call for international assistance. Although the most rigorously compiled and most comprehensive database available at the global scale, this is probably subject to significant under-reporting, so that estimates are likely to be conservative. EM-DAT quotes numbers of people killed, injured and affected. However for this assessment only the numbers of people killed are used as the EM-DAT group (EM-DAT Director, pers. comm.) considers injury numbers for floods to be unreliable, and currently it is not possible to fully characterize the health impact of being affected by flooding. Annual incidence of death attributable to such disasters under baseline climate conditions was estimated as 20-year averages for each region.

Baseline incidence rates alter over time, according to vulnerability. Some factors decrease vulnerability, such as improving flood defences implemented by populations becoming richer, and some increase vulnerability, such as increasing population density in coastal areas. Adjustments were made to account for these effects. Nicholls' model (54) incorporates coastal flooding defences in line with GNP change and population distribution. For inland floods, vulnerability effects are approximated by an analysis for all natural disasters (58). These effects are not specific to inland floods but nevertheless were applied as the specific relationship has not been modelled. There is some evidence that young children and women are more vulnerable to acute impacts of natural disasters from earthquakes (59) and famines (60). This information is considered insufficient to apply to these estimates, equal impacts for all age and sex groups therefore are assumed.

TABLE 7.7 Range of estimates for the relative risks of flood deaths attributable to climate change in 2030.

Region	Inland floods			Coastal floods		
	Unmitigated emissions	S570	S550	Unmitigated emissions	S570	S550
African region	(1.00–2.27)	(1.00–2.65)	(1.00–3.16)	(1.20–1.79)	(1.15–1.59)	(1.13–1.55)
Eastern Mediterranean region	(1.00–6.83)	(1.00–6.69)	(1.00–3.16)	(2.16–5.61)	(1.86–4.46)	(1.80–1.55)
Latin American and Caribbean region	(1.00–4.24)	(1.00–4.43)	(1.00–3.74)	(1.80–4.20)	(1.61–3.43)	(1.57–3.28)
South-East Asian region	(1.00–1.75)	(1.00–2.39)	(1.00–2.49)	(1.06–1.21)	(1.04–1.15)	(1.04–1.15)
Western Pacific region[a]	(1.00–3.13)	(1.00–2.70)	(1.00–2.50)	(1.03–1.10)	(1.02–1.08)	(1.02–1.07)
Developed countries[b]	(1.00–8.79)	(1.00–8.69)	(1.00–7.73)	(1.32–2.27)	(1.34–2.36)	(1.45–2.81)

[a]   without developed countries.
[b]   and Cuba.

Uncertainty of these estimates of course is related to the frequency of extreme weather events as modelled by the various climate scenarios and models, and to evolving protection over time due to projected increases in GNP. Results for coastal flooding are more reliable; they are driven by changes in sea level rise that are relatively consistent across climate models. The estimates are much more uncertain for inland flooding, as precipitation predictions vary considerably between climate models and scenarios. In addition, while the models do account for changes in protection proportional to GNP, individual responses to risk have not been quantified (61). As it can be expected that individual response acts as protection, the results are considered as an upper limit. Mid-estimates are assumed as 50% of the upper limit, the lower estimate assumes that 90% of the projected impacts would be avoided. For inland flooding estimates, the upper and lower estimates are expanded to include a relative risk of 1 (i.e. no change) to 50% greater exposure and no adaptation, to take account of the greater uncertainty inherent in the precipitation estimates.

The ranges of estimates for relative risks of floods in different regions are presented in Table 7.7.

Future research

The link between extreme weather events and the health impacts of the resulting disasters are surprisingly poorly researched. Substantial improvements could be made by improved investigation of:

- current health impacts from natural disasters, particularly in developing countries
- more detailed description of disasters
- analysis of health impacts versus intensity of precipitation at higher temporal and spatial resolution
- formal sensitivity analyses for each model parameter
- longer-term health effects: particularly those resulting from population displacement or drought periods and their effects on food production.

Such research would improve the accuracy of estimates and the inclusion of probably more important health effects.

Falciparum *malaria*

Strength of evidence

Vector-borne diseases are among the most important causes of global ill-health, particularly in tropical regions (2). As described in chapter 6, substantial laboratory (62, 63) and field evidence (64) indicate that both vectors and the pathogens they transmit are highly sensitive to climate conditions, and therefore likely to be affected by future climate change. There is, however, considerable debate over the degree to which potential climate-driven increases in geographical distributions and rates of disease will be prevented by modifying factors (availability of sufficient rainfall or suitable habitat) and the effects of control programmes, socioeconomic developments and population immunity (11, 65–68).

Although climate change is likely to have some effect on all climate-sensitive diseases, only a few have been investigated at the global scale. This assessment is restricted to *falciparum* malaria, which has been subjected to more detailed study, by more independent research groups, than other diseases.

Exposure distribution and exposure-response relationships

The main parameters affecting vector-borne diseases include temperature, rainfall, and absolute humidity. These were mapped for each considered scenario as described above. Quantified relationships between climate, vector population biology and disease incidence have not been described in generalized models, as they depend upon a variety of modifying factors also described above. In addition, the complexity of immune response of populations to changing exposure to infection is difficult to predict (69, 70). The only global models available to date predict changes in geographical and temporal distributions, and therefore populations at risk, rather than incidence of disease. This analysis assumes that relative changes in disease incidence are proportional to changes in the population at risk.

Of the various models that investigate the relationship between climate and malaria, only two have been validated directly to test how well they explain the current distribution of the disease over wide areas. The MARA climate model (Mapping Malaria Risk in Africa) (71, 72) is based on observed effects of climate variables on vector and parasite population biology and malaria distributions in local field studies. This information is used to define areas that are climatically suitable for *falciparum* malaria transmission, and therefore the population at risk, throughout Africa. Predictive distribution maps generated from the model show a close fit to the observed margins of the distribution in Africa, based on a detailed historical database, independent of the data used to create the original model. The major disadvantages of the model for this exercise are that the validation by visual rather than statistical comparison of the predicted and observed maps, and the distribution limits, are assumed to be constrained only by climate rather than by control or other socioeconomic factors. While the validation indicates that this is a reasonable assumption for Africa, it may be less appropriate for other regions.

The other validated model is that of Rogers and Randolph (66) which uses a direct statistical correlation between climate variables and observed disease distributions to give a highly significant and reasonably accurate fit to the current global distribution of all malaria. This model has the significant advantages of not making *a priori* assumptions about climate–disease relationships, and being tested directly against observed data. However, the quality of the available distribution data (relatively coarse maps of the distribution of both *falciparum* and *vivax* malaria) means that the model can be validated only against a subset of the original data used for model building, rather than a completely independent data set. Neither is it clear what effect the combination of distributions of different parasites, with different climate sensitivities, may have on model sensitivity to future climate changes.

As both models are informative but imperfect descriptions of climate-malaria relationships, and have not been directly compared with one another, the results of both are considered in this assessment. Relative risks presented here are the ratios of the population at risk in each region, relative to the population at risk under the 1961–1990 climate, according to the MARA model. "Population at risk" is considered as the population living in areas climatically suitable for more than one month of malaria transmission per year. In order to estimate disease burdens, these relative risks are multiplied with the baseline incidences of malaria for each region. This method is conservative, as it accounts only for malaria in the additional population at risk and not for increasing incidence within already endemic populations. An additional conservative assumption built into the model is that climate change will not cause expansion of the disease into developed regions, even if they become climatically suitable. We are therefore estimating climate-driven changes in the population at risk within those regions where current and predicted future socioeconomic conditions are suitable for malaria transmission.

Possible sources of uncertainty may include:

- results based on different climate projections, as for the other factors
- the degree to which the model validated for Africa applies to other regions
- the relationship between the increase of the population at risk and the incidence of disease for each region
- the influence of control mechanisms.

These uncertainties are likely to be considerable, but have not been formally quantified. As the other model validated for field data (66) predicts practically no increase in the population at risk even under relatively severe climate change, the lower uncertainty estimate assumes no effect. The upper range is estimated as a doubling of the mid-range estimate.

TABLE 7.8 Range of estimates for the relative risks of malaria attributable to climate change in 2030, under the alternative exposure scenarios.

Region	Relative risks		
	Unmitigated emissions	S570	S550
African region	(1.00–1.17)	(1.00–1.11)	(1.00–1.09)
Eastern Mediterranean region	(1.00–1.43)	(1.00–1.27)	(1.00–1.09)
Latin American and Caribbean region	(1.00–1.28)	(1.00–1.18)	(1.00–1.15)
South-East Asian region	(1.00–1.02)	(1.00–1.01)	(1.00–1.01)
Western Pacific region[a]	(1.00–1.83)	(1.00–1.53)	(1.00–1.43)
Developed countries[b]	(1.00–1.27)	(1.00–1.33)	(1.00–1.52)

[a] without developed countries.
[b] and Cuba.

Future research

Additional information on the following would contribute to improvements in quantitative predictions of vector-borne disease frequency caused by climate change:

- models relating climate parameters to disease incidence rather than areas and populations at risk
- relationships between climate and other vector-borne diseases
- effects of population vulnerability
- model validation with past and current data on climate parameters and disease frequency
- effects of climate variability rather than change in average values alone.

Aggregated estimates for 2000

Projections of DALYs for specific diseases are required in order to convert relative risks into estimates of burden of disease. While DALY projections for the period to 2030 will shortly be released by WHO, currently they are available for 2000 alone. The application of the relative-risk models described above may give a better estimate of the current health impacts of climate change than directly measuring long-term changes in health states and correlating them against long-term changes in climate (see chapter 10). Although it is perhaps counter-intuitive and somewhat unsatisfactory to use models rather than direct observation to estimate current disease, it is a necessary consequence of both the poor surveillance data that is available for monitoring long-term trends, and the difficulties of separating out the contributions of climatic and non-climatic factors.

Relative risks for 2000 have been estimated as described above, and applied to the disease burden estimates for that year, with the exception of the effects of extreme temperatures on cardiovascular disease, for the reasons described above (Table 7.9). While the resulting estimates are clearly of limited value in informing policies related to future GHG emissions, they do address two purposes. Firstly, illustrating the approximate magnitude of the burden of disease that already may be caused by climate change, if current understanding of climate-health relationships is correct. Secondly, serving to highlight both the specific diseases (particularly malnutrition, diarrhoea and malaria) and the geographical regions (particularly those made up of developing countries) that are likely to make the greatest contribution to the future burden of climate-change associated disease.

TABLE 7.9 Estimates for the impact of climate change in 2000 in thousands of DALYs, given by applying the relative risk estimates for 2000 to the DALY burdens for specific diseases quoted in the World Health Report (2002) (2).

	Malnutrition	Diarrhoea	Malaria	Floods	Total	Total DALYs/ million population
African region	616	414	860	4	**1894**	3071.5
Eastern Mediterranean region	313	291	112	52	**768**	1586.5
Latin American and Caribbean region	0	17	3	72	**92**	188.5
South-East Asian region	1918	640	0	14	**2572**	1703.5
Western Pacific region[a]	0	89	43	37	**169**	111.4
Developed countries[b]	0	0	0	8	**8**	8.9
World	2847	1460	1018	192	**5517**	920.3

[a] without developed countries.
[b] and Cuba.

Conclusions

Attempts to predict the future health impacts of any risk factor are necessarily uncertain. They rely on a reasonable projection of future exposures to the risk factor, unbiased measurement of the relationship between the exposure and health impacts, and the assumption that this relationship will either hold constant, or change in a predictable manner.

Climate change differs from other health risk factors in that considerable effort has been devoted to generating and evaluating formal models to forecast future climate in response to likely trajectories of atmospheric gaseous compositional change. Arguably we therefore have better information on future climate than for most health exposures. Substantial knowledge also has been accumulated on the relationship between climate variations (either over short time periods or geographically) and a series of important health impacts. Although this information is far from complete, it provides a basis for a first approximation of the likely scale of climate change effects on a range of impacts.

The health impacts of climate change were estimated for the disease outcomes that (1) are of global importance, (2) the IPCC concludes are most likely to be affected by climate change, and (3) for which sufficient information for global modelling was available.

Climate change is expected to affect the distribution of deaths from the direct physiological effects of exposure to high or low temperatures (i.e. reduced mortality in winter, especially in high latitude countries, but increases in summer mortality, especially in low latitudes). However, the overall global effect on mortality is likely to be more or less neutral. The effect on the total burden of disease has not been estimated, as it is unclear to what extent deaths in heat extremes are simply advancing deaths that would have occurred soon in any case.

It is estimated that in 2030 the risk of diarrhoea will be up to 10% higher in some regions than if no climate change occurred. Uncertainties around these estimates mainly relate to the very few studies that have characterized the exposure-response relationship.

Estimated effects on malnutrition vary markedly across regions. By 2030, the relative risks for unmitigated emissions relative to no climate change vary from a significant increase in the south-east Asia region, to a small decrease in the western Pacific region. There is no consistent pattern of reduction in relative risks with intermediate levels of climate change stabilization. Although these estimates appear somewhat unstable due to the high sensitivity to regional variation in precipitation, they are large and relate to a major disease burden.

Proportional changes in the numbers of people killed in coastal floods are very large, but induce a low disease burden in terms of people immediately killed and injured. Impacts of inland floods are predicted to increase by a similar order of magnitude and generally cause a greater acute disease burden. In contrast to most other impacts, the relative increase in risks tends to be similar in developed and developing regions. However, these apply to baseline rates that are much higher in developing than developed countries. Estimates are subject to uncertainty around the likely effectiveness of adaptation measures, and around the quantitative relationships between changes in precipitation, the frequency of flooding and associated health impacts. The suggestion of a trend towards decreasing incidence with increasing GHG emissions in some regions most probably is due to the uncertainties inherent in predicting precipitation trends.

Relatively large changes in relative risk are estimated for *falciparum* malaria in regions bordering current endemic zones. Relative changes are much smaller in areas that already are highly endemic, mainly because increases in transmission in already endemic zones are not considered in this analysis. Most temperate regions are predicted to remain unsuitable for transmission, either because they remain climatically unsuitable (most of Europe), and/or socioeconomic conditions are likely to remain unsuitable for reinvasion (e.g. the southern United States). The principal uncertainties relate to the reliability of extrapolations made between regions, and the relationship between changes in the population at risk of these diseases and disease incidence.

Application of the models derived above to the disease estimates for the present (i.e. 2000) suggest that, if the understanding of broad relationships between climate and disease is realistic, then climate change already may be having some impacts on health. This shows the advantages of using the DALY system to take into account not only the proportional change in each impact, but also the size of the disease burden. Although proportional changes in impacts such as diarrhoea and malnutrition are quite modest (compared to floods for example) they are likely to be extremely important in public health because they relate to such a large burden of disease. Similarly, such analyses emphasise that the impacts are likely to be much larger in the poorest regions of the world. Unfortunately, the relatively poor health surveillance systems that operate in many of the areas likely to be most affected by climate change, coupled with the difficulties of separating climatic and non-climatic influences, make it extremely difficult to test directly whether the modest expected changes have occurred or been prevented by non-climatic modifying factors. Improvements in models, and particularly in the collection of health surveillance data, will be essential for improving the reliability and usefulness of such assessments.

The total estimated burden for the present is small in comparison to other major risk factors for health measured under the same framework. Tobacco consumption, for example, is estimated to cause over ten times as many DALYs (3). It should be emphasised, however, that in contrast to many risk factors for health, exposure to climate change and its associated risks are increasing rather than decreasing over time.

All of the above models are based on the most comprehensive currently available data on the quantitative relationships between climate and disease. However, other factors clearly affect rates of all of these diseases and in many cases interact with climatic effects. As far as possible, the effect of non-climatic factors (both current and future) has been included in these analyses. Understanding of the interactions between climate and non-climatic effects remains far from perfect, and the degree to which population adaptation (physiological, behavioural or societal) may absorb climate-driven changes in risk represents the greatest degree of uncertainty in our projections. Research on these interactions clearly is necessary, and should greatly improve the accuracy of future estimates, as well as indicating how best to adapt to climate change.

In every assessment of disease burden at global level, a model relying on a number of hypotheses needs to be constructed, as only a fraction of the necessary data is ever available. While these results still bear considerable uncertainty, the international climate research community (represented by the UN IPCC) concludes that anthropogenic climate change has occurred already, will continue to occur and will adversely affect human health. This first global assessment, based on a comparable and internally consistent method, provides the opportunity to explore the diverse and potentially large health impacts anticipated.

This assessment serves not only to generate the best estimates possible given current knowledge, but also to highlight the most important knowledge gaps that should be addressed in order to improve future assessments. A very large part of possible health effects were not included in this assessment, either because of insufficient baseline data on health and climate or because the exposure-response relationships have been inadequately researched for quantifying those impacts. No indirect (air pollution and then disease), synergistic (poverty), or longer-term effects (displacement of populations) have been considered in this analysis. In addition the projections are made only until 2030, which is somewhat unsatisfactory for a health exposure that accumulates gradually and perhaps irreversibly. For these reasons the estimates should be considered not as a full accounting of health impacts but as a guide to the likely magnitude of some health impacts of climate change, in the near future.

References

1. Murray, C.J.L. & Lopez, A.D. *The global burden of disease: a comprehensive assessment of mortality and disability from diseases, injuries, and risk factors in 1990 and projected to 2020.* Cambridge, UK, Harvard University Press, 1996.
2. World Health Organization (WHO). *The world health report 2002.* Geneva, Switzerland, World Health Organization, 2002.
3. Ezzati, M. et al. Selected major risk factors and global and regional burden of disease. *Lancet* 360(9343): 1347–60 (2002).
4. Murray, C.J.L. Quantifying the burden of disease—the technical basis for disability-adjusted life years. *Bulletin of the World Health Organization* 72(3): 429–445 (1994).
5. Last, J.M. *A dictionary of epidemiology.* 2nd edition. New York, USA, Oxford University Press, 2001.
6. McMichael, A.J. et al. Climate Change. In: Comparative quantification of health risks: Global and regional burden of disease due to selected major risk factors. Ezzati, M., Lopez, A.D., Rodgers, A., Murray, C.J.L. eds. Geneva, Switzerland, World Health Organization, 2003.
7. Martens, P. & McMichael, A.J. *Environmental change, climate and health* Cambridge, UK, Cambridge University Press, 2002.
8. Intergovernmental Panel on Climate Change (IPCC). *Climate change 2001: the scientific basis.* Contribution of Working Group I to the Third Assessment Report of the Intergovernmental Panel on Climate Change. Cambridge, UK, Cambridge University Press, 2001.
9. Intergovernmental Panel on Climate Change (IPCC). *Climate change 1995: the science of climate change.* Contribution of Working Group I to the Second Assessment Report of the Intergovernmental Panel on Climate Change. Houghton, J.T., et al. eds. Cambridge, UK & New York, USA, Cambridge University Press, 1996.
10. Hadley Centre. *Climate change and its impacts: stabilisation of CO_2 in the atmosphere 1999:* Hadley Centre, UK, 1999.
11. Johns, T.C. et al. Correlations between patterns of 19th and 20th century surface temperature change and HadCM2 climate model ensembles. *Geophysical Research Letters* 28(6): 1007–1010 (2001).
12. McMichael, A.J. & Githeko, A. Human Health. In: *Climate change 2001: impacts, adaptation and vulnerability.* McCarthy, J.J. et al. eds. Cambridge, UK, Cambridge University Press: 451–485 (2001).

13. National Research Council NAoS. *Under the weather: climate, ecosystems, and infectious disease.* Washington, DC, USA, National Academy Press, 2001.
14. Shindell, D.T. et al. Increased polar stratospheric ozone losses and delayed eventual recovery owing to increasing greenhouse-gas concentrations. *Nature* 392(6676): 589–592 (1998).
15. Intergovernmental Panel on Climate Change (IPCC). *Climate change 2001: impacts, adaptation and vulnerability.* Contribution of Working Group II to the Third Assessment Report. Cambridge, UK, Cambridge University Press, 2001.
16. Alderson, M.R. Season and mortality. *Health Trends* 17: 87–96 (1985).
17. Green, M.S. et al. Excess winter-mortality from ischaemic heart disease and stroke during colder and warmer years in Israel. *European Journal of Public Health* 4: 3–11 (1994).
18. Hajat, S. & Haines, A. Associations of cold temperatures with GP consultations for respiratory and cardiovascular disease amongst the elderly in London. *International Journal of Epidemiology,* 31(4): 825–830 (2002).
19. Hajat, S. et al. Association between air pollution and daily consultations with general practitioners for allergic rhinitis in London, United Kingdom. *American Journal of Epidemiology,* 153(7): 704–714 (2001).
20. Keatinge, W.R. et al. Increases in platelet and red-cell counts, blood-viscosity, and arterial-pressure during mild surface cooling—factors in mortality from coronary and cerebral thrombosis in winter. *British Medical Journal* 289(6456): 1405–1408 (1984).
21. Pan, W.H. et al. Temperature extremes and mortality from coronary heart disease and cerebral infarction in elderly Chinese. *Lancet* 345: 353–355 (1995).
22. Schanning, J. et al. Effects of cold air inhalation combined with prolonged submaximal exercise on airway function in healthy young males. *European Journal of Respiratory Diseases* 68(Suppl.143): 74–77 (1986).
23. Australian Bureau of Meteorology (BOM) Climate zones for urban design. 2001. http://www.bom.gov.au/climate/environ/design/climzone.shtml.
24. Kunst, A. et al. Outdoor air temperature and mortality in the Netherlands—a time series analysis. *American Journal of Epidemiology* 137(3): 331–341 (1993).
25. ISOTHURM. ISOTHURM Study Group: International study of temperature and heatwaves on urban mortality in low and middle income countries. *Lancet* submitted (2003).
26. Braga, A.L. et al. The time course of weather-related deaths. *Epidemiology* 12(6): 662–667 (2001).
27. Braga, A.L. et al. The effect of weather on respiratory and cardiovascular deaths in 12 U.S. cities. *Environmental Health Perspectives* 110(9): 859–863 (2002).
28. Semenza, J.C. et al. Heat-related deaths during the July 1995 heat wave in Chicago. *New England Journal of Medicine* 335(2): 84–90 (1996).
29. Gouveia, N. et al. Socio-economic differentials in the temperature-mortality relationship in São Paulo, Brazil. *Epidemiology* 12(4): 413 (2001).
30. Drasar, B.S. et al. Seasonal aspects of diarrhoeal disease. *Seasonal dimensions to rural poverty.* University of Sussex, UK, 1978.
31. Blaser, M.J. et al. eds. *Infections of the gastrointestinal tract.* New York, USA, Raven Press, 1995.
32. Checkley, W. et al. Effects of El Niño and ambient temperature on hospital admissions for diarrhoeal diseases in Peruvian children. *Lancet* 355(9202): 442–450 (2000).
33. Curriero, F.C. et al. The association between extreme precipitation and waterborne disease outbreaks in the United States, 1948–1994. *American Journal of Public Health* 91(8): 1194–1199 (2001).
34. Singh, R.B.K. et al. The influence of climate variation and change on diarrhoeal disease in the Pacific Islands. *Environmental Health Perspectives* 109(2): 155–159 (2001).

35. Robins-Browne, R.M. Seasonal and racial incidence of infantile gastroenteritis in South Africa. *American Journal of Epidemiology* 119(3): 350–355 (1984).

36. Rosenzweig, C. & Parry, M.L. Potential impact of climate-change on world food supply. *Nature* 367(6459): 133–138 (1994).

37. Intergovernmental Panel on Climate Change (IPCC). Climate change 1995: impacts, adaptations and mitigation of climate change. Contribution of Working Group II. In: *Second Assessment Report of the Intergovernmental Panel on Climate Change*. Watson, R.T. et al. eds. Cambridge, UK, & New York, USA, Cambridge University Press, 1996.

38. Parry, M. et al. Climate change and world food security: a new assessment. *Global Environmental Change-Human and Policy Dimensions* 9: S51–S67 (1999).

39. Otter-Nacke, S. et al. *Testing and validating the CERES-Wheat model in diverse environments*. Houston, USA, Johnson Space Center, 1986. (AGGRISTARS YM-15–00407).

40. Waterlow, J. et al. *Feeding a world population of more than eight billion people*. Oxford, UK, Oxford University Press, 1998.

41. Dyson, T. Prospects for feeding the world. *British Medical Journal* 319(7215): 988–990 (1999).

42. International Benchmark Sites Network for Agrotechnology Transfer (IBSNAT). Decision Support System for Agrotechnology Transfer Version 2.1. Honolulu, Department of Agronomy and Soil Science, College of Tropical Agriculture and Human Resources, University of Hawaii, 1989. (DSSAT V2.1).

43. Fischer, G. et al. Climate-change and world food-supply, demand and trade—who benefits, who loses. *Global Environmental Change-Human and Policy Dimensions* 4(1): 7–23 (1994).

44. Fischer, G. et al. Linked National Models. *A tool for international food policy analysis*. Dordrecht: Kluwer, 1988.

45. Bos, E.T. et al. *World population projections 1994–1995: estimates and projections with related demographic statistics*. World Bank. New York, USA, The Johns Hopkins University Press, 1994.

46. Energy Modelling Forum (EMF). *Second round study design for EMF14*. Energy Modelling Forum, 1995.

47. United Nations Food and Agriculture Organization (FAO). *Fifth World Food Survey*. Rome, Italy, United Nations Food and Agriculture Organization, 1987.

48. World Health Organization (WHO). Global Database on Child Growth and Malnutrition, 2002. http://www.who.int/nutgrowthdb/

49. Noji, E.K. *The public health consequences of disasters*. New York, USA, Oxford University Press, 1997.

50. International Federation of Red Cross and Red Crescent Societies (IFRC). *World Disaster Report 2001*. Oxford, UK & New York, USA, Oxford University Press, 2001.

51. Bouma, M.J. et al. Global assessment of El Niño's disaster burden. *Lancet* 350(9089): 1435–1438 (1997).

52. Kovats, R.S. et al. *El Niño and Health*: World Health Organization, 1999.

53. Hoozemans, F.M.J. & Hulsburgen, C.H. Sea-level rise: a worldwide assessment of risk and protection costs. In: *Climate change: impact on coastal habitation*. Eisma, D. ed. London, UK, Lewis Publishers, pp.137–163, 1995.

54. Nicholls, R.J. et al. Increasing flood risk and wetland losses due to global sea-level rise: regional and global analyses. *Global Environmental Change-Human and Policy Dimensions* 9: S69–S87 (1999).

55. Pielke, R.A. Nine fallacies of floods. *Climatic Change* 42(2): 413–438 (1999).

56. Kundzewicz, Z.W. & Kaczmarek, Z. Coping with hydrological extremes. *Water International* 25(1): 66–75 (2000).

57. OFDA/CRED. EM-DAT: The International Disaster Database. Brussels, Belgium, Université Catholique de Louvain, 2001. www.cred.be/emdat.

58. Yohe, G. & Tol, R.S.J. Indicators for social and economic coping capacity—moving

toward a working definition of adaptive capacity. *Global Environmental Change* 12: 25–40 (2002).

59. Beinin, C. An examination of health data following two major earthquakes in Russia. *Disasters* 5(2): 142–146 (1981).

60. Rivers, J.P.W. Women and children last: An essay on sex discrimination in disasters. *Disasters* 6(4): 256–267 (1982).

61. Hoozemans, F.M.J. et al. *A global vulnerability analysis: vulnerability assessment for population, coastal wetlands and rice production on a global scale*, 2nd Edition, the Netherlands, Delft Hydraulics, 1993.

62. Martens, W.J. *Health and climate change: modelling the impacts of global warming and ozone depletion.* London, UK, Earthscan, 1998.

63. Massad, E. & Forattini, O.P. Modelling the temperature sensitivity of some physiological parameters of epidemiologic significance. *Ecosystem Health* 4(2): 119–129 (1998).

64. Kovats, R.S. et al. *Climate and vector-borne disease: an assessment of the role of climate in changing disease patterns*: United Nations Environment Programme, 2000.

65. Sutherst, R.W. et al. Global change and vector-borne diseases. *Parasitology Today* 14: 297–299 (1998).

66. Rogers, D.J. & Randolph, S.E. The global spread of malaria in a future, warmer world. *Science* 289(5485): 1763–1766 (2000).

67. Mouchet, J. & Manguin, S. Global warming and malaria expansion. *Annales de la Société Entomologique de France* 35: 549–555 (1999).

68. Reiter, P. Climate change and mosquito-borne disease. *Environmental Health Perspectives* 109: 141–161 (2001).

69. Snow, R.W. et al. Relation between severe malaria morbidity in children and level of *Plasmodium falciparum* transmission in Africa. *Lancet* 349(9066): 1650–1654 (1997).

70. Coleman, P.G. et al. Endemic stability—a veterinary idea applied to human public health. *Lancet* 357: 1284–1286 (2001).

71. Craig, M.H. et al. A climate-based distribution model of malaria transmission in sub-Saharan Africa. *Parasitology Today* 15(3): 105–111 (1999).

72. Tanser, F.C. et al. Malaria seasonality and the potential impacts of climate change in Africa. *Lancet* (in press).

CHAPTER 8

Stratospheric ozone depletion, ultraviolet radiation and health

A.J. McMichael,[1] R. Lucas,[1] A.-L. Ponsonby,[1] S.J. Edwards[2]

Introduction

To our forebears the sky was the realm of the gods, inaccessible to mere humans. Only 100 years ago, the few scientists studying environmental problems would have been incredulous at suggestions that, by the late twentieth century, humankind would have begun to change the composition and function of the stratosphere. Yet this has happened. After 8000 generations of *Homo sapiens*, this generation has witnessed the onset of the remarkable process of human-induced depletion of stratospheric ozone.

By the usual definition, stratospheric ozone depletion is not an integral part of the process of "global climate change". The latter process results from the accrual of greenhouse gases in the troposphere, physically separate from the stratosphere. The stratosphere extends from around 10 to 50 km altitude (see Figure 8.1). It is distinguishable from the lower atmosphere (troposphere) and the outer atmosphere (mesosphere and thermosphere). In particular, most of the atmosphere's ozone resides within the stratosphere. The ozone layer absorbs much of the incoming solar ultraviolet radiation (UVR) and thus offers substantial protection from this radiation to all organisms living at, or near to, Earth's surface.

Intriguingly, atmospheric ozone is not part of the planet's original system but a product of life on Earth, which began around 3.5 billion years ago. Until a half billion years ago, living organisms could not inhabit the land surface. Life was confined to the world's oceans and waterways, relatively protected from the intense unfiltered solar ultraviolet radiation. About 2 billion years ago as photo-synthesising organisms emitted oxygen (O_2), a waste gas (ozone–O_3) gradually began to form within the atmosphere (*1*). From around 400 million years ago aqueous plants were able to migrate onto the now-protected land and evolve into terrestrial plants, followed by animal life that ate the plants. So the succession has evolved, via several evolutionary paths, through herbivorous and car-nivorous dinosaurs, mammals and omnivorous humans. Today, terrestrial species are shielded by Earth's recently acquired mantle of ozone in the stratosphere that absorbs much of the solar ultraviolet.

Unintentionally, the human species has now reversed some of that strato-spheric ozone accumulation. Surprisingly, various industrial halogenated chemi-cals such as the chlorofluorocarbons (CFCs, used in refrigeration, insulated

---

[1] National Centre for Epidemiology and Population Health, The Australian National University, Canberra, Australia.
[2] London School of Hygiene and Tropical Medicine, London, England.

packaging and spray-can propellants), inert at ambient temperatures, react with ozone in the extreme cold of the polar stratospheric late winter and early spring. This time of year combines cold stratospheric temperatures with the "polar dawn", as solar ultraviolet radiation begins to reach the polar stratosphere, where it causes photolytic destruction of human-made gases in the stratosphere, such as the CFCs, methyl bromide and nitrous oxide. This, in turn, generates reactive "free radicals" that destroy stratospheric ozone.

The Montreal Protocol—noticing and responding to ozone depletion

Colour-enhanced pictures of the winter-spring polar "ozone hole" on the United States NASA web-site depict an overall loss which had crept up to around one-third of total Antarctic ozone, by the late 1990s, relative to the pre-1975 figure. Winter-spring losses in the Arctic are smaller because local stratospheric temperatures are less cold than in the Antarctic. During the 1980s and 1990s at northern mid-latitudes (such as Europe), the average year-round ozone concentration declined by around 4% per decade: over the southern regions of Australia, New Zealand, Argentina and South Africa, the figure has approximated 6–7%. Long-term decreases in summertime ozone over New Zealand have been associated with significant increases in ground level UVR, particularly in the DNA-damaging waveband (2). Ozone depletion is one of several factors, including cloud cover and solar elevation, which affect ground level UV radiation. An examination of atmospheric changes in Australia from 1979 to 1992 has shown that the deseasonalised time series of UVR exposures were a linear function of ozone and cloud cover anomalies. In tropical Australia a trend analysis indicated a significant increase in UVR, estimated from satellite observations, of 10% per decade in summer associated with reduced ozone (1–2% per decade) and reduced cloud cover (15–30% per decade). In southern regions, a significant trend for UVR over time was not observed, partially due to increased cloud cover. Thus, in Tasmania, despite a significant ozone reduction of 2.1% per decade, measures of ground level UVR have not increased (3).

Estimating the resultant changes in actual ground-level ultraviolet radiation remains technically complex. Further, the methods and equipment used mostly have not been standardised either over place or time. While there is good agreement between similarly calibrated spectroradiometers, this may not be true when comparing different types of instruments—spectroradiometers, broad-band meters, filter radiometers. There is little or no reliable evidence on levels of UV radiation prior to concerns related to ozone depletion (pre-1980s) due to maintenance and calibration difficulties with these older instruments. The advent of satellite measuring systems allowed reliable measurement of UVR. However, satellite measurements may not accurately reflect ground level UVR due to failure to take adequate account of lower atmospheric changes. For example, satellite estimates suggest that the difference in summertime erythemal UV irradiances between northern and southern hemispheres is around 10–15%. However, ground level measurements indicate that this difference may be even higher, probably due to lesser atmospheric pollution in the southern hemisphere.

It is clear that under cloud-free skies there is a strong correlation between ground level erythemal UV radiation and levels of atmospheric ozone (4). Yet the effects of clouds, increasing tropospheric ozone and aerosol pollution of the lower atmosphere modify this relationship making the detection of long-term trends in UVR related to ozone depletion difficult to elucidate. Long-term predictions are uncertain since they involve assumptions about not only future ozone levels but also future variations in cloud cover, tropospheric ozone and lower atmospheric pollution. However, exposures at northern mid-latitudes are projected to peak around 2020, entailing an estimated 10% increase in effective ultraviolet radiation relative to 1980s levels (5).

Fears of ozone depletion due to human activities first emerged in the late 1960s. A decade of denial and debate followed with eventual acceptance by scientists and policy-makers that ozone depletion was likely to occur and would represent a global environmental crisis. In the mid-1980s governments responded with alacrity to the emerging problem of ozone destruction. The Montreal Protocol of 1987 was adopted, widely ratified and the phasing out of major ozone-destroying gases began. The protocol was tightened further in the 1990s. At first sight, the solution to this particular global environmental change appears to be unusually simple: a substitution of particular industrial and agricultural gases for others. However, the problem has not yet been definitely solved. First, there is a large range of human-made ozone-destroying gases, including some of those chemicals developed to replace the early CFCs. Second, compliance with the international agreement remains patchy. Third, scientists did not foresee the interplay (see below) between a warming lower atmosphere and an ozone-depleted stratosphere. Nevertheless, scientists anticipate that there will be slow but near-complete recovery of stratospheric ozone during the middle third of the twenty-first century.

Difference between stratospheric ozone depletion and human-enhanced greenhouse effect

Stratospheric ozone destruction is an essentially separate process from greenhouse gas (GHG) accumulation in the lower atmosphere (see Figure 8.1), although there are several important and interesting connections. First, several of the anthropogenic greenhouse gases (e.g. CFCs and N_2O) are also ozone-depleting gases. Second, tropospheric warming apparently induces stratospheric cooling that exacerbates ozone destruction (6, 7). As more of Earth's radiant heat is trapped in the lower atmosphere, the stratosphere cools further, enhancing the catalytic destruction of ozone. Further, that loss of ozone itself augments the cooling of the stratosphere. Interactions between climate change and stratospheric ozone may delay recovery of the ozone layer by 15–20 years (5).

Third, depletion of stratospheric ozone and global warming due to the build-up of greenhouse gases interact to alter UVR related effects on health. In a warmer world, patterns of personal exposure to solar radiation (e.g. sun-bathing in temperate climates) are likely to change, resulting in increased UVR exposure. This may be offset by changes in cloud cover and cloud optical thickness as a result of global climate change. Predictions of future UVR exposures based on ozone depletion, behavioural changes and climate change are uncertain. A recent analysis of trends in Europe reports a likely increase of 5–10% in yearly UV doses received over the past two decades (5).

Stratospheric ozone depletion has further indirect health effects. One important effect is that ozone depletion in the stratosphere increases the formation of photochemical smog, including ozone accumulation, in the lower troposphere. That is, ozone depletion in the upper atmosphere will allow more ultraviolet radiation to reach the troposphere where photochemical smog forms via a UVR-mediated breakdown of nitrogen dioxide (a common fossil fuel pollutant) and

FIGURE 8.1 Layers of the Earth's atmosphere. *Source: reproduced from reference 8.*

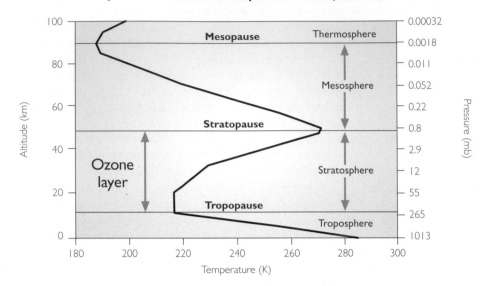

other products. Photochemical smog is a complex chemical mixture containing nitric acid (HNO_3); peroxyacyl nitrates (PANs), aldehydes (e.g. formaldehyde) ozone (O_3) and other substances. It has been estimated that the concentration of tropospheric ozone has increased from 10 ppb 100 years ago to 20–30 ppb in some locations today, with peaks of >100 ppb reported in some centres (9). The ozone component of photochemical smog acts as a respiratory irritant, causing oxidant damage to the respiratory epithelium and possibly enhancing allergen-induced airway inflammation.

Solar UVR measurement

Sunlight consists of solar rays of differing wavelengths. Visible light ranges from 400 nm (violet) to 700 nm (red). Infrared radiation, or heat, has longer wavelengths than visible light; ultraviolet radiation has shorter wavelengths than visible light. UVR is further divided into UVA (315–400 nm), UVB (280–315 nm) and UV-C (<280 nm). Almost all incoming solar UVC and 90% of UVB are absorbed by stratospheric ozone, while most UVA passes through the atmosphere unchanged. Although UVA penetrates human skin more deeply than UVB, the action spectra from biological responses indicate that it is radiation in the UVB range that is absorbed by DNA—subsequent damage to DNA appears to be a key factor in the initiation of the carcinogenic process in skin (*10*).

The amount of ambient UVB experienced by an individual outdoors with skin exposed directly to the sky is dependent on the following:

(i)   stratospheric ozone levels
(ii) solar elevation
(iii) regional pollution
(iv) altitude of the individual
(v)   cloud cover
(vi) presence of reflective environmental surfaces such as water, sand or snow.

The amount of received UVR exposure can be measured in terms of the energy of the transmitted photons, often expressed as energy per unit area irradiated (e.g. joules per square metre).

To examine the health effects of solar UVR, it is necessary also to consider measurement in the biological dimension. Hence, UVR also is described in units of erythemal (sunburn) efficacy. To this end, exposure is spectrally weighted over the relevant wavelengths according to erythemal impact (using the Commission Internationale de l'Éclairage {CIE} (11) erythemal standard action spectrum (12)). Thus, standard erythemal doses (SEDs) can be defined (13) by which daily, monthly or annual UV exposures can be quantified. A UV index also has been defined to express the daily maximum in biologically effective UVR, reached around midday.

Main types of health impacts

There is a range of certain or possible health impacts of stratospheric ozone depletion. These are listed in Table 8.1.

Many epidemiological studies have implicated solar radiation as a cause of skin cancer (melanoma and other types) in fair-skinned humans (14, 15). The most recent assessment by the United Nations Environment Program (1998) projected significant increases in skin cancer incidence due to stratospheric ozone depletion (16). The assessment anticipates that for at least the first half of the twenty-first century (and subject to changes in individual behaviours) additional ultraviolet radiation exposure will augment the severity of sunburn and incidence of skin cancer.

High intensity UVR also damages the eye's outer tissues causing "snow blindness", the ocular equivalent of sunburn. Chronic exposure to UVR is linked to conditions such as pterygium (17). UVB's role in cataract formation is complex but some subtypes, especially cortical and subcapsular cataracts, appear to be associated with UVR exposure while others (nuclear cataracts) do not.

In humans and experimental animals, UVR exposure causes both local and whole-body immunosuppression (16). Cellular immunity is affected by variation in the ambient dose of UVR (18). UVR-induced immunosuppression therefore could influence patterns of infectious disease and may also influence the occurrence and progression of various autoimmune diseases. Nevertheless, little direct evidence exists for such effects in humans, and uncertainties remain about the underlying biological processes.

Finally, there is an ecological dimension to consider. Ultraviolet radiation impairs the molecular chemistry of photosynthesis both on land (terrestrial plants) and at sea (phytoplankton). This could affect world food production, at least marginally, and thus contribute to nutritional and health problems in food-insecure populations. However, as yet there is little information about this less direct impact pathway.

Disorders of the skin

Since the 1850s it has been known that excessive exposure to sunlight can cause skin damage. Observation of boatmen, fishermen, lightermen, agricultural labourers and farmers revealed that skin cancer developed on areas most frequently exposed (e.g. hands, neck and face) (19). The exact process by which

TABLE 8.1 Summary of possible effects of solar ultraviolet radiation on the health of human beings.

Effects on skin
- Malignant melanoma
- Non-melanocytic skin cancer—basal cell carcinoma, squamous cell carcinoma
- Sunburn
- Chronic sun damage
- Photodermatoses.

Effects on the eye
- Acute photokeratitis and photoconjunctivitis
- Climatic droplet keratopathy
- Pterygium
- Cancer of the cornea and conjunctiva
- Lens opacity (cataract)—cortical, posterior subcapsular
- Uveal melanoma
- Acute solar retinopathy
- Macular degeneration.

Effect on immunity and infection
- Suppression of cell mediated immunity
- Increased susceptibility to infection
- Impairment of prophylactic immunization
- Activation of latent virus infection.

Other effects
- Cutaneous vitamin D production
  - prevention of rickets, osteomalacia and osteoporosis
  - possible benefit for hypertension, ischaemic heart disease and tuberculosis
  - possible decreased risk for schizophrenia, breast cancer, prostate cancer
  - possible prevention of Type 1 (usually insulin dependent) diabetes
- Non-Hodgkin's lymphoma
- Altered general well-being
  - sleep/wake cycles
  - seasonal affective disorder
  - mood.

Indirect effects
- Effects on climate, food supply, infectious disease vectors, air pollution, etc.

exposure to sunlight causes skin cancer was not understood until relatively recently.

The incidence of skin cancer, especially cutaneous malignant melanoma, has been increasing steadily in white populations over the past few decades (*20*). This is particularly evident in areas of high UVR exposure such as South Africa, Australia and New Zealand. Human skin pigmentation has evolved over hundreds of thousands of years, probably to meet the competing demands of protection from the deleterious effects of UVR and maximization of the beneficial effects of UVR. Skin pigmentation shows a clear, though imperfect, latitudinal gradient in indigenous populations (*21*). Over the last few hundred years, however, there has been rapid migration of predominantly European populations away from their traditional habitats into areas where there is a mismatch of pigmentation and UVR. The groups most vulnerable to skin cancer are white Caucasians, especially those of Celtic descent (see Box 8.1) living in areas of high UVR. Further, behavioural changes particularly in fair-skinned populations, have led to much higher UV exposure through sun-bathing and skin-tanning. The marked increase in skin cancers in these populations over recent decades reflects, predominantly, the combination of post-migration geographical vulnerability and modern behavioural patterns. It remains too early to identify any adverse effect of stratospheric ozone depletion upon skin cancer risk.

UVR and skin cancer

UVR exposure was first linked experimentally to skin cancer in the 1920s (*19*). Using a mercury-vapour lamp as a source of UVR, Findlay exposed mice experimentally to daily doses of UVR over 58 weeks. Malignant tumours developed in four of the six mice that developed tumours, leading to the conclusion that exposure to UVR could result in skin cancer (*19*). Epidemiologists' interest in this association was further stimulated by the possibility of human-induced damage to stratospheric ozone, first theorized in the 1970s. The International Agency for Research on Cancer in 1992 concluded that solar radiation is a cause of skin cancer (*14*). A summary of the evidence appears in Box 8.1.

Within the ultraviolet radiation waveband, the highest risk of skin cancer is related to UVB exposure. UVB is much more effective than UVA at causing biological damage, contributing about 80% towards sunburn while UVA contributes the remaining 20% (*22*). UVB exposure (from both sunlight and artificial sources) has been linked conclusively to cutaneous malignant melanoma (CMM) and non-melanoma skin cancer (NMSC) (*23, 24*). Figure 8.2 shows diagrammatically the UV spectrum and the erythemal effectiveness of solar radiation in humans.

There is a strong relationship between the incidence (and mortality) of all types of skin cancer and latitude, at least within homogeneous populations. Latitude approximately reflects the amount of UVR reaching the earth's surface (*24*).

BOX 8.1 Evidence linking skin cancer to solar radiation

■ Skin cancer—cutaneous malignant melanoma (CMM) and non-melanoma skin cancer (NMSC)—occurs predominantly in white populations. It is uncommon in populations with protective melanin pigmentation of the skin, e.g. Africans, Asians, Hispanics, etc.

■ Especially common in fair complexioned individuals who freckle and sunburn easily, notably those of Celtic ancestry, e.g. Irish, Welsh, etc.

■ Occurs primarily on parts of the body most often exposed to sunlight.

■ Incidence of skin cancer is inversely correlated with latitude and shows a positive relation to estimated or measured levels of UVR.

■ Outdoor workers with chronic sun exposure are at greater risk than indoor workers for NMSC. Indoor workers with intermittent sun intensive exposure appear more prone to CMM.

■ Risk of skin cancer is associated with various measures of solar skin damage.

■ Individuals with certain genetic skin diseases, such as albinism, are prone to skin cancer by virtue of their sensitivity to UVR.

■ Experimental animals develop skin cancer with repeated doses of UVR (especially UVB).

■ Most SCC and BCC have highly specific mutations of the tumour suppressor gene p53 that are characteristic of UV-induced changes in model systems.

Source: adapted from reference 23

FIGURE 8.2 Biologically active UV radiation. Diagrammatic representation of the range of ultraviolet and visible radiation, the relative incidence of different wavelenghts (in nanometres) at Earth's surface, and the predicted source of cancer risk (carcinogenic effectiveness) as the product of both incident radiation and the experimentally-shown "action spectrum" for DNA damage (25). Source: reference 26.

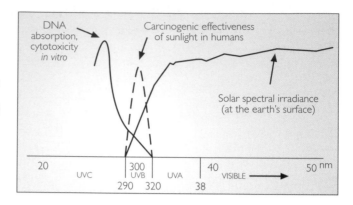

This is due partly to the differing thickness of the ozone layer at different latitudes, and partly to the angle at which solar radiation passes through the atmosphere.

In response to UVB exposure the epidermis thickens via an increase in the number of cell layers (epidermal hyperplasia). This occurs particularly in people who do not tan readily. This thickening reduces the amount of UVB penetration to the basal layer providing partial natural protection against the harmful effects of UVR (27). Animal experiments indicate that despite this epidermal protection, further UVB exposure can act as a potent tumour promoter on damaged basal cells (28).

Ozone depletion and skin cancer

Scientists expect the combined effect of recent stratospheric ozone depletion, and its continuation over the next one to two decades, to be (via the cumulation of additional UVB exposure) an increase in skin cancer incidence in fair-skinned populations living at mid to high latitudes (29).

Future impacts of ozone depletion on skin cancer incidence in European and North American populations have been modelled (30). Figure 8.3 summarizes the estimates for the expected excess skin cancer incidence in the US white population, following three scenarios of ozone depletion. The first entails no restrictions on CFC emissions. The second, reflecting the original Montreal protocol of 1987, entails a 50% reduction in the production of the five most important ozone-destroying chemicals by the end of 1999. In the third scenario, under the Copenhagen amendments to that protocol, the production of 21 ozone-depleting chemicals is reduced to zero by the end of 1995. This (vertically integrated) modelling study estimated that, for the third scenario, by 2050 there would be a park relative increase in total skin cancer incidence of 5–10% in "European" populations living between 40°N and 52°N (based on a 1996 baseline of 2,000 cases of skin cancer per million per year in the United States and 1,000 cases per million per year in northwest Europen). The figure would be higher, if allowing for the ageing of the population. The equivalent estimation for the United States' population is a 10% increase in skin cancer incidence by around 2050.

It must be remembered that all such modelling makes simplifying assumptions and entails a substantial range of uncertainty. Not only is the shape of the UVR-cancer (dose-response) relationship poorly described in human populations, but also there is inevitable uncertainty about actual future gaseous emissions; the physical interaction between human-induced disturbances of the lower and

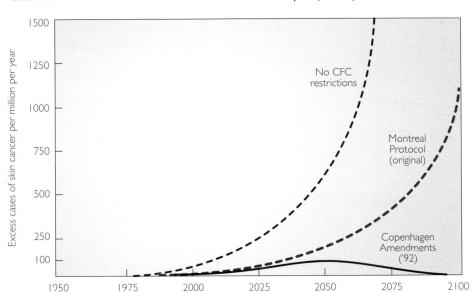

FIGURE 8.3 Estimates of ozone depletion and skin cancer incidence to examine the Montreal Protocol achievements. *Adapted from reference 30.*

middle atmospheres (including changes in cloud cover under conditions of climate change); and future changes in patterns of human exposure-related behaviours.

Eye disorders

Both age related macular degeneration (AMD) and cataract show associations with low or depleted antioxidant status and higher oxidative stress (smoking), suggesting common aetiological factors. Approximately 50% of incident UVA and 3% of UVB penetrates the cornea, where a further 1% of UVB is absorbed by the aqueous humor (31). Remaining UVR is absorbed by the lens, hence the UVR association with lens opacities is the most plausible. There is some evidence that sunlight exposure (possibly the blue light component) may be implicated in macular degeneration (32).

Solar radiation and risk of lens opacities: current level of evidence

The shorter wavelength constituents of solar radiation (notably UVA, UVB and UVC) are more damaging to biological molecules than is visible light. Although UVB is only 3% of the UVR that reaches the earth, it is much more biologically active than UVA.

In vivo and in vitro laboratory studies demonstrate that exposure to UVR, in particular to UVB, in various mammalian species induces lens opacification (33). The actual mechanisms remain unclear but a range of adverse effects is observed as a result of free radical generation from UVR energised electrons. There has been criticism that UVR doses in laboratory studies are much higher than those encountered in natural conditions (34). However, based on ambient UVA and UVB fluxes in the north-eastern United States, it has been estimated that 26 hours of continuous UVA exposure or 245 hours of continuous UVB exposures at those ambient levels would exceed the rabbit lens threshold for lens damage (31). While direct extrapolation from animal studies to humans is not possible it is plausible that in humans, with much longer age spans than laboratory animals, cumulative damage to the lens from UVR could explain the high prevalence of lens opacities in elderly people.

There is mixed evidence for UVR's role in lens opacities in human populations (35). Cataracts are more common in some (but not all) countries with high UVR levels. However, few studies have examined whether UVR can explain differences between populations in the prevalence of lens opacities. One study of cataract surgical rates in the United States' Medicare programme estimated a 3% increase in the occurrence of cataract surgery for each 1° decrease in latitude across the United States (36). However, surgery rates are not a good measure of the prevalence of opacities in the population; they are influenced by service access and differences in the thresholds for eligibility for surgery. Studies in non-Western populations have provided some weak evidence for opacities being higher in areas with greater UVB radiation. These studies based on eye examinations included surveys among Australian Aboriginal populations (37); rural Chinese populations (38); and across areas of Nepal (39). These associations may have been confounded by other unmeasured lifestyle factors, such as diet.

Studies measuring UVR or outdoor exposure in individuals have shown inconsistent results. The strongest evidence is provided by a study of a high UVR-exposed group (fishermen), in the Chesapeake Bay Watermen Study in the United States (40, 41), which showed an association between adult UVR dose and risk of cortical and posterior subcapsular opacities. In general population studies in the United States, UVR exposure was related to cortical opacities in one study (42) but not in another (43), or has been observed in men but not in women (44). Further support for the association with cortical opacities and UVR comes from mannikin studies showing the largest doses of UVR to be received by the lower and inner (nasal) lens—the site where cortical opacities predominate (45). Cortical opacities are rare in the upper lens. However, it has been suggested that the lack of an association between UVR and nuclear opacities may reflect failure to measure exposures occurring in earlier life (46). Since the nuclear material is the oldest in the lens capsule, the most relevant exposures are those that occur in early life. In India, where rates of lens opacities are higher than in Western populations, estimated lifetime sunlight exposure was associated with all types of lens opacities, including nuclear (47).

Few studies have been conducted in European populations. A hospital-based case control study conducted in Parma, northern Italy showed an increased risk of cortical cataracts with a four point scaled estimate of time spent outdoors (48). In a small population based study in the north of Finland, working outdoors was a risk factor for cortical cataracts in women, but not men (49). The POLA (Pathologies Oculaires Liées à l'Age) study showed a significant association between annual ambient solar radiation and cortical and mixed (mainly cortical and nuclear) cataracts, with a modest trend also in nuclear-only cataracts (50). The POLA study was undertaken in a small town in the south of France close to the sea where there were high levels of outdoor professional and leisure activities. The study also showed an excess risk of posterior subcapsular cataracts for people who were professionally exposed to sunlight (eg fishing, agriculture, building industry). Of the European studies, only that of POLA attempted to measure ambient UVR. Such measurements of UVR exposure (i.e. taking account of occupation, leisure and residence) rarely have been made in other European populations.

The Reykjavik Eye Study, a population-based study in Iceland, found a positive relationship between cortical cataract and time spent outside on weekdays (51). In Australia, the Melbourne Visual Impairment Project demonstrated a relationship between UVR and cortical cataract, as well as an interaction between ocular UVB exposure and vitamin E for nuclear cataract (52).

An evaluation of the possible risk from UVR must take account of both confounding factors and factors that may modify the association. Factors that may increase susceptibility to UVR-induced damage include poor nutrition and smoking. Smoking may act as an additional source of oxidative stress and consistently has been shown to increase the risk of cataract. Antioxidant micronutrients may enhance the free radical scavenging defence system of the eye. There is some evidence that low dietary intakes of vitamins C, E and carotenoids increase cataract risk (53, 54).

Solar radiation effects on the cornea and conjunctiva

Acute exposure of the eye to high levels of UVR, particularly in settings of high light reflectance such as snow-covered surroundings, can cause painful inflammation of the cornea or conjunctiva. Commonly called snow-blindness, photokeratitis and photoconjunctivitis are the ocular equivalent of acute sunburn.

Pterygium is a common condition that usually affects the nasal conjunctiva, sometimes with extension to the cornea. It is particularly common in populations in areas of high UVR or high exposure to particulate matter. Studies of the Chesapeake Bay watermen showed a dose-response relationship between history of exposure to UVR and risk of pterygium (55). Others have found measures of UV exposure to be strongly related to pterygium risk (56, 57). In a large population-based study in Melbourne, almost half of the risk of pterygium was attributable to sun exposure (58).

Effects on the retina

Other eye disorders associated with UVR are uncommon but cause significant morbidity to affected individuals. Acute solar retinopathy, or eclipse retinopathy, usually presents to medical attention soon after a solar eclipse when individuals have looked directly at the sun. Effectively this is a solar burn to the retina. Usually the resulting scotoma resolves but there may be permanent minor field defects. Several cases of solar retinopathy in young adults, possibly related to sun-gazing during a period of low stratospheric ozone in the United States, have been described (59, 60).

Immune system function and immune-related disorders

Although most of the available evidence comes from studies of experimental animals, it appears that ultraviolet radiation suppresses components of both local and systemic immune functioning. An increase in ultraviolet radiation exposure therefore may increase the occurrence and severity of infectious diseases and, in contrast, reduce the incidence and severity of various autoimmune disorders. The damping down of the T lymphocyte (helper cell type 1), or "T_H1", component of the immune system may alleviate diseases such as multiple sclerosis, rheumatoid arthritis and insulin-dependent (Type 1) diabetes. Undifferentiated T_H0 cells are immunologically primed to develop into either T_H1 or T_H2 cells; in animals these two groups are thought to be mutually antagonistic (61). Thus UVR exposure theoretically could worsen T_H2-mediated disease by suppressing T_H1 cell function (62), however, more recent work has shed some doubt on this notion. In mice UVR exposure is associated with decreased systemic T_H2 as well as T_H1 immune responses (63). UVR leads to increased secretion of the cytokine, interleukin (IL)-10 (64) appears to suppress T_H1 and T_H2 cytokine responses to external antigens (65). Much remains unknown. Partly in response to questions about the biological impacts of stratospheric ozone depletion, among scientists there is new interest in assessing the influence of ultraviolet radiation upon immune system function, vitamin D metabolism (see Box 8.2) and the consequences for human disease risks.

Recent research suggests that UVR exposure can weaken T_H1-mediated immune responses through several mechanisms:

- UVR can cause local epidermal immunosuppression and a reduction in contact hypersensitivity (CH) and delayed type hypersensitivity (DTH) (62);
- UVR acts to convert urocanic acid (UCA) from the *trans*-UCA form to its isomer, the *cis*-UCA form, within the stratum corneum (64). This process induces changes in epidermal cytokine profiles from a wide range of cell types. UVR-induced DNA damage also alters cytokine profiles, leading to immunosuppression (64). Liposome therapy with a DNA repair enzyme can prevent UVR-induced cytokine alterations such as the upregulation of IL-10 (66). Importantly, subepidermal cytokine signalling alterations also can induce soluble products that can exert systemic immunosuppression (61);
- sunlight suppresses secretion of the hormone melatonin. Activation of melatonin receptors on T helper cells appears to enhance T lymphocyte priming and the release of T_H1 type cytokines such as interferon gamma (67);
- a role for UVR in promoting the secretion of melanocyte stimulating hormone (MSH), which may suppress T_H1 cell activity, also has been proposed (68);
- the active form of vitamin D ($1,25(OH)_2D_3$), derived from UVR-supported biosynthesis has well-documented immunomodulatory effects. Peripheral monocytes and activated T helper cells have vitamin D receptors, vitamin D or its analogues can down-regulate T helper cell activity (69).

Overall, these findings indicate that UVR suppresses T_H1-mediated immune activity. It is important to note that part of this effect occurs independently of vitamin D.

Possible effect on human infectious disease patterns

Higher UVR exposure could suppress the immune responses to infection of the human host (70). The total UVR dose required for immune suppression is likely to be less than that required for skin cancer induction but direct human data are not available. In animals, high UVR exposure has been shown to decrease host resistance to viruses such as influenza and cytomegalovirus, parasites such as malaria and other infections such as *Listeria monocytogenes* and *Trichinella spiralis* (71). However, significant inter-species variation in UVR-induced immune suppression and other differences in host response to infection limit direct extrapolation of these findings to humans.

Recently, data from these animal studies have been used to develop a model to predict the possible changes in infection patterns in humans due to increased UVR resulting from stratosphere ozone depletion (72). Importantly, the model did account for likely inter-species variation in susceptibility to UVR-induced

TABLE 8.2 Predicted effects of stratospheric ozone decreases on the biologically effective ultraviolet irradiance, and hence, on suppression of the specific cellular immune responses to Listeria bacteria (local noon, clear skies, southern Europe).

Latitude (month)	Decrease in ozone (%)	Ozone (dobson units)	Biologically Effective irradiance (W/m^2)	Increase in BE$_{imm}$ %	RAF$_{imm}$	Calculated time (min) for 50% immunosuppression[a]
40°N	0	335.6	0.073	0.0	—	350
January	5	318.8	0.075	3.0	0.60	340
	10	302.0	0.078	6.3	0.63	327
	20	268.5	0.083	13.5	0.68	307
40°N	0	307.9	0.278	0.0	—	92
July	5	292.5	0.285	2.5	0.50	90
	10	277.1	0.292	5.3	0.53	87
	20	246.3	0.310	11.5	0.58	82

Abbreviations: BEI$_{imm}$, biological effective irradiance for immunosupression: RAF$_{imm}$, radiation amplification factor for immunosupression.

[a] Lymphocyte proliferation in response to Listeria bacteria.

Source: adapted from reference 72.

immunosuppression. The theoretical model demonstrated that outdoor UVB exposure levels could affect the cellular immune response to the bacteria *Listeria monocytogenes* in humans. Using a worst-case scenario (sun-sensitive individuals with no UVR adaptation), ninety minutes of noontime solar exposure in mid-summer at 40°N was predicted to lead to a 50% suppression of human host lymphocyte responses against *Listeria monocytogenes*. A 5% decrease in ozone layer thickness might shorten this exposure time by about 2.5% (72) (Table 8.2).

Human epidemiological studies are required to confirm the findings from laboratory or animal studies. They also are needed to provide clearer risk assessments of the adverse immunosuppressive effect of increased UVR exposure. Such studies also should consider the role of vitamin D in host resistance to infection.

Personal UVR exposure in humans has been demonstrated to increase the number and severity of orolabial herpes simplex lesions (i.e. around the mouth). Recent questions have been raised about the potential adverse consequences of UVR-induced immunosuppression for HIV-infected individuals. A 1999 review concluded that despite experimental evidence in laboratory animal studies demonstrating HIV viral activation following UV radiation, there were no data in humans that consistently showed clinically significant immunosuppression in HIV-positive patients receiving UVB or PUVA therapy (73). A small follow-up study of HIV-positive individuals failed to detect any association between sun exposure and HIV disease progression. However, the review concluded that larger follow-up studies were required to assess fully this important issue (73).

Increased UVR exposure: a possible effect to reduce vaccine efficacy?

There has been concern that increased exposure to UVR due to stratospheric ozone depletion could hamper the effectiveness of vaccines, particularly BCG, measles and hepatitis (70). BCG vaccine efficacy has a latitudinal gradient with reduced efficacy at lower latitudes. Seasonal differences in vaccine efficacy have been observed for hepatitis B (74). While this ecological observation may reflect other latitude-related factors, it is also consistent with UVB depressing an effective host response to intradermally administered vaccines (75).

In animal studies, pre-exposure to UVB prior to intradermal vaccination with *Mycobacterium bovis* (BCG) impairs the DTH immune response of the host animal to mycobacterial antigens (62). Local UV irradiation of the skin prior to, and following, inoculation decreases the granulomatous reaction to lepromin in sensitised individuals (76). Overall these studies indicate that a potential health effect of increased UV exposure could be reduced vaccine efficacy particularly for vaccines that require host immune responses to intradermally administered antigens.

Non-Hodgkin's Lymphoma

The incidence of Non-Hodgkin's Lymphoma (NHL) has increased greatly worldwide in recent decades. The reasons for this increase are not known but high personal UVR exposure has been suggested as a possible contributary factor, for the following reasons:

- NHL incidence in England and Wales is positively associated with higher solar UV radiation by region (77);
- patients with NHL also have been noted to have an increased likelihood of non-melanoma skin cancer;
- chronic immunosuppression is an established risk factor for NHL and, as discussed, UVR has immunosuppressive effects on humans.

BOX 8.2 The beneficial role of UVR for Vitamin D synthesis in humans

---

The active metabolite of Vitamin D ($1,25(OH)_2 D_3$) is a human hormone with an important role in calcium and phosphorous regulation in humans. It also has other important roles. In 1822, the link between sunlight deprivation and the bone disease rickets was postulated. This link was confirmed in the early 1900s by experiments that showed that sunlight exposure could cure rickets. More recently, vitamin D has been shown to have an important role in the immune system and also may be important in the growth of neural tissue during early life. Furthermore, vitamin D receptors (VDR) have now been located in a variety of cells (e.g. brain, breast and pancreas).

Sunlight exposure is the primary determinant of vitamin D levels in terrestrial vertebrates, including humans. UVB rays enter the epidermis and release energy that changes a pre-existing cholesterol metabolite to previtamin D and its isomer cholecalciferol. Cholecalciferol ($25(OH)D$) is carried in the blood stream to the liver and then kidney, where, after a series of biological reactions, the active vitamin D hormone ($1,25(OH)D_3$) is formed. Circulating serum $25(OH)D$ concentration provides an integrated assessment of vitamin D intake and stores (79).

The exact dose of UVR exposure for optimal vitamin D levels is not known particularly as the required UVR dose will be influenced by host factors such as skin pigmentation, vitamin D receptor gene allelic status and dietary vitamin D intake. Whole body exposure in a bathing suit to one minimum erythemal dose of UVR is equivalent to ingesting 10 000 international units of vitamin D. It is important to note that while excessive dietary vitamin D can lead to vitamin D toxicity, excessive UVR exposure cannot lead to vitamin D toxicity.

Source: adapted from references 80 and 81

CLIMATE CHANGE AND HUMAN HEALTH

A causal link has not been established (78). Nevertheless, NHL is a disease that should be monitored closely because of its possible increase with any future increases in UVR.

Is UVR exposure beneficial for some autoimmune diseases?

Recent developments in photoimmunology and epidemiology suggest that UVR may have a beneficial role in autoimmune diseases such as multiple sclerosis (MS), type 1 diabetes mellitus (IDDM) and rheumatoid arthritis (RA). Each of these autoimmune diseases is characterized by a breakdown in immunological self-tolerance that may be initiated by an inducing agent such as an infectious micro-organism or a foreign antigen (82). A cross-reactive auto-immune response occurs and a "self-molecule" is no longer self-tolerated by the immune system. At this stage, the host tissue becomes immunogenic, attracting a T helper cell type 1 (T_H1) mediated immune response resulting in chronic inflammation (82). That is, the T_H1 lymphocytes no longer recognise the host tissue as such and instead try to eliminate the host tissue by inflammation.

The well-established gradient of MS increasing with increasing latitude may reflect differential UV-induced immune suppression of autoimmune activity. That is, at lower latitudes where MS prevalence is lower high levels of UVR exposure may dampen down the immune over-activity that occurs in MS. In particular, the autoimmune profile of MS is characterized by disturbances of those T cell-related activities specifically affected by UVB (83). A strong inverse association between UVR exposure and MS has been shown. In Australia the negative correlation between regional UVR and MS prevalence is higher than the magnitude observed for the positive correlation between regional UVR and malignant melanoma (84).

A recent case-control study found that compared to indoor workers living in a low sunlight region, the odds ratios for an outdoor worker dying from MS in low, medium and high residence sunlight were, respectively, 0.89 (with 95% confidence intervals of 0.64 to 1.22), 0.52 (0.38, 0.71) and 0.24 (0.15, 0.38) (85). Thus high residential and occupational solar exposure (in combination) were associated with a reduced likelihood of MS. UVR may affect not only the development of MS but also its clinical course. An ecological study recently has shown a striking inverse correlation between serum 25(OH) D, a metabolite of vitamin D, and high MS lesion activity (86).

For type 1 diabetes, a disease resulting from T cell-mediated inflammation with destruction of pancreatic tissue, the epidemiological evidence also suggests a possible beneficial role for UVR. An increasing disease prevalence gradient with increasing latitude has been noted. In a Finnish birth cohort study, vitamin D supplementation in infancy was inversely associated with subsequent type 1 diabetes (relative risk 0.22 {0.05, 0.89}) (87). Vitamin D receptor gene allelic status has been found to relate to MS and type 1 diabetes in some populations (88). For rheumatoid arthritis, dietary supplementation with vitamin D has been related to lower levels of disease activity (89).

Overall, the epidemiological features of these three autoimmune diseases are consistent with a protective effect for high personal UVR exposure. However, the data are not conclusive and further research work is required.

UVR and other diseases with immune dysfunction

Although the three diseases above are characterized by T_H1 cell over-activity, other immune diseases may be characterized by T_H2 cell over-activity or a mixed

T cell over-activity pattern. Systemic lupus erythematosus (SLE) is characterized by a mixed T_H2/T_H1 disturbance. It has been postulated that the immune dysfunction in SLE begins under the skin where UV-induced keratinocytes produce antigens that are recognized by the body to be foreign (*90*). UVR plays a major role in the induction of lesions of patients with the cutaneous form of lupus disease and photo-aggravation of systemic disease may occur in systemic SLE (*91*).

Atopic eczema, a disease of immune disturbance that includes T_H2 over-activity, appears to be inversely related to UVR. Strong latitudinal gradients for increasing eczema with increasing latitude have been reported in the Northern Hemisphere (*92*). In a clinical trial, narrow-band UVB therapy significantly improved allergic eczema (*93*). Thus, high UVB exposure appears to have a beneficial effect on the immune disorder of atopic eczema even though this disease is not characterized by a purely T_H1 immune over-activity pattern.

Other diseases that could be exacerbated by decreased UVR exposure, particularly if dietary vitamin D sources were inadequate

Although a detailed discussion is beyond the scope of this chapter, it should be noted that inadequate UVR exposure in the absence of adequate dietary D sources, could lead to vitamin D deficiency. This would increase the likelihood of rickets, osteomalacia, osteoporosis, muscle pain and possibly hypertension (*94*) or ischaemic heart disease (*95*). Certain cancers (e.g. prostate and breast) have been linked to vitamin D deficiency, although not conclusively (*80*). Vitamin D deficiency may increase tuberculosis (TB) risk (*96*). Evidence suggests that the explanation for this may reflect the immunological modulation caused by vitamin D. Vitamin D activates one group of white blood cells, the monocytes, thereby increasing their capacity to resist cell infection by the mycobacterium (*96*). Further, a recent case-control study showed that the combination of vitamin D deficiency and the "high-risk" allele of the vitamin D receptor gene was strongly associated with the occurrence of TB (*97*).

During pregnancy, inadequate maternal UVR exposure in the absence of adequate dietary vitamin D sources will lead to low foetal exposure to vitamin D. As vitamin D appears to be important in neural growth this could influence the developing brain of the foetus. In fact, this has been proposed as an explanation for the finding that winter-born babies appear at increased risk of schizophrenia (*98*). Furthermore, inadequate UVR exposure usually is associated with reduced visible light and a reduction in photoperiod. This will alter melatonin levels, a hormone important in maintaining the rhythm of wake/sleep patterns. Changes in photoperiod also have been related to seasonal affective disorder (*99*). Although not well understood, the relationship between solar radiation and mood is important to consider (*100*).

Public health message re UVR exposure

Encouraging total sun avoidance (with the related notion of solar radiation as a "toxic" exposure) is a simplistic response to the hazards of increased ground level UVR exposure due to stratospheric ozone depletion, and should be avoided. Any public health messages concerned with personal UVR exposure should consider the benefits as well as the adverse effects. The notion that UVR is inherently an adverse exposure to be maximally avoided cannot fully be reconciled with evolutionary heritage. It is a reasonable presumption that levels of skin pigmentation in regional populations originally evolved over many millennia to optimise the amount of UVR absorbed by the skin in order to balance biological benefits and risks. The possible benefits and adverse effects of UVR exposure on human health therefore should be assessed concurrently (100).

Many modern infants and young children already receive less solar radiation than children several decades ago. This reflects an increase in indoor living and medical recommendations promoting sun avoidance advice (100). A growing recognition of low winter 25(OH) D levels, particularly among children of Asian origin residing in the United Kingdom, has led to the current United Kingdom recommendation that all pregnant women and children up to age five should have a vitamin D supplement, unless solar and dietary sources are adequate (101). Clear guidelines on the optimal age-appropriate solar radiation dose are not yet available (101) and are difficult to formulate because the recommended level of appropriate solar radiation depends on host factors as described above. However, the lack of clear recommendations could lead to inappropriate personal solar exposure. For example, a recent case report of severe rickets in a Caucasian child residing in Toronto, Canada, highlighted the possible adverse effects of inadequate UVR exposure in childhood (102). The child went outdoors in summer but was always covered by potent sunscreen. The child's rickets subsequently responded well to dietary vitamin D. This case highlights the importance of considering UVR as an exposure that requires titration rather than avoidance.

Although measured UVR exposures are proportional to ambient UVR for similar population groups, there is a wide variation in inter-personal UVR exposure within each of these groups. Some individuals may have only one-tenth of the population average for UVR exposure, others may have an individual UVR exposure of ten times the average (12). The factors affecting this large inter-individual variation are not well understood. In addition to the difficulty in quantifying both UVR exposure and inter-individual variation in susceptibility to UVR, this large variation in sun exposure behaviour makes difficult the correct titration of UVR at a population level.

To negate the adverse effects of increased UVR exposure due to ozone depletion, an alternative response to that of careful titration of sun exposure dose is, in theory, to recommend total sun avoidance and large-scale vitamin D supplementation. However this approach:

- runs the possible risk of hypervitaminosis D with resultant hypercalcaemia, a documented cause of infant mortality 40 years ago;
- neglects the possible benefits of UVR exposure than are not mediated via vitamin D;
- neglects that other beneficial factors such as visible light exposure are correlated with UVR exposure.

Conclusions

The occurrence of stratospheric ozone depletion over the past quarter-century, and its anticipated continuation for at least the next several decades, has focused attention on questions about the impact of UVR on human biology and disease risks. This has coincided with a growth of knowledge about some of the basic biological pathways via which UVR affects human biology. In particular, it is evident that a change in levels of UVR exposure will affect the incidence of skin cancer, and is likely to affect the incidence of several ocular disorders, including cataract, and various immune-related diseases and disorders.

Uncertainties remain about the extent to which the loss of stratospheric ozone to date has resulted in increases in ground-level UVR. Environmental monitoring systems often have been unstandardized, non-spectral, and suboptimally located. Climate-related changes in cloud cover appear to have compounded the relationship between ozone depletion and ground-level UVR.

The majority of the known health consequences of increased UVR exposure are detrimental. However, UVR exposure also has some beneficial effects. Therefore, while excessive solar exposure should be avoided—the more so during the current and foreseeable period of stratospheric depletion—so should excessive sun avoidance. Future public health advice about solar exposure should take account of the changing ambient UVR environment and the available knowledge about the health risks and benefits of UVR exposure.

References

1. McMichael, A.J. *Planetary overload: global environmental change and the health of the human species.* Cambridge, UK, Cambridge University Press, 1993.
2. McKenzie, R. et al. Increased summertime UV radiation in New Zealand in response to ozone loss. *Science* 285(5434): 1709–1711 (1999).
3. Udelhofen, P.M. et al. Surface UV radiation over Australia, 1979–1992: effects of ozone and cloud cover changes on variations of UV radiation. *Journal of Geophysical Research* 104(D16): 19,135–19,159 (1999).
4. Madronich, S. et al. Changes in biologically active ultraviolet radiation reaching the Earth's surface. *Journal of Photochemistry and Photobiology B: Biology* 46(1–3): 5–19 (1998).
5. Kelfkens, G. et al. *Ozone layer-climate change interactions. Influence on UV levels and UV related effects.* Dutch National Research Programme on Global Air Pollution and Climate Change. Report no. 410 200 112.
6. Shindell, D.T. et al. Increased polar stratospheric ozone losses and delayed eventual recovery owing to increasing greenhouse gas concentrations. *Nature* 392(6676): 589–592 (1998).
7. Kirk-Davidoff, D.B. et al. The effect of climate change on ozone depletion through changes in stratospheric water vapour. *Nature* 402(6760): 399–401 (1999).
8. Jacobson, M.Z. *Fundamentals of atmospheric modelling.* Cambridge, UK, Cambridge University Press, p. 656, 1999.
9. Ashmore, M. Human exposure to air pollutants. *Clinical and Experimental Allergy* 25(3): 12–22 (1995).
10. Horneck, G. Quantification of the biological effectiveness of environmental UV radiation. *Journal of Photochemistry and Photobiology B: Biology* 31(1–2): 43–49 (1995).
11. CIE Standard Erythema reference action spectrum and standard erythema dose. CIE S 007/E-1998. Vienna: Commission International de l'Éclairage, 1998.
12. Gies, P. et al. Ambient solar UVR, personal exposure and protection. *Journal of Epidemiology* 9(6 Suppl): S115–S122 (1999).
13. Diffey, B.L. Sources and measurement of ultraviolet radiation. *Methods* 28: 4–13 (2002).
14. International Agency for Research on Cancer (IARC). *Solar and ultraviolet radiation. IARC monographs on the evaluation of carcinogenic risks to humans.* Vol. 55. Lyon, France, International Agency for Research on Cancer, 1992.
15. World Health Organization (WHO). *Environmental health criteria 160: ultraviolet radiation.* Geneva, Switzerland: World Health Organization, p. 352, 1994a.
16. United Nations Environment Program (UNEP) *Environmental effects of ozone depletion: 1998 assessment.* Nairobi, Kenya, United Nations Environment Program, 1998.

17. World Health Organization (WHO). *The effects of solar radiation on the eye.* Geneva, Switzerland, World Health Organization Programme for the Prevention of Blindness, 1994b.

18. Garssen, J. et al. Estimation of the effect of increasing UVB exposure on the human immune system and related resistance to infectious disease and tumours. *Journal of Photochemistry and Photobiology B: Biology* 42(3): 167–179 (1998).

19. Findlay, G.M. Ultraviolet light and skin cancer. *Lancet* 2: 1070–1073 (1928).

20. Armstrong, B.K. & Kricker, A. Cutaneous melanoma. *Cancer Surveys* 19–20: 219–240.

21. Jablonski, N.G. & Chaplin, G. The evolution of human skin coloration. *Journal of Human Evolution* 39(1): 57–106 (2000).

22. International Agency for Research on Cancer (IARC). *Handbooks on cancer prevention: sunscreens* Volume 5. Lyon, France, International Agency for Research on Cancer, 2001.

23. Scotto, J. et al. Nonmelanoma skin cancer. In: *Cancer epidemiology and prevention.* Schottenfeld, D. & Fraumeni, J.F. eds. New York, USA, Oxford University Press, pp. 1313–1330, 1996a.

24. Scotto, J. et al. Solar radiation. In: *Cancer epidemiology and prevention.* Schottenfeld, D. & Fraumeni, J.F. eds. New York, USA, Oxford University Press, pp. 355–372, 1996b.

25. McKinlay, A.F. & Diffey, B.L. A reference action spectrum for ultra-violet induced erythema in human skin. In: *Human exposure to ultraviolet radiation: risks and regulations.* Passchler, W.R. & Bosnajakovic, B.F.M. eds. pp. 83–87, Elsevier, Amsterdam, 1987.

26. Tyrrell, R.M. The molecular and cellular pathology of solar ultraviolet radiation. *Molecular Aspects of Medicine* 15: 1–77 (1994).

27. Gonzalez, S. et al. Development of cutaneous tolerance to ultraviolet B during ultraviolet B phototherapy for psoriasis. *Photodermatology, Photoimmunology, Photomedicine* 12(2): 73–78 (1996).

28. Mitchell, D.L. et al. Identification of a non-dividing subpopulation of mouse and human epidermal cells exhibiting high levels of persistent ultraviolet photodamage. *Journal of Investigative Dermatology* 117(3): 590–595 (2001).

29. Madronich, S. & de Gruijl, F.R. Skin cancer and UV radiation. *Nature* 366(6450): 23 (1993).

30. Slaper, H. et al. Estimates of ozone depletion and skin cancer incidence to examine the Vienna Convention achievements. *Nature* 384(6606): 256–258 (1996).

31. Zigman, S. Environmental near-UV radiation and cataracts. *Optometry and Vision Science* 72(12): 899–901 (1995).

32. Taylor, H.R. et al. The long-term effects of visible light on the eye. *Archives of Ophthalmology* 110(1): 99–104 (1992).

33. Young, R.W. The family of sunlight-related eye diseases. *Optometry and Vision Science* 71(2): 125–144 (1994).

34. Harding, J.J. The untenability of the sunlight hypothesis of cataractogenesis. *Documenta Opthalmologica* 88(3–4): 345–349 (1994–1995).

35. Dolin, P.J. Ultraviolet radiation and cataract: a review of the epidemiological evidence. *British Journal of Ophthalmology* 78: 478–482 (1994).

36. Javitt, J.C. & Taylor, H.R. Cataract and latitude. *Documenta Ophthamologica* 88(3–4): 307–325 (1995).

37. Hollows, F. & Moran, D. Cataract—the ultraviolet risk factor. *Lancet* 2(8258): 1249–1250 (1981).

38. Mao, W.S. & Hu, T.S. An epidemiological survey of senile cataract in China. *Chinese Medical Journal* 95(11): 813–818 (1982).

39. Brilliant, L.B. et al. (1983). Associations among cataract prevalence, sunlight hours and altitude on the Himalayas. *American Journal of Epidemiology* 118(2): 250–264 (1982).

40. Taylor, H.R. et al. Effect of ultraviolet radiation on cataract formation. *New England Journal of Medicine* 319(22): 1429–1433 (1988).

41. Bochow, T.W. et al. Ultraviolet light exposure and risk of posterior subcapsular cataracts. *Archives of Ophthalmology* 107(3): 369–372 (1989).

42. West, S.K. et al. Sunlight exposure and risk of lens opacities in a population based study. The Salisbury eye evaluation project. *Journal of the American Medical Association* 280(8): 714–718 (1998).

43. Leske, M.C. et al. The lens opacities case-control study: risk factors for cataract. *Archives of Ophthalmology* 109(2): 244–251 (1991).

44. Cruickshanks, K.J. et al. Ultraviolet light exposure and lens opacities: the Beaver Dam eye study. *American Journal of Public Health* 82(12): 1658–1662 (1992).

45. Merriam, J.C. The concentration of light in the human lens. *Transactions of the American Ophthalmological Society* 94: 803–918 (1996).

46. Christen, W.G. Sunlight and age-related cataracts. *Annals of Epidemiology* 4(4): 338–339 (1994).

47. Mohan, M. et al. India-US case-control study of age-related cataracts. *Archives of Ophthalmology* 107(5): 670–676 (1989).

48. Rosmini, F. et al. A dose-response effect between a sunlight index and age-related cataracts. Italian-American Cataract Study Group. *Annals of Epidemiology* 4(4): 266–270 (1994).

49. Hirvela, H. et al. Prevalence and risk factors of lens opacities in the elderly in Finland. A population-based study. *Ophthalmology* 102(1): 108–117 (1995).

50. Delcourt, C. et al. Light exposure and the risk of cortical, nuclear and posterior subcapsular cataracts. *Archives of Ophthalmology* 118(3): 385–392 (2000).

51. Katoh, N. et al. Cortical lens opacities in Iceland. Risk factor analysis—Reykjavik eye study. *Acta Ophthalmologica Scandinavica* 79(2): 154–159 (2001).

52. McCarty, C.A. et al. The epidemiology of cataract in Australia. *American Journal of Ophthalmolog*, 128(4): 446–465 (1999).

53. Sarma, U. et al. Nutrition and the epidemiology of cataract and age related maculopathy. *European Journal of Clinical Nutrition* 48(1): 1–8 (1994).

54. Christen, W.G. et al. Antioxidants and age-related eye disease. Current and future perspectives. *Annals of Epidemiology* 6(1): 60–66 (1996).

55. Taylor, H.R. et al. Corneal changes associated with chronic UV irradiation. *Archives of Ophthalmology* 107(10): 1481–1484 (1989).

56. Mackenzie, F.D. et al. Risk analysis in the development of pterygia. *Ophthalmology* 99(7): 1056–1061 (1992).

57. Threlfall, T.J. & English, D.R. Sun exposure and pterygium of the eye: a dose-response curve. *American Journal of Ophthalmology* 128(3): 280–287 (1999).

58. McCarty, C.A. et al. Epidemiology of pterygium in Victoria, Australia. *British Journal of Ophthalmology* 84(3): 289–292 (2000).

59. Ehrt, O. et al. Microperimetry and reading saccades in retinopathia solaris. Follow-up with the scanning laser ophthalmoscope. *Ophthalmologe* 96(5): 325–331 (1999).

60. Yannuzzi, L.A. et al. Solar retinopathy. A photobiologic and geophysical analysis. *Retina* 9(1): 28–43 (1989).

61. Holt, P.G. A potential vaccine strategy for asthma and allied atopic diseases during early childhood. *Lancet* 344(8920): 456–458 (1994).

62. Kripke, M.L. Ultraviolet radiation and immunology: something new under the sun—presidential address. *Cancer Research* 54(23): 6102–6105 (1994).

63. Van Loveren, H. et al. UV exposure alters respiratory allergic responses in mice. *Photochemistry and Photobiology* 72(2): 253–259 (2000).

64. Duthie, M.S. et al. The effects of ultraviolet radiation on the human immune system. *British Journal of Dermatology* 140: 995–1009 (1999).

65. Akdis, C.A. & Blaser, K. Mechanisms of interleukin-10-mediated immune suppression. *Immunology* 103:131–136 (2001).

66. Wolf, P. et al. Topical treatment with liposomes containing T4 endonuclease V protects human skin *in vivo* from ultraviolet-induced upregulation of interleukin-10 and tumour necrosis factor-α. *Journal of Investigative Dermatology* 114(1): 149–156 (2000).

67. Liebmann, P.M. et al. Melatonin and the immune system. *International Archives of Allergy and Immunology* 112(3): 203–211 (1997).

68. Constantinescu, C.S. Melanin, melatonin, melanocyte-stimulating hormone and the susceptibility to autoimmune demyelination: a rationale for light therapy in multiple sclerosis. *Medical Hypotheses* 45(5): 455–458 (1995).

69. Hayes, C.E. et al. Vitamin D and multiple sclerosis. *Proceedings of the Society for Experimental Biology and Medicine* 216(1): 21–27 (1997).

70. Selgrade, M.K. et al. Ultraviolet radiation-induced immune modulation: potential consequences for infectious, allergic, and autoimmune disease. *Environmental Health Perspectives* 105(3): 332–334 (1997).

71. Norval, M. et al. UV-induced changes in the immune response to microbial infections in human subjects and animal models. *Journal of Epidemiology* 9(6 Suppl): S84–92 (1999).

72. Goettsch, W. et al. Risk assessment for the harmful effects of UVB radiation on the immunological resistance to infectious diseases. *Environmental Health Perspectives* 106(2): 71–77 (1998).

73. Akaraphanth, R. & Lim, H.W. HIV, UV and immunosuppression. *Photodermatology Photoimmunology Photomedicine* 15(1): 28–31 (1999).

74. Termorshuizen, F. et al. Influence of season on antibody response to high dose recombinant Hepatitis B vaccine: effect of exposure to solar UVR? *Hepatology* 32(4): 1657 (2000).

75. Fine, P.E. Variation in protection by BCG: implications of and for heterologous immunity. *Lancet* 346(8986): 1339–1345 (1995).

76. Cestari, T.F. et al. Ultraviolet radiation decreases the granulomatous response to lepromin in humans. *Journal of Investigative Dermatology* 105(1): 8–13 (1995).

77. Bentham, G. Association between incidence of non-Hodgkin's lymphoma and solar ultraviolet radiation in England and Wales. *British Medical Journal* 312(7039): 1128–1131 (1996).

78. Zheng, T. & Owens, P.H. Sunlight and non-Hodgkin's lymphoma. *International Journal of Cancer* 87(6): 884–886 (2000).

79. Utiger, R.D. The need for more Vitamin D. *New England Journal of Medicine* 338(12): 828–829 (1998).

80. Holick, M.F. Sunlight dilemma: risk of skin cancer or bone disease and muscle weakness. *Lancet* 357(9249): 4–6 (2001).

81. Holick, M.F. McCollum Award Lecture, 1994: vitamin D—new horizons for the 21st century. *American Journal of Clinical Nutrition* 60(4): 619–630 (1994).

82. Mackay, I.R. Science, medicine, and the future: tolerance and autoimmunity. *British Medical Journal* 321(7253): 93–96 (2000).

83. McMichael, A.J. & Hall, A.J. Does immunosuppressive ultraviolet radiation explain the latitude gradient for multiple sclerosis? *Epidemiology* 8(6): 642–645 (1997).

84. van der Mei, I.A. et al. Regional variation in multiple sclerosis prevalence in Australia and its association with ambient ultraviolet radiation. *Neuroepidemiology* 20(3): 168–174 (2001).

85. Freedman, D. et al. Mortality from multiple sclerosis and exposure to residential and occupational solar radiation: a case-control study based on death certificates. *Occupational and Environmental Medicine* 57(6): 418–421 (2000).

86. Embry, A.F. et al. Vitamin D and seasonal fluctuations of gadolinium-enhancing magnetic resonance imaging lesions in multiple sclerosis. *Annals of Neurology* 48(2): 271–272 (2000).

87. Hyponnen, E. et al. Intake of vitamin D and risk of type 1 diabetes: a birth-cohort study. *Lancet* 358(1): 1500–1503 (2001).

88. Zmuda, J.M. et al. Molecular epidemiology of vitamin D receptor gene variants. *Epidemiologic Reviews* 22(9): 203–217 (2000).

89. Oelzner, P. et al. Relationship between disease activity and serum levels of vitamin D metabolites and PTH in rheumatoid arthritis. *Calcified Tissue International* 62(3): 193–198 (1998).

90. Lee, L.A. & Farris, A.D. Photosensitivity diseases: cutaneous lupus erythematosus. *Journal of Investigative Dermatology Symposium Proceedings* 4(1): 73–78 (1999).

91. Millard, T.P. & Hawk, J.L. Ultraviolet therapy in lupus. *Lupus* 10(3): 185–187 (2001).

92. McNally, N.J. et al. Is there a geographical variation in eczema prevalence in the UK? Evidence from the 1958 British Birth Cohort Study. *British Journal of Dermatology* 142(4): 712–720 (2000).

93. Reynolds, N.J. et al. Narrow-band ultraviolet B and broad-band ultraviolet A phototherapy in adult atopic eczema: a randomised controlled trial. *Lancet* 357(9273): 2012–2016 (2001).

94. Rostand, S.G. Ultraviolet light may contribute to geographic and racial blood pressure differences. *Hypertension* 30(2, pt 1): 150–156 (1997).

95. Pell, J.P. & Cobbe, S.M. Seasonal variations in coronary heart disease. *Quarterly Journal of Medicine* 92(12): 689–696 (1999).

96. Bellamy, R. Evidence of gene-environment interaction in development of tuberculosis. *Lancet* 355(9204): 588–589 (2000).

97. Wilkinson, R.J. et al. Influence of vitamin D deficiency and vitamin D receptor polymorphisms on tuberculosis among Gujarati Asians in west London: a case-control study. *Lancet* 355(9204): 618–621 (2000).

98. McGrath, J. Hypothesis: is low prenatal vitamin D a risk-modifying factor for schizophrenia? *Schizophrenia Research* 40(3): 173–177 (1999).

99. Mersch, P.P. et al. Seasonal affective disorder and latitude: a review of the literature. *Journal of Affective Disorders* 53(1): 35–48 (1998).

100. Ness, A.R. et al. Are we really dying for a tan? *British Medical Journal* 319(7202): 114–116 (1999).

101. Wharton, B.A. Low plasma vitamin D in Asian toddlers in Britain. *British Medical Journal* 318(7175): 2–3 (1999).

102. Zlotkin, S. Vitamin D concentrations in Asian children living in England. Limited vitamin D intake and use of sunscreens may lead to rickets. *British Medical Journal* 318(7195): 1417 (1999).

CHAPTER 9

National assessments of health impacts of climate change: a review

R.S. Kovats,[1] B. Menne,[2] M.J. Ahern,[1] J.A. Patz[3]

Introduction

Previous chapters have shown that climate change represents a serious environmental threat over the coming century. The public health community has a responsibility to provide policy-makers with evidence of the potential impacts of climate change on human population health. Policy-makers are obliged to respond to this risk even in the face of scientific uncertainties. The public health community has established methods for assessing the risks to health for a population. WHO defines health impact assessment as "a combination of procedures, methods and tools by which a policy, project or hazard may be judged as to its potential effects on the health of a population, and the distribution of those effects within the population" (1, 2). Despite recent advances in the methodology for health impact assessment, the greater goal of integration into mainstream policy-making has yet to be achieved. This objective is a long way off in both developed and developing countries.

Global climate change presents many unique problems for health impact assessment. It is a highly diffuse global exposure for which very limited information (in the form of climate projections) is available at local or national level. For many health impacts this is not an immediate problem but one that will develop over decades or longer. Action is needed now to avert the worst impacts through the reduction of greenhouse gas emissions. Further, guidance is required now on policies to enhance the capacity to deal with climate change (chapter 12). Health impact assessments typically refer to impacts in the next 10 to 20 years, rather than the 50 to 100 year time-scale of climate change projections and the assessments of impacts in other sectors.

Global assessments have been undertaken by the Intergovernmental Panel on Climate Change (IPCC) in the Second and Third Assessment Reports (3, 4). Such global assessments make general statements about the types of impacts that climate change may have upon human health outcomes. In theory, national assessments should provide the global assessments with important information about regional and local vulnerability. In practice, this has proved difficult to achieve for a variety of reasons that will be discussed below. Assessments should be country driven and reflect local environmental and health priorities. Where sufficient resources have been available, there has been a preference to include

[1] London School of Hygiene and Tropical Medicine, London, England.
[2] World Health Organization Regional Office for Europe, European Centre for Environment and Health, Rome, Italy.
[3] Johns Hopkins University, Baltimore, MD, USA.

some quantification of health impacts in future decades. However many assessments rely on a qualitative assessment of the available literature.

Several agencies have produced guidelines on assessing the impacts of climate change (5, 6). These are seen as inadequate for addressing impacts in social systems and outcomes such as human health that are more complex and context-specific. Health impact assessments must use the best available scientific evidence to inform policy decisions. In 1999 the European Ministerial Conference on Environment and Health recommended countries to:

- develop capacities to undertake national health impact assessments with the aim of identifying the vulnerability of populations and subgroups. Ensure the necessary transfer of know-how among countries;
- carry out ongoing reviews of the social, economic and technical prevention, mitigation and adaptation options available to reduce the adverse impacts of climate change and stratospheric ozone depletion on human health.

This chapter reviews national assessments that have addressed climate change impacts on human population health in some detail, with particular reference to:

- methods and tools used
- main findings
- integration with other climate assessments (in non-health sectors and other geographical regions)
- assessments of adaptation (what is likely and what is recommended).

Health impact assessments: key concepts and methods

Health impact assessments (HIA) are undertaken for a variety of purposes and under a variety of circumstances. The purpose of the health impact assessment will therefore determine its scope, form and content. Health impact assessments are multi-disciplinary by necessity and bring together a range of methods and tools: policy appraisal, evidence-based risk assessment and environmental impact assessment (7). The common elements of health impact assessments include:

- integrated assessment of impacts, i.e. not concentrating on single risk factors and disease outcomes (a holistic view of health);
- relate to policies and projects outside the health sector;
- multidisciplinary process;
- provide information for decision-makers, therefore designed with needs of decision-makers in mind;
- quantification of the expected health burden due to an environmental exposure in a specific population.

There is growing consensus that systematic assessments of health effects are needed to inform the development of policies and to include health in the agendas of other sectors—such as water, food, housing, trade, etc. Legislation and legally binding agreements at the international level now make provision for HIA in policy-making. Anthropogenic climate change as an exposure is different to other types of hazard and new methods and tools need to be developed.

Bernard and Ebi (8) identify a critical distinction between climate change impact assessment and traditional environmental health quantitative risk assessment (QRA). The primary assumptions underlying QRA are no longer applicable. Climate change is not a defined exposure to a specific agent (the pollutant) that causes an adverse health outcome to identifiable exposed populations. Climate change is associated with a range of climate and weather exposures that are mediated in complex fashion through a range of mechanisms. The challenges that this poses for scientific assessment are described in detail in chapter four. If the impacts of climate change are to be measured quantitatively, then existing population-based methodologies such as Comparative Risk Assessment (CRA) have to be adapted and applied in a flexible manner, with appropriate description of uncertainties (see chapter seven).

Health impact assessment often is part of a wider prospective environmental impact assessment of a specific project or development (mezzosocial focus) for which specific tools are available. Climate change assessments operate at the macrosocial level because the impacts on an entire population (country or region) are considered. This would be similar to the evaluation of a particular policy, e.g. energy policy. Such assessments should include reports of current health status using standard epidemiological indicators, e.g. mortality rates, etc. Health impact assessment of government policy has been implemented in Canada (7). The methods used a framework of questions to guide decision-makers in considering factors that influence population health for the following key areas: employment and economy; education and skills; environment and safety; programmes and services. Other examples of broad policy related assessments are the quantification of the health impacts of traffic related air pollution in three European countries (9), and the estimation of health costs for different fuel cycles—such as coal, nuclear (ExternE project (10)).

Methods for climate-change impact assessments

Approaches to climate impact assessment have evolved rapidly since climate change became a policy concern in the 1980s. Methods and tools have been developed primarily for climate assessments for biophysical impacts for either economic (agriculture, forestry) or intrinsic value (biodiversity). Impact assessment in social systems has addressed the impact of weather disasters and the implications for the insurance industry. The IPCC has produced methodological guidelines in order to ensure standardization in methods across sectors and disciplines (5, 6, 11), but these have limited use for assessing impacts in socioeconomic or human systems. The IPCC guidelines focus on impacts rather than adaptive measures (or adaptive capacity) as adaptive responses vary greatly between countries and are less easily described in terms of generic methods (12).

Figure 9.1 illustrates the framework first described in the IPCC guidelines (5) and expanded upon by Parry and Carter (6). The methodological approach is rigidly top down. That is, scenarios of climate change (generated by global climate models) are used as input to large-scale biophysical models. Current methods now incorporate future projections of populations and GDP together with storylines regarding the future worlds in which the impacts of climate change will be experienced (11, 13).

The first specific guidelines for health appeared in the UNEP Handbook on Methods for Climate Change Impact Assessment and Adaptation Strategies (14). The aim of the UNEP country studies programme was to improve the methods for assessing climate change impacts in developing countries or countries with economies in transition. One of the stated aims was to test the methods described in the Handbook on Methods for Climate Change Impact and Adaptation

FIGURE 9.1 Steps in climate change impact and adaptation assessment. *Source: reproduced from reference 6.*

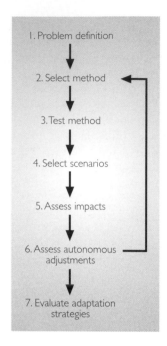

Assessment. The UNEP country studies programme ended in 2000 with completed studies on four countries: Antigua and Barbuda, Cameroon, Estonia and Pakistan (*15*). Only the Cameroon and Antigua and Barbuda studies addressed health. Several problems were found with the guidelines: in particular, the 100-year time frame for climate change projections had little practical relevance. Further, the methods described in the Handbook were found to rely too much on health (and other) data that were not available in either Cameroon or Antigua and Barbuda. The lack of data with which to generate and validate the climate-health models is cited often as a significant limitation. However, even if such data are available, the development and interpretation of such models creates many difficulties.

There are limitations with the scenario driven predictive modelling approach to assessing health impacts at the national level. A major problem is the mismatch between the spatial and temporal scale of environmental factors that affect health (local concentrations of air pollutants, focal vector distributions) and the scenarios of future climate change (global climate model grid boxes). Given the limits of climate scenarios (*16*), it may be appropriate to use analogue scenarios of changes in temperature and precipitation that can provide important information regarding the relative contribution of climate and non-climate factors to the burden of disease.

Health impact assessments include an evaluation of the epidemiological evidence base for the causal relationships on which to base future projections. Some discussion is necessary of the current knowledge of the local relationships between climate variability and disease, with particular attention to the plausible mechanisms by which climate/weather affect health; such relationships are generally population specific. If such information is not available in the published literature, this type of analysis could be part of the assessment.

For the United States of America's health impact assessment, the target questions were widened to address current vulnerability to climate-sensitive diseases and weather extremes:

CLIMATE CHANGE AND HUMAN HEALTH

- what are current environmental stresses and issues for the United States that will form a backdrop for potential additional impacts of climate change?
- how might climate variability and change exacerbate or ameliorate existing problems?
- what are the priority research and information needs that can better prepare policy-makers to reach wise decisions related to climate variability and change?
- what research is the most important over the short-term? Over the long term?
- what coping options exist that can build resilience into current environmental stresses, and also possibly lessen the impacts of climate change?

The United States' assessment was able to make use of a considerable amount of published literature on health relationships. Other countries, particularly developing countries, are unlikely to have the background information available. Assessments have thus used a variety of methods to estimate future impacts on health. These are discussed in more detail below.

Review of national health assessments

This chapter reviews national assessments on the potential health impacts of climate change that were published before mid 2002. There is some difficulty in defining what constitutes a national impact assessment. Many countries have addressed health impacts in their national communications to the United Nations Framework Convention on Climate Change (UNFCCC), however, few of these constitute a formal impact assessment. Our criteria for including a report are:

- assessment at national level or below (excluded international assessments such as IPCC reports)
- undertaken for the Ministry of Health or Environment, i.e. central or regional government
- explicitly addresses global climate change and human health
- involves some formal methods of assessment, e.g. systematic literature review or modelling.

Table 9.1 lists the national assessments, the majority from developed countries, which fit these criteria. A few large studies have been undertaken as part of a comprehensive, multi-sectoral assessment, such as in the United States and Canada. In contrast, European countries have conducted few assessments: only the United Kingdom and Portugal have conducted comprehensive assessments. Assessments in developing countries have been undertaken only under the auspices of donor-funded capacity building initiatives. It is possible that sub-national or local assessments have been undertaken for climate change impacts that address health but the authors have been unable to find them. The majority of these studies are in the grey literature and not widely available.

Several types of assessments have been undertaken. At one level there is a basic review of the types of potential impacts with little evaluation of evidence that they are likely to occur. Such assessments can be undertaken with few resources but provide very limited information. At another level, comprehensive well-funded and well-supported assessments are undertaken. For example, in the United States assessment, health was one of 5 sectoral assessments, and included in 16 regional assessments and the overall synthesis report. This assessment involved stakeholder participation and extensive consultation and peer review (8).

TABLE 9.1 National Climate Health Impact Assessments.

Country	Reference	Methods	Areas of concern for climate change
Antigua and Barbuda	O'Marde & Michael, 2000 (17)	Forecasting by analogy.	Coastal flooding due to sea level rise, impacts of hurricanes, increases in dengue transmission.
Australia	NHMRC, 1991 (18)	Expert judgement.	None specified, focuses on adaptation.
Australia	McMichael et al., 2002 (19)	Predictive modelling.	No impact on heat stress. Changes in distribution of malaria vectors. Increases in diarrhoeal disease.
Cameroon	UNEP/ Ministry of Environment and Forestry, Cameroon, 1998 (20)	Expert judgement.	Possible increases in cholera, malaria, yellow fever, meningitis, malnutrition.
Canada	Duncan et al., 1997 (21)	Literature review, expert judgement, predictive modelling.	Increases in heat-related deaths, risk that vector-borne diseases may extend north into Canada, environmental refugees.
Fiji	de Wet & Hales, 2000 (22)	Literature review, predictive modelling.	Increases in dengue, diarrhoeal disease.
Japan	Ando, 1993 (23)	Literature review.	Heat stress and photochemical air pollution may increase. Malaria.
Japan	Ando et al. 1998 (24)	Literature review, expert judgement, predictive modelling.	Heat stress. Malaria.
Kiribati	Taeuea, T. et al. 2000 (25)	Literature review, predictive modelling.	Increases in dengue, diarrhoeal disease and ciguatera fish poisoning.
Netherlands	Martens, 1996 (26)	Literature review, expert judgement.	Heat stress, flooding.
New Zealand	Woodward et al. 2001 (27)	Literature review, predictive modelling.	Risk of introduction of mosquito vector species into north island.
Panama	Sempris, E. & Lopez, R. eds. 2001.—ANAM/UNDP (28)	Quantitative assessment of climate with diarrhoeal illness and influenza.	Potential increases in diarrhoeal disease in vulnerable areas.
Portugal	Casimiro,E. & Calheiros, J.M. 2002 (29)	Extensive literature review, predictive modelling, identified populations at risk.	Heat-related deaths, food and water borne diseases, air pollution effects, vector and rodent borne disease.
Sri Lanka	Ratnasari, J. 1998 (30)	Expert judgement, predictive model.	Malaria.
St Lucia	St Lucia National Communication (31)	Assessment of current burden of climate sensitive diseases.	Drought and diarrhoeal disease.
United Kingdom	CCIRG. 1996 (32)	Literature review.	Heat-related deaths and extreme events.
United States	Dept of Health, 2002 (33)	Expert judgement, predictive modelling	Thermal stress, air pollution, flooding, water borne disease.
	USEPA, 1989 (34)	Literature review, some modelling.	Increases in vector-borne diseases. See table 9.3
	Patz et al., 2000 (35) +various documents	Literature review, expert judgement.	
Zambia	Phiri, J. & Msiska, D. 1998 (36)	Expert judgement.	Malaria, schistosomiasis, water-borne diseases and malnutrition.

CLIMATE CHANGE AND HUMAN HEALTH

The assessments do not consistently report their conclusions regarding future risks to health. The outcomes listed in Table 9.1 refer to likely impacts of climate change on that country's health addressed within the report. In general, the confidence or uncertainty surrounding these estimates is not described. It can be seen that vector-borne diseases, particularly malaria, are consistently addressed. Other impacts that may have greater effects, such as weather disasters, are less well addressed.

It is important to distinguish between the process of conducting the assessment and the product. HIA is a policy tool therefore the process of conducting assessments, particularly the involvement of stakeholders, is very important. Yet information on the process aspects of these assessments is not widely available, and is not reviewed here. Stakeholder participation and evaluation are essential activities for a policy-orientated assessment process, discussed in chapter ten and elsewhere (37).

Developed countries

The most important environmental health problem in developed countries is generally considered to be outdoor air pollution. For these countries, the focus of impacts of climate change was also the potential increase in heat stress and heatwaves. There was relatively little discussion of weather disasters except for the United Kingdom (33). The risk of vector-borne disease was considered important even if the diseases were not currently present in that particular country. The effect of climate and weather on food and water-borne diseases was also addressed, perhaps without sufficient attention.

The United States study involved a comprehensive review of the epidemiological literature for a number of climate sensitive diseases and exposures. An expert panel was formed, their conclusions summarised in Table 9.2. The assessment was well disseminated and summary documents for the health specific chapters were published as a special issue of a health journal (8, 35). The United States assessment health chapter team took the decision not to use modelling to quantify future estimates. It was decided that the primary objective was to produce a consensus document to serve as a foundation for future quantitative assessments. In contrast, the United Kingdom assessment was focussed totally on delivering quantitative results for the following outcomes (33) for three time periods and four climate scenarios:

- heat and cold related deaths and hospital admissions
- cases of food poisoning
- changes in distribution of *Plasmodium falciparum* malaria (global) and tick-borne encephalitis (Europe), and seasonal transmission of *P. vivax* malaria (UK)
- cases of skin cancer due to stratospheric ozone depletion.

The large uncertainty surrounding these estimates was acknowledged, the main source of uncertainty being the United Kingdom's capacity to control these diseases in the future. For example, cases of salmonella and levels of air pollutants are generally declining, therefore the climate-change attributable effects will decline also. The main conclusions of the report were the impact of increases in river and coastal flooding, and severe winter gales. This report also clearly addressed the balance between the potential benefits and adverse impacts of climate change: the potential decline in winter deaths due to milder winters is much larger than the potential increase in heat-related deaths. Climate change

TABLE 9.2 Summary of the health sector assessment for the United States. *Source: reproduced from reference (35).*

Potential health impacts	Weather factors of interest[a]	Direction of possible change in health impact	Priority research areas
Heat-related illnesses and deaths	extreme heat and stagnant air masses	↑	Improved prediction, warning and response Urban design and energy systems Exposure assessment Weather relationship to influenza and other causes of winter mortality
Winter deaths	extreme cold snow ice	↓	
Extreme weather events-related health effects storms	precipitation variability (heavy rainfall events)[b]	↑	Improved prediction, warning and response Improved surveillance Investigation of past impacts and effectiveness of warnings
Air pollution-related health effects	temperature stagnant air masses	↑	Relationships between weather and air pollution concentrations Combined effects of temperature/humidity on air pollution Effect of weather on vegetative emissions and allergens (e.g., pollen)
Water- and food-borne diseases	precipitation estuary water temperatures	↑	Improved monitoring effects of weather/environment on marine-related diseases Land use impacts on water quality (watershed protection) Enhanced monitoring/mapping of fate and transport of contaminants
Vector- and rodent-borne diseases	temperature precipitation variability relative humidity	↕	Rapid diagnostic tests Improved surveillance Climate-related disease transmission dynamic studies

[a] Based on projections provided by the National Assessment Synthesis Team. Other scenarios might yield different changes.
[b] Projected change in frequency of hurricanes and tornadoes is unknown.

also is anticipated to lead to a decline in air pollution-related illnesses and deaths, except for those associated with tropospheric ozone.

The Portuguese national assessment made several methodological advances (*29*). The following approach was used:

- assessment of current health status in Portugal
- identification of populations most vulnerable to climate change
- identification of mechanisms by which projected climate changes may affect health
- assessment of strategies that may reduce potential impacts on health
- identification of knowledge gaps.

The main findings of the Portuguese assessment were that climate change would lead to:

- an increase in heat-related deaths in Lisbon (cold related deaths were not addressed) even with full acclimatization;
- decline in meteorological conditions suitable for high nitrogen dioxide levels in Lisbon, but an increase in conditions suitable for high ozone levels;

- increased transmission of water and food-borne diseases due to higher temperatures (not quantified);
- increase in number of days per year suitable for malaria (*P. falciparum* and *P. vivax*), West Nile virus fever, and dengue transmission (based on temperature threshold model). However, the actual (rather than potential) risk of transmission was either low or none using a qualitative risk assessment method (Table 9.3);
- actual risk of increases in leishmaniasis and Mediterranean spotted fever were assessed to be medium to high;
- increased risk of leptospirosis transmission due to increased flooding.

TABLE 9.3a Scenarios used in vector borne disease assessment.

	Assuming current knowledge of vector and parasite prevalence in Portugal	Assuming the introduction of a small population of parasite infected vectors into Portugal.
Current climate	Scenario 1	Scenario 2
Climate change (2x CO_2)	Scenario 3	Scenario 4

TABLE 9.3b Potential risk of mosquito borne diseases in Portugal (Casimiro and Calheiros, 2002) (29).

Disease	Scenario	Suitable Vector	Parasite	Risk Level
Vivax malaria	1	Widespread distribution	Imported cases only	Very low
	2	Focal distribution (new vector)	Low –> high prevalence	Low
	3	Widespread distribution	Imported cases only	Very low
	4	Focal –> potentially regional distribution (new vector)	High focal prevalence –> high prevalence regional distribution	Low-Medium
Falciparum malaria	1	None present	Imported cases only	None
	2	Focal distribution	Low –> high prevalence	Low
	3	None present	Imported cases only	None
	4	Focal v potentially regional distribution	High focal prevalence –> high prevalence regional distribution	Low-medium
Dengue	1	None present	Imported cases only	None
	2	Focal distribution	Low –> high prevalence	Low
	3	None present	Imported cases only	None
	4	Focal –> potentially regional distribution	High focal prevalence –> high prevalence regional distribution.	Low-medium
Yellow Fever	1	Widespread distribution	Imported cases only	Very low—None
	2	Focal distribution (new vector)	Low –> high prevalence, focally distributed	Low
	3	Widespread distribution	Imported cases only	Very low
	4	Focal –> potentially regional distribution (new vector)	Low prevalence, widespread distribution	Low medium
West Nile fever	1	Widespread distribution	Low prevalence, focally distributed	Low
	2	Focal distribution (new vector)	Low –> high prevalence, focally distributed	Low
	3	Widespread distribution	Low –> high prevalence, regionally distributed	Low medium
	4	Focal distribution (new vector)	Low –> high prevalence, focally distributed	Low

Developed country assessments were more likely to identify research gaps and include an adaptation assessment. In all cases the adaptation assessment referred to planned measures or strategies (see Table 9.2).

Developing countries

Assessments of adaptation (i.e. responses to the impacts of climate change) are linked to the development status of developing countries and those with economies in transition. Several initiatives have supported assessments in developing countries. The following programmes have provided financial and technical support, with a main focus on capacity building within the scientific and stakeholder communities:

- United States Country Studies Programme managed by USEPA (*38*)
- Country Studies Programme managed by UNEP, funded through Global Environment Facility (GEF) (*15*)
- Dutch Country Studies Programme
- Pacific Islands Climate Change Assistance Programme (PICCAP).

The Cameroon UNEP Country Study was restricted to looking at the impacts in only two regions: the coastal zone, the most densely populated area; and the Sudano-Sahelian zone, the poorest region in Cameroon (*20*). The burden of infectious diseases is high in this latter region (cholera, yellow fever, malaria). Water projects constructed for climate change response could increase malaria in dry regions, despite less rain being projected. Modelled projections for malaria and schistosomiasis were undertaken. Increased drought may lead to extension of the meningococcal meningitis season although there is little epidemiological literature to support this conclusion. Cholera is associated with flooding in the lowlands and therefore any increased flooding associated with climate change was anticipated to lead to an increase in cholera. Although Cameroon has an early warning system for cholera and meningitis epidemics, it was acknowledged as not fully operational due to lack of trained personnel. Major constraints in the health sector were identified as insufficient amounts of human, material and financial resources.

The United States country studies programme supported climate change studies in 49 developing countries and those with economies in transition, enabling these countries to develop emissions inventories, assess vulnerabilities to climate change, and evaluate response strategies for mitigation and adaptation. Zambia and Sri Lanka completed health assessments under this programme. Zambia qualitatively addressed the implications of climate change for infectious diseases such as malaria, schistosomiasis, cholera, dysentery, bubonic plague, and malnutrition (*36*). No modelling was undertaken and the assessment was limited by the lack of health data (*38*). Based on expert judgement, the study found that existing environmental health problems (due to environmental degradation) are likely to be exacerbated by climate change. Sri Lanka also studied the potential effects of climate change on malaria and concluded that this could become prevalent in areas that are currently clear. The effect of changes in population, income or quality of health care in these countries was not assessed.

The general conclusion from the developing country studies is that new innovative methods are needed, given the lack of data. It was also suggested that basic indicators of vulnerability be developed. The United States country studies programme recognised that some of the methods proposed were not applicable to developing country situations.

Small island developing states

Small Island Developing States (SIDS) have more incentive than most to undertake vulnerability and adaptation assessments. The effects of climate change in the Caribbean will be felt through increasingly severe tropical storms and likely increases in the severity and frequency of low rainfall events and droughts in all areas (39). The UNEP country study report for Antigua and Barbuda found droughts and hurricanes to be of particular concern. The methods described in the UNEP handbook were followed. However, only one vector-borne disease was considered (dengue) because this disease is currently present on the islands. A dengue model (40) was used to assess changes in the seasonal pattern of epidemic potential (EP). An assessment of the local relationship between dengue and climate, using local records on the Infestation Index (an index of dengue risk), was inconclusive. Antigua's assessment was very useful as it provided a comprehensive assessment of the current health system and its adaptive capacity in relation to specific diseases, such as the cholera and food safety plans.

The Fijian assessment's health chapter (22) addresses current health status in the context of health services and other provision. Fiji's main concerns were dengue fever (recent epidemic in 1998), diarrhoeal disease and nutrition related illness. The islands are malaria free and a mosquito vector (*Anopheles*) population has not been established despite a suitable climate. The risk of introduction and establishment of malaria and other mosquito borne diseases due to climate change was considered to be very low therefore. Filariasis, an important vector-borne disease on the islands, is likely to be affected by climate change due to higher temperatures. The distribution of the vector (*Aedes polynesiensis*) may be affected by sea level rise as it breeds in brackish water.

Climate change was anticipated to increase the rates of diarrhoeal disease in Fiji and Kiribati due to decreases in rainfall and increases in temperature. Evidence was not presented to show current association between flooding or heavy rainfall and cases of diarrhoea. However, the 1997/98 drought (associated with El Niño) had widespread impact, including malnutrition and micronutrient deficiency in children and infants (41).

The Kiribati assessment quantified potential impacts of climate change on cases of ciguatera fish poisoning, dengue fever and diarrhoeal diseases (25). Kiribati has the highest rates of ciguatera in the Pacific. A linear relationship between sea surface temperature anomalies and annual cases of ciguatera (42) was extrapolated to estimate future reported cases and incidence using air surface temperature as a parameter of future changes in sea surface temperature. The study suggests an increase in cases of ciguatera under climate change. The many caveats of this approach were clearly stated, including the extrapolation of a relationship beyond the temperature range observed in the original study. Further, this model does not take account of other factors, such as human behaviour or the effect of climate warming on the reef ecosystems.

For both Kiribati and Fiji, a dengue transmission model was incorporated into PACCLIM, a climate impacts model developed for the Pacific Islands (DenSIM) (40). The model estimates changes in EP (epidemic potential), a relative indicator of potential transmission based on temperature (assuming other factors remain constant) for selected population centres: Nadi, Suva (Fiji), South Tarawa

(Kiribati). Dengue is primarily an urban disease (see chapter 6 on vector-borne disease). The outcome measures EP was related to categories of risk of epidemics, however it had not been validated for these populations. The modelling indicates that climate change may extend the transmission season and geographical distribution in Fiji (not examined in Kiribati). In the Kiribati assessment, absolute values and changes in EP are reported for decadal means that are difficult to interpret.

Recommendations for developing methods and tools

Assessments of the potential health impacts of climate change have used a variety of methods and tools. Both qualitative and quantitative approaches may be appropriate depending on the level and type of knowledge. The outcome of an assessment need not be quantitative for it to be useful to stakeholders. An integrated approach is likely to be an informative approach as climate impacts are likely to transcend traditional sector and regional boundaries. Impacts in one sector may affect the capacity to respond of another sector or region.

It is important to distinguish between epidemiological methods and health impact assessment methods. Current epidemiological research methods are best able to deal with the health impacts of short-term (daily, weekly, monthly) variability, which require only a few years of continuous health data (*43*). In contrast, health impact assessment methods address the application of epidemiological functions to a population to estimate the burden of disease. Attributable burdens can only be estimated for those weather-disease relationships for which epidemiological studies have been conducted. The available evidence indicates that weather-disease relationships are highly context specific and vary between populations, therefore such models need to be derived from data from relevant populations.

The IPCC has developed formal methods of reporting uncertainty in assessments (*44*). Figure 9.2 illustrates the importance of communicating to policy-makers the scientific evidence behind particular estimates. Further, the IPCC authors were encouraged to apply probabilities to statements regarding future impacts. To ensure that these probabilities were reported consistently a quantitative scale was agreed, for example, "high" confidence referred to probabilities of 67–95%, and "very high" confidence referred to probabilities greater than 95%.

These methods are suitable for health impact assessment and were applied in the Portuguese assessment (*29*). The standard epidemiological approach to quantifying uncertainty relates to the use of confidence intervals around estimates. It is not possible to apply these to the results of scenario-based health risk assessment when biological or processed based models are used. However, new approaches to quantifying uncertainty that apply Bayesian methods have been developed. It is important to specify the likely range of uncertainties and the magnitude and direction of errors.

Literature reviews

Only the well-funded assessments (e.g. Canada, United States) have been able to include comprehensive literature reviews. Reviewing the literature and consulting experts are both processes open to bias and inaccuracies (*45*). Robust and transparent methods are needed. There is a problem with a lack of published

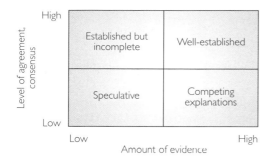

FIGURE 9.2 Uncertainty assessment.
Source: reproduced from reference 44.

studies on climate/weather and health outcomes, and the limited scope for extrapolation given that such relationships are highly context-specific. An additional problem has been that some relevant literature comes from non-health disciplines (e.g. air quality studies) and therefore difficult to detect and easy to overlook. This underscores the need for a multi-disciplinary writing team. No assessment so far has published search protocols or specified criteria for inclusion/exclusion, or made formal attempts to assess the quality of individual studies.

Considerable time, effort and resources are required to undertake a systematic and comprehensive literature review, usually beyond the means available. Many assessments rely too heavily on the IPCC Assessment Reports that necessarily lack geographical specificity. There is considerable opportunity for the assessments to address more relevant problems of the population in question. In most cases such opportunities have been lost.

The literature review should focus on the following evidence:

- mechanisms by which climate change may affect health;
- estimates of current burden of climate-sensitive diseases;
- estimates of future burden of climate-sensitive diseases using scenario based modelling;
- descriptions of future vulnerability to climate change, particularly identification of most vulnerable groups/populations/locations;
- studies that address early health effects of climate change;
- how to identify and evaluate impacts of climate variability on health.

Predictive modelling

Quantified scenario-driven modelling was undertaken in several assessments using a range of methods (see chapter 6 on vector-borne disease). It is likely that climate change will affect vector-borne disease. The majority of predictive modelling studies have estimated future changes in the distribution of vectors and/or measures of disease risk within existing, or predicting newly endemic, areas (see review in chapter 7).

The model used in Fiji and Kiribati to estimate changes in dengue transmission was not derived for that area, but was a biological model that relied on generalised assumptions about dengue transmission. This had limited value for a local study, its application is therefore unknown, and ideally requires validation with local data. Most model outcomes were assessed for long time periods (2020s, 2050s and 2080s) rather than periods of more relevance to public health. This was primarily because the climate scenarios were available only for those time periods and did not address the needs of the health sector.

Use of climate scenarios

National climate scenarios were available for several of the assessments (United Kingdom, United States, Portugal, Fiji, Kiribati). As noted above, there is great uncertainty on future changes in climate at local or national level. In order to address such uncertainty around future climate projections, impacts studies should consider a range of emission scenarios and a range of climate models. In practice this is not always feasible, climate scenarios are expensive to produce and impact groups must use what is available. The United Kingdom study was able to address four climate scenarios based on assumptions of future emissions (high, medium-high, medium-low and low estimates), and the Portuguese study compared the output from two different regional climate models (PROMES and HadRM2).

Climate information was not available for all assessments. Antigua and Barbuda found that the recommended climate scenarios did not even resolve the islands from the ocean. Analogue scenarios were used in this assessment (46) (see chapter 2 on climate). The Pacific Islands used PACCLIM, a climate change integrated assessment model for the Pacific region. In addition to analogue scenarios, simple climate models (e.g. COSMIC and MAGIC) can be used to give a range of climate scenarios based on a range of climate models and sensitivities, and emissions scenarios. COSMIC (Country Specific Model for Intertemporal Climate) contains simplified versions of the 14 different climate models used by the IPCC, and gives output at the country level (47).

When discussing (or quantifying) the range of future impacts it is important to use a range of scenarios in order to address the uncertainty surrounding future emissions that drive climate change, and inherent within the climate models. However, when planning adaptation measures, it may be most appropriate to address the upper limits of the projections. A report on impacts in small islands recommended planning for the worst-case scenario—i.e. the top range of the climate scenarios—5 or 6 °C temperature rise by 2100, and sea level rise of 0.9 m (39).

Integrated assessment

In order to optimise their use to decision-makers, assessments should integrate across all relevant sectors. This also involves stakeholder needs assessment (discussed in chapter 12) due to the multi objective nature (or tradeoffs) inherent within any policy responses to climate change health risks. Within an integrated assessment framework, physical, biological and societal system responses and assumptions can be simulated. This, in turn, affords improved guidance in decision evaluation when comparing policy tradeoffs in terms of potential risks or benefits.

Integration between sectors

All major economic sectors are likely to be affected by climate change. These impacts should be included in the assessments because of the:

- direct impacts of climate change and health implications, e.g. loss of food supply
- impacts of responses to climate change (adaptation and mitigation strategies). This assessment is within the traditional role of health impact assessment i.e. that specific projects should be evaluated.

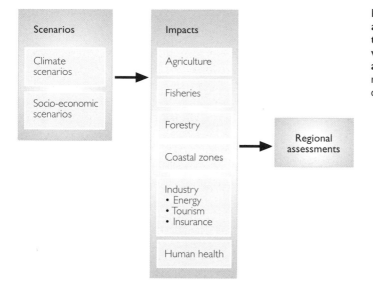

FIGURE 9.3 Multi-sectoral assessments: example of typical sectors addressed within a national assessment. Each box represents an expert group and chapter within final report.

Figure 9.3. describes the sectors typically included in climate change impact and vulnerability assessments. There was a consistent lack of integration between the sectors in the multi-sectoral assessments. By integration, we mean that the water assessment provides information for the health assessment and vice versa. This has been identified consistently as a problem for climate change assessments. Health risk assessment emphasises the need for monitoring of human diseases in relation to climate and environmental factors.

The United Kingdom Climate Impacts Programme (UKCIP) supports climate impact assessments in the United Kingdom although no specific guidelines have yet been developed. UKCIP promotes the need for a cross-sectoral and integrated approach to impact assessments as they are unlikely to respect sectoral or regional boundaries. In particular, impacts on health will depend on the impacts of climate change on water resources management, transport, and coastal and flood plain infrastructure, as well as the responses to those impacts. The Portuguese assessment was able to make use of the results from other sectoral assessments, particularly for future flood risk and water availability.

There was some consideration of cross-sectoral impacts in the developing country studies. In Antigua and Barbuda, it was recommended that the government consider an incentive scheme to encourage the public to construct larger water storage tanks to collect rainwater. However, the health sector made a recommendation to reduce the storage of water because such tanks are a known breeding site for the dengue vector.

Integration across a region

Countries in close proximity will generally experience similar climatic conditions and—if socioeconomically alike—similar vulnerabilities to climate change. The regional scale approach can offer a useful organisational unit on which to co-ordinate and evaluate research cognisant of socioeconomic needs and geophysical and jurisdictional boundaries. In this setting assessment of critical climate-sensitive issues can augment nation-specific assessments, affording a more complete assessment of risk and vulnerabilities in a region commensurate with the design and support of effective adaptation responses (discussed below). Region-based (multi-national) assessment frameworks are encouraged where common climate exposures and health outcomes occur (e.g. Pacific Island or Southern African nations).

Adaptation assessment

In most assessments adaptation typically is addressed by recommendations for measures or interventions (Table 9.4). That is, what should be done now to reduce the potential health impacts of climate change. The autonomous responses to climate change that will tend to reduce impacts are often poorly described: the non-climate factors that affect health are not fully considered. The relative contributions of climate and non-climate factors should be discussed in the context of current sensitivity and future vulnerability.

Climate adaptation and sustainability goals can be advanced jointly by changes in policy that lessen pressure on resources, improve management and environmental risks, and enhance adaptive capacity. An analysis of the IPCC guidelines for coastal zone management found greater emphasis on assessing impacts rather than adaptation and that impact studies include poor assessment of autonomous adjustments (adaptations "likely" to occur without additional policy incentives) (48). For health impact assessment, autonomous adjustments would include the physiological and behavioural acclimatization of populations.

The literature distinguishes between specific adaptations (discussion of specific measures or health interventions) and general adaptive capacity. Adaptive capacity has been defined as the enabling environment and relates to institutional and environmental issues, often clearly linked to sustainable development. Strengthening and maintenance of the public health infrastructure will increase adaptive capacity.

Specific adaptation assessment may be addressed in four increasingly complex ways:

1. List of adaptation options/strategies/policies with no evaluation
2. Estimation of the health benefit or effectiveness of specific strategies
3. Evaluation of specific strategies e.g. cost-effectiveness analysis
4. Policy analysis that addresses the feasibility of implementation of specific strategies or policies

Only step 1 has been seen in the assessments reviewed so far. The UNEP country studies placed heavy emphasis on the analysis of adaptation strategies for climate change—with a focus on dealing with current vulnerability to climate variability and extremes—the win-win strategies (15). For example, Samoa advises a no-regrets approach, defined as strategies that benefit both society and the environment in the long-term in spite of initial economic costs. A range of sectoral adaptation measures were qualitatively assessed, based on economic and environmental costs, cultural suitability and practicability (49). The UNEP Country Studies Assessment concluded that technology transfer was not required but that there needed to be an increase in institutional capacity to regulate environmental issues—e.g. settlement patterns or over-use of resources (15).

Review of National Communications to the United Nations Framework Convention on Climate Change

The UNFCCC is the international legal mechanism under which national governments are responsible for reducing greenhouse gas emissions, with the aim of avoiding, postponing or reducing the environmental, economic, and social impacts of climate change. Assessments of vulnerability and adaptation address a country's response to climate change. Article 4(f) of the UNFCCC states:

> "All Parties [. . .] shall: Take climate change considerations into account, to the extent feasible, in their relevant social, economic and environmental policies and actions, and employ appropriate methods, for example impact assessments, formulated and determined nationally, with a view to minimising adverse effects on the economy, on public health and on the quality of the environment, of projects or measures undertaken by them to mitigate or adapt to climate change."

Whether and how an impact assessment is conducted is based on a country's voluntary decision. Although several agencies have produced guidelines on how to assess the impacts of climate change, no standard approach has been developed with regard to human health and a comparative mechanism is not in place.

These reports (referred to as National Communications) are submitted to the UNFCCC which provides guidelines to promote consistent reporting. There are few references to human health within these guidelines. Countries decide what additional information they want to submit to the secretariat. Most Annex I parties (most developed countries) submitted their first report in 1994 or 1995 and their second in 1997. Most parties with economies that were in transition submitted their second communications in 1998 (according to the longer timeframe granted to them). The third national communication for Annex I parties was due in November 2001.

All national communications available online from the UNFCCC website, and in English, in March 2002 were reviewed to see if they addressed health. Most Annex I countries do not address health impacts. Some countries address health but conclude that climate change impacts on health are of "no great concern" (e.g. Austria, Denmark). Where health is mentioned, a range of impacts is discussed, primarily based on the IPCC conclusions. Italy and Japan address health in some detail. The potential impacts on health were not well addressed in the assessments of non-Annex I countries, except for the small island states that have conducted extensive vulnerability assessments. These populations are considered extremely vulnerable to climate change and also contribute significantly to the UNFCCC process. Table 9.4 summarises the conclusions of National Communications in non-Annex I countries (i.e. developing countries) on strategies to reduce the impacts on human health. The range of impacts is similar as countries identify familiar concerns of increases in vector-borne disease, food and water-borne disease, heat stress, air pollution and impacts from natural disasters (such as floods and droughts).

The National Communications are important and address a range of issues that support the Framework Convention, primarily for national inventories of greenhouse emissions. The potential impacts of climate change in economic sectors (e.g. agriculture) are typically well-addressed (50). National Communications provide invaluable information about vulnerability to the impacts of climate change. Many governments recognise that there are important gaps in knowledge of the relationship between human health, social change and

TABLE 9.4 Non-Annex I countries: summary of conclusions of National Communications for adaptation to reduce health impacts of climate change, as supplied to the United Nations Framework Convention on Climate Change.

Country	Date of submission	Adaptation strategies
Bolivia	16/11/00	• Environmental Care • Sanitary Education • Reservoir Control • Decreasing vector/human contact • Epidemic/climate warning • Biological control • Chemical Control
Colombia	18/12/01	Strengthen prevention and control of malaria and dengue
Cook Islands	30/10/99 revised edition March 2000	Specific capacity building needs • Regular access to workshops, conferences, meetings and internet services • Education and awareness • Training and equipment • Integrated pest management systems • Legislation
Indonesia	27/10/99	• promote use of environmentally friendly fuels and healthy transportation system • promotion of healthy environment housing • promotion of emergency response system for sporadic climate change disaster
Malaysia	22/08/00	Strengthening existing emergency preparedness and disaster management programme for international health surveillance and monitoring systems. Multidisciplinary approach and collaboration with other agencies—such as, agricultural, meteorological, environmental and planning agencies—will be intensified to ensure adequate weight given to health impacts due to climate change in Malaysia
Marshall Islands	24/11/00	Develop comprehensive suite of human health policies to address water borne diseases and other sicknesses related to climate-induced change, including those arising from poor water quality and nutrition
Niue	02/10/01	• Health education and promotion programmes to incorporate health impacts of climate change on infectious diseases • Future health services delivered on Niue to acknowledge emerging infectious diseases with an adaptive perspective on human health impacts • Preventative health programmes and projects within Public Health division to be strengthened and supported, with emphasis on community involvement in the projects • Policies for disaster preparedness with adaptive strategies to be formulated and implemented • Database and information system to be established for accurate monitoring and data collation
Papua New Guinea	27/02/02	Adaptation measures
Saint Kitts & Nevis	30/11/01	• Development of Health Forecast System for acute respiratory, cardiovascular • and many other diseases • Strengthening of data collection and reporting systems • Vaccination campaigns for all possible diseases • Sustained and improved sanitary conditions in human settlements • Sustained and improved disease vector control • Educational and promotional health related public campaigns

TABLE 9.4 *Continued*

Country	Date of submission	Adaptation strategies
Saint Lucia	30/11/01	• Public awareness • Surveillance and monitoring • Infrastructure development • Engineering and technological responses • Medical interventions
Samoa	30/10/99	• Health education and awareness need to be implemented at a community level. Regular cleaning campaigns need to be conducted for sites and places where mosquito vector is abundant • Conduct research programmes on the use of biological control. • Encourage development of proper waste disposal methods to minimise existence of vector breeding habitats. • Reliable and safe drinking water supply is essential.
Seychelles	15/11/00	In view of considerable lack of data and expertise on the nature and magnitude of these impacts, it is important to undertake research both at population and individual levels so as to provide a solid basis for the formulation of adaptation strategies.
Sri Lanka	06/11/00	Increase awareness of climate change in health sector Detailed responses with respect to following impacts • Natural disaster preparedness • Heat stress, heat-related illnesses and disorders • Spread of infectious diseases • Food shortage and nutritional disorders
Thailand	13/11/00	Chemical control of outbreak of malaria may not be appropriate due to disease resistance and ecological effects. More research and development of alternative approaches to control possible malaria outbreaks are required.
Vanuatu	30/10/99	• Promotion of hygienic waste disposal methods will help to prevent contamination with disease pathogens in the event of cyclones and floods. • Management of surface water catchments will help to maintain quality of domestic water and continuity of water supply.
Zimbabwe	25/05/98	Health impacts, i.e. from malaria, will require investments in education and prevention techniques such as netting, repellents, and low-cost anti-malarial drugs.

environmental problems arising from climate-induced change. To date only limited attention has been given to understanding specific ways in which changes in population size and settlement density, economy and traditional practices are creating heightened vulnerability to health problems arising from climatic variations. There is a need to determine development policies to lessen the health impacts arising from vulnerability to climate change.

Conclusions

Health impacts have not been well addressed in the climate change impact assessments, which have followed the climate assessment methods rather than health specific approaches. Assessments should be driven by region and country priorities in order to determine which health impacts are considered. No single set of guidelines can cover all health and institutional situations.

The most effective HIA is a prospective activity. A further distinction can be made between interventions where health changes are an explicit objective (e.g. vaccination programmes) and where they are not explicitly part of the objectives (e.g. energy policy). As with other areas of environmental health impact assessment, there is a need to create awareness of the problem. There is also concern that health impact assessments are not incorporated into the UNFCCC National Communications and other climate change assessments, strategies or action plans.

Assessments should set an agenda for future research. Nearly all the assessments identify research gaps and often specify narrow research questions that should be answered. Assessment also should be linked to follow up activities such as monitoring and revised reports. A major shortcoming of many climate change impact assessments has been the superficial treatment of the adaptive capacities and options of diverse populations (6). Strategies to enhance population adaptation should promote measures that are not only appropriate for current conditions but also build the capacity to identify and respond to unexpected future developments. The restoration and improvement of general public health infrastructure will reduce vulnerability to the health impacts of climate change. In the longer-term and more fundamentally, improvements in the social and material conditions of life and the reduction of inequalities within and between populations are required for sustained reduction in vulnerability to global environmental change.

The development of guidelines would improve methods used in assessments, allow for some standardization and the development of a set of indicators (51). Health Canada has prepared an initial framework (52). There are three distinct phases:

1. Scoping: to identify the climate change problem (concerns of vulnerable groups) and its context, describe the current situation (health burdens and risks) and identify key partners and issues for the assessment.
2. Assessment: estimations of future impacts and adaptive capacity, and evaluation of adaptation plans, policies and programmes.
3. Risk management: to minimize the impacts on health and follow-up assessment process and health risk management actions.

Health impacts assessment requires guidelines that fit in with the larger HIA framework of WHO and other international agencies. There is a need to move beyond the climate change environmental policy domain and into the public health arena. Guidance is needed on capacity and needs assessment. In most countries the policy environment does not encourage intersectoral collaboration. Policies for resource allocation in the health sector aim at dealing with problems of the present with the highest burden of disease. Unfortunately, the medical perspective (i.e. provision of health services in clinics/hospitals) prevails over the public health perspective (i.e. activities to prevent disease) in much policy-making.

References

1. World Health Organization (WHO). *Health impact assessment as a tool for intersectoral health policy.* WHO European Centre for Environment and Health/European Centre for Health Policy, 1999.

2. BMA. Health and environmental health impact assessment. London, UK, Earth-Scan, 1999.

3. McMichael, A.J. et al. Human population health. In: *Climate change 1995: impacts, adaptations, and mitigation of climate change: scientific-technical analyses*. Contribution of Working Group II to the Second Assessment Report of the Intergovernmental Panel on Climate Change. Cambridge, UK, Cambridge University Press: 561–584, 1996.

4. McMichael, A.J. & Githeko, A. Human health. In: *Climate change 2001: impacts, adaptation, and vulnerability*. Contribution of Working Group II to the Third Assessment Report of the Intergovernmental Panel on Climate Change. McCarthy, J.J. et al. eds. New York, USA, Cambridge University Press, pp. 451–485, 2001.

5. Carter, T.R. et al. *IPCC Technical guidelines for assessing climate change impacts and adaptations*. London, UK, University College, 1994.

6. Parry, M.L. & Carter, T. *Climate impact and adaptation assessment*. London, UK, Earth-Scan, 1998.

7. Lock, K. Health impact assessment. *British Medical Journal* 320: 1395–1398 (2000).

8. Bernard, S. & Ebi, K.L. Comments on the process and product of the health impacts assessment component of the national assessment of the potential consequences of climate variability and change for the United States. *Environmental Health Perspectives* 109: supplement 2: 1877–1884 (2001).

9. Kunzli, N. et al. Public health impact of outdoor and traffic related air pollution: a European assessment: *Lancet*, 356: 795–801 (2000).

10. Berry, J.E. *Power generation and the environment: the UK perspective*. Vol. 1. Abingdon, UK: AEA Technology, 1998.

11. IPCC-TGCIA *Guidelines on the use of scenario data for climate impact and adaptation assessment. Version 1*. Prepared by Carter, T. et al. Intergovernmental Panel on Climate Change, Task Group on Scenarios for Climate Impact Assessment, pp. 69, 1999. http://ipcc-ddc.cru.uea.ac.uk/cru_data/.

12. Smit, B. et al. An anatomy of adaptation to climate change and variability. *Climatic Change* 45: 223–251 (2000).

13. Nakicenovic, N. & Swart, R. *Emissions Scenarios*. A Special Report of Working Group III of the Intergovernmental Panel on Climate Change. New York, USA, Cambridge University Press, 2001.

14. Balbus, J.M. et al. Human health. In: *Handbook on methods of climate change impact assessment and adaptation strategies*, version 2.0. Feenstra, J.F. et al. eds. Nairobi:UNEP/ Amsterdam, Institute for Environmental Studies, 1998.

15. O'Brien, K. Developing strategies for climate change: The UNEP Country Studies on Climate Change Impact and Adaptation Assessment. *CICERO Report 2000:2*. Oslo, Norway, University of Oslo, 2000. http://www.cicero.uio.no/media/314.pdf.

16. Carter, T.R. & La Rovere, E. Developing and applying scenarios. In: *Climate change 2001: impacts, adaptation, and vulnerability*. Contribution of Working Group II to the Third Assessment Report of the Intergovernmental Panel on Climate Change. Eds. McCarthy, J.J. et al. New York, USA, Cambridge University Press, pp. 145–190, 2001.

17. O'Marde, D. & Michael, L. Human Health. In: *Antigua and Barbuda climate change impacts and adaptation assessments*. UNEP Country Studies Programme, 2000.

18. NHMRC. *Health implications of long term climatic change*. Canberra, Australia, Australian Government Publishing Service, 1991.

19. McMichael, A.J. et al. *Risk assessment regarding human health and climate change in Australasia*. Draft report submitted to Australian Government (2002).

20. UNEP/Ministry of Environment and Forestry, Cameroon. *Country case study on climate change impacts and adaptation assessment. Volume 2*. Cameroon: UNEP, Nairobi, and Ministry of Environment and Forestry, Cameroon, 1998.

21. Duncan, K. et al. Health Sector, Canada Country Study: *Impacts and adaptation: Environment Canada*, pp. 520–580, 1997.

22. de Wet, N. & Hales, S. Human health. In: *Climate change vulnerability and adaptation assessment for Fiji.* Prepared for World Bank Group by International Global Change Institute, in partnership with South Pacific Regional Environment Programme (SPREP) and PICCAP, 2000.

23. Ando, M. Health. In: *The potential effects of climate change in Japan.* Nishioka, S. et al. eds. Tsukuba, Japan, Center for Global Environmental Research/National Institute for Environmental Studies, pp. 87–93, 1993.

24. Ando, M. et al. Health. In: *Global warming: the potential impact in Japan.* Nishioka, S. & Harasawa, eds. Tokyo, Japan, Springer Verlag, pp. 203–214, 1998.

25. Taeuea, T. et al. Human health. In: *Climate change vulnerability and adaptation assessment for Kiribati.* Prepared for World Bank Group by International Global Change Institute, in partnership with South Pacific Regional Environment Programme (SPREP) and PICCAP, 2000.

26. Martens, W.J.M. ed. *Vulnerability of human population health to climate change: state-of-knowledge and future research directions.* Bilthoven, Netherlands, Dutch National Research Programme on Global Air Pollution and Climate Change, Report No. 410200004, 1996.

27. Woodward, A. et al. *Climate change: potential effects on human health in New Zealand.* A report prepared for the Ministry for the Environment as part of the New Zealand Climate Change Programme. Wellington, New Zealand, Ministry for the Environment, 2001.

28. Sempris, E. & Lopez, R. eds. *Vulnerability and adaptation to the adverse impacts of climate change in the health sector.* National Program on Climate Change, National Environment Authority ANAM/UNDP (2001).

29. Casimiro, E. & Calheiros, J.M. Human Health. In: *Climate Change in Portugal: scenarios, impacts and adaptation measures—SIAM project.* Santos, F.D. et al. eds. Lisbon, Portugal, Gradiva, 2002.

30. Ratnasari, J. ed. *Final report of the Sri Lanka climate change country study: March 1998.* Colombo, Sri Lanka, Ministry of Forestry and Environment—Environment Division, 1998.

31. St Lucia climate change and vulnerability assessment. St Lucia's initial national communication to the UNFCCC, Chapter 4, 2002. http://unfccc.int/resource/docs/natc/lucnc1.pdf.

32. Climate Change Impacts Review Group (CCIRG). *Review of the potential effects of climate change in the United Kingdom.* London, UK: HMSO, 1996.

33. Department of Health. *The Health Effects of Climate Change in the UK.* London, UK: HMSO, 2002.

34. USEPA. *The potential effects of global climate change on the United States* Washington DC, USA: USEPA, Office of Policy, Planning and Evaluation, EPA 230-05-89-057, 1989.

35. Patz, J.A. et al. The potential health impacts of climate variability and change for the United States: executive summary of the report of the health sector of the US National Assessment. *Environmental Health Perspectives* 108: 367–376 (2000).

36. Phiri, J.S. & Msiska, D.B. *Vulnerability and adaptation studies: health impact assessments.* Environment Council of Zambia, 1998.

37. Scheraga, J.D. & Furlow, J. From assessment to policy: lessons learned from the US national assessment. *Human and Ecological Risk Assessment* 7: 1227–1246 (2001).

38. Smith, J.B. & Lazo, J.K. *A summary of climate change impacts assessments from the US Country Studies program.* Unpublished document. Boulder: Stratus Consulting (2001).

39. Sear, C. et al. *The impacts of global climate change on the UK overseas territories: technical report and stakeholder survey.* Natural Resources Institute/Tyndall Centre of Climate Change Research, UK, 2001.

40. Focks, D.A. et al. A simulation model of epidemiology of urban dengue fever: literature analysis, model development, preliminary validation, and samples of

simulation results. *American Journal of Tropical Medicine and Hygiene* 53(5): 489–506 (1995).

41. OCHA. *UNDAC Mission Report Fiji Drought.* UN Office for Co-ordination of Humanitarian Affairs, 1998.

42. Hales, S. et al. Ciguatera fish poisoning, El Niño and Pacific sea surface temperatures. *Ecosystem Health* 5: 20–25 (1999).

43. Kovats, R.S. & Bouma, M.J. Retrospective studies: analogue approaches to describing climate variability and health. In: *Environmental change, climate and health: issues and research methods.* Martens, P & McMichael, A.J. eds. Cambridge, UK, Cambridge University Press, 2002.

44. Schneider, S. & Moss, R. Uncertainties in the IPCC TAR. Recommendations to lead authors for more consistent assessment and reporting. Unpublished document (1999).

45. Parry, J. & Stevens, A. Prospective health impact assessment: pitfalls, problems and possible ways forward. *British Medical Journal* 323: 1177–1182 (2001).

46. Ebi, K.L. & Patz, J.A. Epidemiologic and impact assessment methods. In: *Health impacts of global environmental change: concepts and methods.* Martens, P. & McMichael, A.J. eds. Cambridge University Press, 2002.

47. Schlesinger, M.E. & Williams, L.J. COSMIC (Country Specific Model for Intertemporal Climate) 1997.

48. Klein, R.J.T. *Adaptation to climate change in German official development assistance: an inventory of activities and opportunities with a special focus on Africa.* Eschborn: Deutsche Gesellschaft fur Technicshe Zusammenarbeit, 2001.

49. World Health Organization (WHO). *Climate variability and change and their health effects in Pacific Island countries.* Report of a Workshop in Apia, Samoa, 25–28 July 2000. Geneva, Switzerland, WHO, 2001 (WHO/SDE/OEH/01.1).

50. Intergovernmental Panel on Climate Change (IPCC). *Climate Change 1995: Impacts, adaptations, and mitigation of climate change: scientific-technical analyses.* Contribution of Working Group II to the Second Assessment Report of the Intergovernmental Panel on Climate Change. Cambridge, UK, Cambridge University Press, 1996.

51. World Health Organization (WHO). *Environmental Health Indicators: Framework and methodologies.* Geneva, Switzerland: WHO, 1999. [WHO/SDE/OEH/99.10].

52. Health Canada (2002) National Health Impact and Adaptation Assessment Framework and Tools. Climate Change and Health Office, Health Canada, Ottawa.

CHAPTER 10

Monitoring the health effects of climate change

P. Wilkinson,[1] D.H. Campbell-Lendrum,[1] C.L. Bartlett[2]

Introduction

Detection and measurement of health effects of climate change are necessary to provide evidence on which to base national and international policies relating to control and mitigation measures.

Unequivocal evidence of health effects, and accurate measurements of their size, can come only from hard data. However, climate varies naturally as well as through human influences and in turn is only one of many determinants of health. There must be careful consideration of how best to collect and analyse information that will provide secure evidence of climate change impacts.

In this chapter, we consider how monitoring may help provide evidence of early health impacts, examining the principles on which monitoring should be based, potential sources of monitoring data, and discussing issues in the analysis and interpretation of such data. There is a complex relationship between climate-change and health. Detecting and quantifying impacts on health will be gained only through broad scientific effort rather than individual monitoring studies.

Methodological considerations

Monitoring has been defined as "the performance and analysis of routine measurements aimed at detecting changes in the environment or health of populations" (1). Thus, it encompasses the notion of continuous or repeated observation using consistent and comparable methods to detect changes in some parameter relating to health or the determinants of health.

In many investigations in public health, it is possible to measure changes in a defined health impact and attribute this trend to changes in a directly related risk factor. This is not so with the health effects of climate change. As climate varies naturally over time (and is one of many determinants of disease rates) modelling is required to identify the climate-attributable part of disease and the long-term trends. This lack of direct connection makes monitoring of climate change impacts on health more complex than many other forms of health monitoring. There are three main issues:

Evidence of climate change

When concerned with health effects resulting from climate change, it is necessary to clarify the extent to which the putative driving factor (the climate) itself

---

[1] London School of Hygiene and Tropical Medicine, London, England.
[2] Centre for Infectious Disease Epidemiology, University College London, England.

has changed at the location where monitoring has been undertaken. This may seem an obvious point, but it is worth remembering that many markers of health show seasonal and inter-annual fluctuation: the demonstration of this provides no direct evidence of health impacts relating to climate *change*—merely that these diseases exhibit a form of seasonal or climatic dependence. An excess of heat-related deaths in a particularly hot summer, or even a succession of hot summers, is evidence of the *potential* for climate change to increase mortality. It does not constitute sufficient evidence that mortality has increased as a result of climate change. That would require additional evidence of a change in the baseline, i.e. that the hot summers were exceptional in historical terms and a consequence of climate change rather than random variation. Indeed, not only would it have to show that the climate had changed, but also quantify how much, so that its contribution to changes in health could be estimated. Chapter 2 summarizes the evidence for being in a period of global climate change. Interpretation of the reason for change in monitored health data at a given location also requires specific evidence of climatic change at that location. Thus, monitoring of climate change impacts on health is not simply a matter of recording certain health outcomes over time. It also requires an analytical process to quantify the component of change in those health outcomes that can be ascribed to measured change in the underlying climatic conditions.

Given this, should the goal of monitoring be to yield direct evidence of changes in health resulting from climate change, or to show that diseases are altered in the short-term by meteorological conditions? If the latter, separate evidence of a changing climate arguably would provide sufficient grounds for assuming an impact on human health. However, some researchers and policy-makers may not be content to make this assumption. Most would prefer to have data from one setting which could both establish that a change in climate had occurred and show that there had been linked changes in health. The difficulty is that very long time series of data, perhaps several decades or longer, may be needed to achieve these aims, yet the quality and consistency of data recording over the long term often are uncertain especially when past data are used.

Attribution

The second key methodological issue in monitoring climate change impacts is attribution. It is clear that climate is but one of many influences on health, and its influence needs to be separated from that of other factors. For example, heat-related deaths require some form of time series analysis, usually at daily or weekly time resolution. This permits researchers to quantify the relationship between mortality and temperature (or other meteorological variables) independently of season, secular trends, and time varying factors. In most cases, the principal focus of these analyses is entirely on within-year fluctuation, i.e. the extent to which daily or weekly variation in mortality can be explained by fluctuation in meteorological conditions. This is the climate-attributable part of the particular health measure. But it is important to realise that there are various choices in the method of model construction, and these analyses can yield varying estimates of attributable cases depending on such factors as the method of adjustment for season, assumptions about the lag between exposure and health effect, and the functional form used to characterize the temperature-health relationship. Those choices will influence the estimates of climate-related health impact in individual years, and hence the assessment of trend in health impacts over the long term.

With some diseases (e.g. vector-borne diseases such as malaria), it is possible to investigate not only temporal changes but also changes in the geographical distribution of vectors and disease over years. Even so, similar questions of attribution arise: the influence of meteorological conditions has to be separated from that of change in other environmental conditions. This requires some form of analysis of the determinants of geographical or short-term temporal variation in disease. The interpretation of these analyses can be more difficult than that of analyses of fluctuation in many other climate-sensitive diseases.

Effect modification

Finally, the issue of effect modification must be acknowledged. Our interest is in changes in weather-related morbidity or mortality over decades, which is the time-scale when change in underlying climatic conditions becomes apparent. Specifically, the aim is to demonstrate a gradual increase or decrease from baseline in the annual climate-attributable part of a range of marker diseases. But over such long time-scales changes in relevant non-climate factors also may occur. The population vulnerability to meteorological influences may alter such that more or fewer climate-related cases of disease may occur even without any change in prevailing climatic conditions. For example, vulnerability to extreme weather events, including floods and storms, will depend on where and how residential housing is built, what flood protection measures are introduced, how land use is changed. Similarly, susceptibility to heat deaths will be affected by the age of the population (vulnerability rises with age), the underlying prevalence of cardio-respiratory morbidity and quality of housing (2), among other factors.

A central question for public health is the extent to which populations are able to adapt to changing climatic conditions, as this determines their vulnerability to future climate change impacts. That adaptation may include physiological acclimatization (the fact that, in a biological sense, people gradually 'get used to' a warmer, wetter, cooler or windier environment) as well as structural adaptation, such as the extension of flood protection measures, introduction of air-conditioning systems in buildings and transport systems, and the implementation of eradication programmes for vector-borne diseases.

Thus, proper interpretation of a long-term trend in a climate sensitive disease needs to pay close attention to these various potential modifiers so that there is sufficient confidence that any observed change in health is the result of a climate effect rather than alteration in population susceptibility. Further, it could be said that the priority for monitoring should be not just to detect change but also to quantify the effect on disease burdens of adaptation and changes in population susceptibility.

This provides a strong case for a broad and sophisticated monitoring programme to enable vital questions about emerging trends in climate-related disease, and the extent of future vulnerability, to be addressed. It also suggests that effective monitoring cannot be simply a matter of sequential recording of markers of climate-sensitive diseases. It also must entail parallel measurements of population and environmental data to allow study of potential modifying influences, accompanied by methodological development to improve methods of modelling the disease impacts attributable to climate change. The data for such modelling must come from observations over periods of decades in multiple settings.

General principles

The objective of detecting health effects of climate change will require similar levels of data collection and analysis to those needed to address broader questions about climate and health relationships. Given that only limited resources will be available to implement new (or revise existing) monitoring systems, priorities must be identified. We suggest that the principal criteria for selecting diseases and settings for monitoring should include the following:

Evidence of climate sensitivity

This will be demonstrated through either observed health effects of temporal or geographical climate variation, or evidence of climate effects on components of the disease transmission process in the field or laboratory. For infectious diseases, detailed knowledge of transmission cycles is essential in identifying the major threats. Climate change effects are likely to be most profound for diseases caused by organisms which replicate outside of human hosts (where they will be subject to ambient conditions), and will be less important and/or more difficult to detect for those where human to human transmission is common.

Public health burden

Monitoring also should be preferentially targeted towards significant threats to public health. These may be diseases with a high current prevalence and/or severity (3) resulting in a large loss of Disability Adjusted Life Years (DALYs), or considered likely to become prevalent under conditions of climate change. For example, diarrhoeal illness is a major contributor to the global burden of disease, and is in part related to meteorological conditions. Spread of malaria, dengue and other important vector-borne diseases to new areas and populations also would cause a substantial burden for those populations.

Practicality

Logistical considerations are important given that monitoring requires dependable and consistent long-term recording of data of health-related indices and other environmental parameters. Monitoring sites must be chosen where change is most likely to occur, but appropriate structures and capacity for reliable measurement will be essential. In some cases, retrospective data series that could be a suitable basis for forward data collection may be available.

Such considerations have been used to identify the most important health issues for both research and monitoring on a global scale. These priorities include:

- direct effects of exposure to low and high temperatures
- health impacts of extreme weather events (floods, high winds etc)
- increased frequency of food and water-borne disease
- geographical change and altered transmission frequency of vector-borne disease, principally malaria, dengue, filariasis, sleeping sickness (African trypanosomiasis), leishmaniasis, shistosomiasis, and Chagas' disease (American trypanosomiasis).

Lower priority threats include:

- other vector-borne diseases, including tick-borne encephalitis, Lyme disease, Toscana virus
- aero-allergens, particularly pollen
- rodent-borne diseases, including hantavirus and leptospirosis.
- harmful algal blooms & biotoxins.

Priorities will vary between regions with differences in current climate, level of socioeconomic development and spectrum of disease (4). Altered food productivity, changes in air pollution, and social, economic and demographic dislocations due to effects on economic infrastructure and resource supply are additional climate-related concerns that may be of high priority in some settings (5).

Monitoring systems should take account of these local needs and be alert to the appearance of potential new health concerns. Systems such as the WHO Global Outbreak and Response Network Programme for Monitoring Emerging Diseases (PROMED) (6) provide a valuable resource in identifying new health concerns. Such emergent concerns may include new geographical distributions of diseases or changes in the frequency of established diseases in particular populations; but also include the possibility of the emergence of new threats. By definition, such occurrence is unpredictable and identifying such climate-related health impacts will require vigilance by the scientific and public health communities. A possible climate-related issue that has emerged within recent years is that of harmful algal blooms.

Data requirements and data sources

A broad range of data is needed to monitor climate effects on health. Where possible, monitoring systems should assemble data on all components required for statistical analysis (including assessment of effect modification) or process-based/biological models. Many relevant variables already are recorded by existing systems and may require only access and cross-referencing with other data sources. For others, new monitoring systems or radical changes to existing systems may be necessary. Relevant measurements fall into the following broad classes:

Meteorology

Various meteorological factors influence health processes either directly or indirectly. Temperature, relative humidity, rainfall and wind-speed are perhaps the most important and all are predicted (with a greater or lesser degree of certainty) to be affected by future climate change (4, 7, 8, 9). There are few difficulties in obtaining reliable series of daily measurements of these variables for representative sites, given that there are extensive networks of meteorological monitoring stations throughout nearly all regions of the world. These measurements also can be interpolated to give estimates for the entire globe, at spatial resolutions as high as 30 km by 30 km (10). In addition, satellite based sensors record proxy measurements of temperature, rainfall and humidity with true global coverage. Data from either source can be geographically and temporally cross-referenced to health and environmental monitoring data in a geographical information system (GIS), providing powerful tools for broad scale analysis of associations between climate and disease.

The main difficulties associated with the use of climate data in monitoring health impacts lie in linking climate and health data at a suitably high resolution. Measurements are either discontinuous point data or averages over an area, and may not describe either local variation in climate (e.g. temperature in city centres vs. nearby rural areas), or microclimatic conditions in specific important environments (e.g. resting sites of adult mosquitoes). Climate measurements at a local level and in important microclimates (where these can be identified) should be recorded in intensively studied sites to test whether they provide closer correlation with health outcomes. As it is not feasible to take such measurements at all sites, their relationship with data from monitoring systems with global coverage should be modelled in order to allow scaling-up across large areas.

Health markers

Analysis of climate effects on health ultimately depends on reliable recording of health status, at suitable temporal and spatial resolution. Such monitoring should be sensitive and specific enough to quantify changes in the intensity and temporal and geographical distribution of climate-sensitive health impacts.

One way to address the complex causality of most health outcomes is carefully to select indicators that are highly sensitive to climate changes but relatively insensitive to other influences. This approach already has given clear evidence of climate-change driven effects in other ecological systems. For example, changes in the seasonality of bird egg-laying (11) and the gradual pole-wards shift of insect and bird populations (12, 13) is most plausibly explained by the gradual warming observed in the study regions. By analogy, long-term monitoring of health-relevant properties such as the seasonal pattern of insect vector abundance in areas without control programmes may demonstrate changes that can be confidently attributed to climate-change.

Human health differs from other systems, however: there is not only an interest in demonstrating that climate change is having some effect, but also in quantifying the effect and making an adaptive response. The data requirements for attributing and measuring impacts may be quite different and the utility of routinely collected data varies markedly, depending on the health issue and region. For studies of direct effects of heat and cold the essential requirement is daily series of counts of death or morbidity subdivided by age and cause. Date of death usually is routinely recorded on death certificates and daily statistics often are collated at regional and national levels. Studies of all-cause mortality therefore are feasible in many settings, though even mortality counts can be difficult to acquire for many low-income countries, especially if cause-specific breakdown also is required.

The difficulties of data assembly become more challenging where the intention is to look at disease morbidity or health effects resulting from complex ecological processes, such as infectious diseases transmitted through food, water or vectors. In these circumstances, case definition, completeness of case ascertainment and reporting efficiency are key issues, as are temporal and geographical variation in diagnosis. Future monitoring must aim to address these limitations. In some cases this may be achieved through revision of existing health data sets and linkage to climate records. For many climate-sensitive diseases, however, the coverage or quality of available data precludes this approach, and improved monitoring systems are necessary to monitor tends (both climate-dependent and independent).

Other explanatory factors

It has been stated that monitoring will need to measure more than just climate and health. Data on time varying risk factors and changes in potential population susceptibility are equally important in order to begin to assess the climatic contribution to any observed change in health status. This is particularly true of infectious diseases transmitted by water, food or vectors. Often these are highly sensitive to climate, but human infections are only the end product of a complex chain of environmental processes (14).

For such diseases climate and disease monitoring ideally should be linked with parallel monitoring of intermediate stages in the transmission cycle (e.g. parasites, vectors and reservoir hosts) and the wider ecosystem (e.g. distribution of habitats suitable for vectors or reservoirs). Collection of accurate data on these variables would help to improve the explanatory value of models in purely statistical terms, and by identifying the key biological processes underlying changes in human disease. For example, such integrated monitoring would help to give a secure answer to the question of whether variation in the incidence of tick-borne encephalitis (TBE) in Europe is mainly due to climate effects on tick population dynamics, changes in the abundance of reservoir hosts, changes in the distribution of habitats where ticks and reservoirs come into contact, or changes in disease control and reporting systems.

In some situations, it may be informative to define the relationship between climate and environmental intermediates, even in the absence of the disease itself. For example, although competent vectors for both dengue and malaria are present in many areas of western Europe, there is no active parasite transmission. However, mosquito population density, biting rate and adult longevity (all known to be affected by temperature) are important determinants of these vector populations' capacity to transmit the relevant pathogens (15). Therefore it is relevant to public health to assess the link between climate and vector populations so as to assess changes in the probability of establishment of autochthonous transmission when parasites are imported to currently non-endemic areas. The specific parameters that should be recorded vary depending on the transmission cycle of specific diseases.

Recording of population and environmental changes is necessary for interpretation of changes in all climate-sensitive diseases. The choice of variables to be measured will depend on the specific disease, but the principal categories of confounding or modifying factors include:

- age structure of the population at risk;
- underlying rates of disease, especially cardiovascular and respiratory disease and diarrhoeal illness;
- level of socioeconomic development;
- environmental conditions e.g. land-use, air pollutant concentrations, housing quality;
- quality of health care;
- specific control measures: e.g. vector control programmes.

Data on the first two of these usually is available at aggregate level from routine sources of demographic and health statistics. Socioeconomic indicators often are available in fairly crude form, but it is less easy to obtain markers that have direct bearing on the health impact of interest. Poverty and living conditions will influence exposure to almost all health threats (from direct thermal effects and natural disasters to infectious diseases) and are likely to increase the harmful effects of such exposures. Other socioeconomic factors may affect a smaller range of health outcomes, such as exposure to some pathogens in occupational environments. Changes in such socioeconomic factors potentially are of great importance for disease changes over periods of decades.

An increasingly useful source of data on land-use and other environmental parameters is remote sensing data from satellites. Such data are not always easy to calibrate and interpret, and may not be available at the required level of spatial resolution or with the desired detail of ground conditions. However, their great advantage is that they are becoming available for most areas of the world, providing the opportunity to capitalize on spatial as well as temporal contrasts. Statistical methods can be used to compare distribution of disease frequency in relation to measured land-use patterns as well as climatic variation, providing valuable insights into the determinants of distribution of malaria and other vector-borne diseases, for example (*16*).

Disease control programmes (e.g. insecticide spraying against insect vectors or chlorination of water supplies) and health care factors are problematic, not because of the difficulty of obtaining data (though this may well be the case) but because it is difficult to quantify their influence on the occurrence of climate-sensitive diseases.

In general, monitoring should aim to record data on all these confounding and modifying factors so that they can be included in analytical models. The difficulty of data acquisition, and proportionate increase in the amount of data with the number of factors being assessed, makes this infeasible in many settings. Indeed, it is impossible even to identify all confounding effects. An alternative approach may be to restrict monitoring to areas where non-climatic factors are unlikely to exert a strong effect. For example, monitoring of climate effects on mosquito populations might be carried out in areas where there are no vector control activities. In this case, climate-health models can be generalized to other areas only under the assumption that the derived relationships hold true elsewhere (e.g. temperature will affect mosquito population dynamics in the same way in habitats which are different from the original study site).

Examples

Table 10.1 summarizes the principal health impacts of climate change, sources of data, methods of data acquisition and the types of climate and other data needed to support their analysis. Where possible, monitoring should be based on existing data acquisition systems for reasons of cost-effectiveness and sustainability, and to ensure fullest use of existing retrospective data series. Particular issues of monitoring arise in different settings for the various health effects. Examples of studies that have entailed direct measurement of health in relation to meteorological parameters are given in Boxes 10.1 and 10.2 below. Other examples recently have been outlined by Patz (*17*).

Data sources and methods of monitoring will depend on the precise purpose of the monitoring, the type of existing data acquisition systems and available resources. There are differences between the sorts of data and analysis needed to quantify long-term trends in the burden of heat-related mortality and those needed for an early warning system that provides the basis for public health action (heatwave warning systems). The former requires systems to link health data to temperature measurements and other factors over a matter of decades;

TABLE 10.1 Principal health impacts of climate change and the sources of monitoring data.

	Principal health outcomes	Which populations/locations to monitor	Sources and methods for acquiring health data	Meteorological data	Other variables
Thermal extremes	Daily mortality; hospital admissions; clinic/emergency room attendance;	Urban populations	National and sub-national death registries (e.g. city specific data)	Daily temperatures (min/max or mean) & humidity	Confounders: influenza & other respiratory infections; air pollution Modifiers: housing conditions (e.g. household/workplace air conditioning), availability of water supplies
Extreme weather events (floods, high winds, droughts)	Attributed deaths; hospital admissions; infectious disease surveillance data; mental state; nutritional status	All regions	Use of sub-national death registries; local public health records	Meteorological event data: extent, timing & severity	Disruption/contamination of food & water supplies; disruption of transportation Population displacement The above parameters will have an indirect impact on health
Asthma and allergy	Changes in seasonal patterns of disease	Sentinel populations in various locations	Primary care data; emergency room attendance; hospital admissions; survey data	Daily/weekly temperature and rainfall; pollen counts	Air quality; influenza & other respiratory infections;
Food- & water-borne disease	Relevant infectious disease deaths & morbidity;	All regions	Death registries; national & sub-national surveillance notifications	Weekly/daily temperature; rainfall for water-borne disease, including cryptosporidium	Long term trends dominated by host-agent interactions (e.g. S enteritidis in poultry) whose effects are difficult to quantify. Indicators may be based on examination of seasonal patterns.
Vector-borne disease	Vector populations; disease notifications; temporal and geographical distributions	Margins of geographical distribution (for changes in altitude, latitude) and within endemic areas (for changes in temporal patterns)	Local field surveys; routine surveillance data (variable availability)	Weekly/daily temperature, humidity and rainfall	Land use Surface configurations of freshwater

the latter would be based on a short-term weather forecast (coupled with evidence from analysis of comparatively short periods of daily data defining the temperature-health relationship).

On the other hand (see Box 10.1) longitudinal data on the inter-annual variation in malaria, which may serve as an analogue of climate change impacts, also have the potential to predict high and low-risk years of malaria to inform preventative action. In this instance, monitoring to detect changes in the frequency of disease over time also may serve a more immediate and practical public health function. On the whole it may be said that early warning systems usually require shorter-term meteorological data together with systems for forecasting, forecast dissemination and initiation of public health action. Monitoring to detect evidence of early impacts of climate change requires much longer series of meteorology, health and other data, together with an analytical framework that allows impacts attributable to climate change to be quantified separately from those of other time varying factors.

Box 10.1 Vector-borne disease

One of the earliest health impacts of climate change may be altered distribution and incidence of vector-borne diseases, already a major cause of illness and death, particularly in tropical and subtropical countries (3). There is substantial evidence for the effects of climate variables on both vector and pathogen population biology in the laboratory and geographical and seasonal patterns of transmission intensity in the field (see chapter 6) (18).

This group of infectious diseases also is affected by factors such as human population density and movement; control programmes; forest clearance and land-use patterns; surface configurations of fresh water; and the population density of insectivorous predators. A recent review of vector-borne diseases concluded that the literature to date does not include unequivocal evidence of an impact of climate change, though the authors suggested that this should be seen as 'absence of evidence' rather than evidence of no effect. They refer to a lack of good quality long-term data on disease and vector distributions in areas where climate change has been observed and where a response is most likely to have occurred (19).

Monitoring for research and early detection would benefit from the creation of enhanced datasets in such areas with prospective data collection of climate, disease and vector populations. As a prerequisite current vector monitoring systems must be developed to improve their reliability. Data should be collected on demographic, socioeconomic and environmental change, including land use. For the present, however, data collected for other purposes are used to investigate effects of climate change on vectors and vector-borne diseases.

An example of the sort of data that could be provided is shown in a study by Bouma et al. (20) in the Northwest Frontier Province (NWFP) of Pakistan. These researchers showed that interannual variations in the proportion of all slides tested which were positive for *Plasmodium falciparum*, and the proportion of all positive slides which were identified as *P. falciparum* (rather than the more cold-tolerant and less pathogenic *P. vivax*), correlated with variation in temperature, rainfall and humidity during the transmission season. They also showed, separately, that all of these parameters have shown an increasing trend in this region since the late 1870s. In the absence of any evidence that other determinants of malaria incidence (e.g. control activities) changed significantly over the study period, it is therefore suggested that the recent increase in the severity of autumnal outbreaks of *P. falciparum* could most probably be explained by the trend towards more favourable climatic conditions.

This study illustrates the advantages of (i) selecting an appropriate indicator (i.e. proportion of *P. falciparum* cases, rather than total number, which would be more sensitive to variation in surveillance

Continued

FIGURE 10.1a Annual proportion of *Plasmodium falciparum* infections in the districts of NWFP, against the estimated average October—December temperature difference from a reference point (Peshawar).

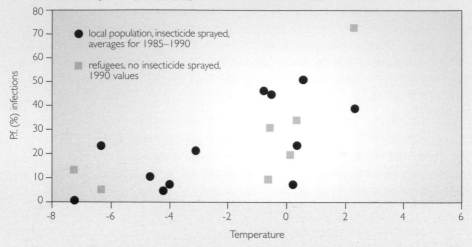

FIGURE 10.1b Mean temperatures in °C with a linear regression line since 1876, for Peshawar station. *Source: reproduced from reference 20*

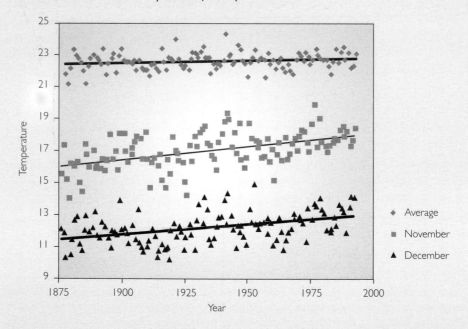

effort), and (ii) defining climate/disease associations on the basis of short-term (i.e. inter-annual) variations, and using this relationship to interpret a long-term trend in climate and disease. It should be noted that even with this longitudinal series, it was not considered informative to correlate the gradual trend in the outcome against the trend in the climatic predictors, as it is very difficult to exclude completely the possibility that unmeasured factors also may contribute to the observed increase in *P. falciparum* frequency.

Aside from the difficulties of interpreting past changes, there is additional uncertainty in predicting how malaria and other vector-borne diseases will respond to the much larger climatic changes that

CLIMATE CHANGE AND HUMAN HEALTH

are expected over the coming century. The principal difficulty lies in limited understanding of the modifying influences of socioeconomic development, control programmes, and the ability of human immunity to absorb increases in transmission intensity, all of which may alter the apparently simple relationship between climate and disease (16, 22, 23, 24, 25). Understanding of these influences will be fundamental to proper assessment of both measured and projected changes in disease.

The IPCC (26) has concluded that climate change is likely to expand the geographical distribution of several vector-borne diseases to higher altitudes and extend the transmission seasons in some locations. There is less confidence that these diseases will expand into higher latitudes, or that decreases in transmission may occur through reductions in rainfall or increases in temperature above a threshold for vector survival. Thus, if monitoring is to find evidence of change, it will be important to ensure appropriate siting of monitoring stations that collect data relatively frequently. This might include frequent and long-term sampling along transects to monitor the full longitudinal and altitudinal range of specific vector species and their seasonal patterns.

Such studies may be the most cost-effective and robust methods of detecting directly the first health-relevant effects of this climate change impact. Unfortunately, current vector monitoring systems often are unable to provide reliable measurement of changes even in the limited number of parameters suggested. The design of surveillance programmes therefore may need to be adapted specifically to monitor sensitive aspects of climate change on vector-borne disease over the coming decades. These should be targeted at areas where the population at risk is large and adaptive capacity is low.

Box 10.2 Diarrhoeal illness

The relative importance of different pathogens and modes of transmission (e.g. via water, food, insects or human-human contact) varies between areas, and is influenced by levels of sanitation (27). As pathogens are known to vary in their response to climate (28, 29) this is likely to cause geographical variation in temperature relationships, depending on level of development. The quantitative relationship between climate and overall diarrhoea incidence (i.e. due to all pathogens) only rarely has been explicitly quantified.

A study by Checkley and colleagues provides evidence of the meteorological dependence of diarrhoeal illness and a possible analogue for longer-term climate change impacts (30). These workers reported a time series study of the relationship between temperature and relative humidity and daily hospital admissions at a single paediatric diarrhoeal disease clinic in Lima, Peru (Figure 10.2). Analyses based on 57 331 admissions over a period of just under 6 years revealed a 4% increase in admissions for each 1 °C increase in temperature during the hotter months, and 12% increase per 1 °C increase in the cooler months. During the 1997–98 El Niño event, there was an additional increase in admissions expected on the basis of pre-El Niño temperature relationships. The time series methods used in this study independently controlled for seasonal variations, other climatic factors and long-term trend, so that the variation in diarrhoea rates can be attributed confidently to variations in temperature. The positive correlation is biologically plausible also, as a high proportion of diarrhoea cases in Peru, as in many tropical developing countries, are caused by bacteria, entamoeba and protozoa (27) which are favoured by high temperatures. Very long term data gathering is necessary to provide clear evidence of changes in disease burdens in relation to longer-term changes in climate. In the present example, ideally this would cover not just one but multiple ENSO cycles.

Daily temporal resolution may not be essential for such monitoring. Singh et al. (31) used similar time series methods to correlate monthly reported incidence of diarrhoea throughout Fiji, 1978–1998, against variations in temperature and rainfall. The reported incidence increased by approximately 3%

Continued

FIGURE 10.2 Daily time series, 1 January 1993–15 November 1998, of admissions for diarrhoea, mean ambient temperature, and relative humidity in Lima, Peru. *Source: reproduced from reference 30*

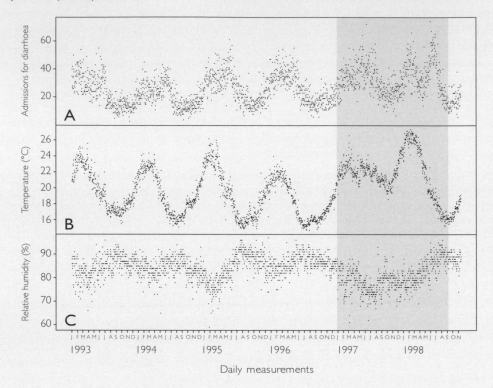

(95% CI 1.2 to 5.0%) for each 1 °C increase in temperature, and also increased significantly with unusually high or low rainfall. These findings are supported by a positive geographical correlation between temperature and diarrhoea incidence in 18 Pacific Island countries. However, adjustment for non-meteorological seasonal factors can be a difficulty especially when the temporal resolution of the data is fairly crude.

Diarrhoeal illness is already of major importance for tropical developing countries because of its large contribution to the burden of ill-health (*32*). Although that burden is much more a consequence of poor sanitation and nutrition than of climatic conditions, the demonstration of climate sensitivity suggests that climate change is likely to contribute to an increase in morbidity unless counteracted by increasing standards of living and improved public health.

Conclusions

Monitoring the health impacts of climate change is an important task by which the public health community provides much needed scientific and policy evidence related to global warming. There are many methodological challenges in carrying out such monitoring but a body of analytical expertise has been acquired through the research efforts of recent years.

It must be recognized that the process of climate change is gradual and detectable only over decades. The impact of such climate change on health will therefore similarly be slow to evolve. Over long periods changes occur also in non-climatic risk factors and in disease detection and recording. These factors often render it difficult to assess the contribution of climate change to trends in attributable burdens, and thus to detect early health effects of climate change. Monitoring studies therefore should be either designed to allow analysis of potential confounding and modifying influences or established in settings where such influences can be minimized.

For the health effects of thermal extremes, reliable temperature, mortality and morbidity data are available in many countries, and the methodological framework is sufficiently developed to allow assembly, analysis and interpretation of data from long time series in multiple locations. The principal focus of such research should be to characterize modification of the temperature-mortality/morbidity relationship by individual, social and environmental factors (poverty, housing quality, pre-existing morbidity etc).

Various databases already exist (e.g. EM-DAT) for extreme weather events and these could be a key resource for monitoring trends. To maximize their utility, attention is needed to ensure completeness and consistency of reporting of extreme weather events across a wide geographical area, and use of standard definitions of events and the application of comparable methods of attribution. Interpretation of temporal patterns will be aided by work to examine the extent to which human activity and environmental modification protects or increases vulnerability to extreme weather. Similarly, concerns over the consistency of notifications of food and water-borne diseases over the long term impose constraints on the interpretation of long-term monitoring data. Methodological work is needed to characterize variation in the temperature-disease relationships (including duration and amplitude of the seasonal rise) between populations in different geographical areas.

For most vector-borne illnesses (malaria, Leishmaniasis, tick-borne encephalitis, Lyme disease) current monitoring data can provide only very broad quantification of the relationship between climate and human disease. Assessment of the climate contribution to long-term trends requires data on factors such as land use, host abundance and intervention measures. Understanding of these relationships could be improved by obtaining high quality serial data on vector abundance at a modest number of sites within or at the margins of endemic areas. Data from sites along specified transects could provide useful indication of changing vector distributions. The latter could include sites to measure altitudinal shifts (e.g. malaria). Use of geographical comparisons based on remote sensing data may offer additional insights into disease trends.

With all forms of monitoring, interpretation of the evidence will be strengthened by procedures for standardization, training and quality assurance/quality control. Linkage of multiple data sets covering meteorology, health, intermediate stages (e.g. vectors, pathogens) and effect modifiers (land use, control programmes) also will be important. Most will be learned from long time series of health changes in populations with steep climate-disease functions—for example, on the edge of endemic areas for vector-borne diseases. Such monitoring will be made more effective through international collaboration and integration with existing surveillance networks.

References

1. Last, J. *A Dictionary of epidemiology*. 2nd edition. New York, USA, Oxford University Press, 1988.
2. Curriero, F. et al. Temperature and mortality in eleven cities of the eastern United States. *American Journal of Epidemiology* 155: 80–87 (2002).
3. World Health Organization (WHO). *The World Health Report 2002*. Geneva, Switzerland, WHO, 2002.
4. Intergovernmental Panel on Climate Change (IPCC). *Climate change 2001: the scientific basis*. The Contribution of Working Group I to the Third Assessment Report of the Intergovernmental Panel on Climate Change. Houghton, J.T. et al. eds. New York, USA, Cambridge University Press, 2001.
5. McMichael, A.J. et al. eds. *Climate change and human health*: an assessment by a task group on behalf of the World Health Organization, the World Meteorological Organization and the United Nations Environment Programme. Geneva, Switzerland, WHO, 1996 (WHO/EHG/96.7).
6. World Health Organization (WHO). *A framework for global outbreak alert and response*. Geneva, Switzerland, WHO 2000 (Report No.: WHO\CDS\CSR\2000.2).
7. Palmer, T. & Ralsanen, J. Quantifying the risk of extreme seasonal precipitation events in a changing climate. *Nature* 415: 512–514 (2002).
8. Milly, P. et al. Increasing risk of great floods in a changing climate. *Nature* 415: 514–517 (2002).
9. Hulme, M. et al. Climate change scenarios for global impacts studies. *Global Environmental Change-Human and Policy Dimensions* 9: S3-S19 (1999).
10. New, M. et al. Representing twentieth-century space-time climate variability. Part I: Development of a 1961–90 mean monthly terrestrial climatology. *Journal of Climate* 12(3): 829–856 (1999).
11. Crick, H. & Sparks, T. Climate change related to egg laying trends. *Nature* 399: 423–424 (1999).
12. Thomas, C. & Lennon, J. Birds extend their ranges northwards. *Nature* 399: (6733): 213 (1999).
13. Parmesan, C. et al. Poleward shifts in geographical ranges of butterfly species associated with regional warming. *Nature* 399: (6736): 579–583 (1999).
14. Epstein, P.R. Climate and health. *Science* 285: 347–348 (1999).
15. Garrett-Jones, C. The human blood index of malaria vectors in relation to epidemiological assessment. *Bulletin of the World Health Organization* 31: 241–261 (1964).
16. Rogers, D.J. & Randolph, S.E. The global spread of malaria in a future, warmer world. *Science* 289 (5485): 1763–1766 (2000).
17. Patz, J. A human disease indicator for the effects of recent global climate change. *Proceedings of the National Academy of Sciences* 99 (20): 12506–12508 (2002).
18. World Kovats, R.S. et al. *Climate and vector-borne diseases: An assessment of the role of climate in changing disease patterns.*, Maastricht, Netherlands, International Centre for Integrative Studies, 2000 (ICIS/UNEP/LSHTM).
19. Kovats, R.S. et al. Early effects of climate change: do they include changes in vector-borne disease? *Philosophical Transactions of the Royal Society Series B* 356: 1057–1068 (2001).
20. Bouma, M.J. et al. Falciparum malaria and climate change in the northwest frontier province of Pakistan. *American Journal of Tropical Medicine and Hygiene* 55(2): 131–137 (1996).
21. Hay, S.I. et al. Etiology of interepidemic periods of mosquito-borne disease. *Proceedings of the National Academy of Sciences of the United States of America* 97(16): 9335–9339 (2000).
22. Sutherst, R.W. Implications of global change and climate variability for vector-borne diseases: generic approaches to impact assessments. *International Journal for Parasitology* 28(6): 935–945 (1998).

23. Mouchet, J. & Manguin, S. Global warming and malaria expansion. *Annales de la Société Entomologique de France* 35: 549–555 (1999).

24. Reiter, P. Climate change and mosquito-borne disease. *Environmental Health Perspectives* 109: 141–161 (2001).

25. Coleman, P.G. et al. Endemic stability-a veterinary idea applied to human public health. *Lancet* 357(9264): 1284–1286 (2001).

26. Intergovernmental Panel on Climate Change (IPCC). *Climate change 2001: impacts, adaptation and vulnerability.* Contribution of Working Group II to the Third Assessment Report. Cambridge, UK, Cambridge University Press, 2001.

27. Black, R.E. & Lanata, C.F. Epidemiology of diarrhoeal diseases in developing countries. In: *Infections of the Gastrointestinal Tract.* Blaser, M.J. et al. eds. New York, USA, Raven Press pp. 13–36, 1995.

28. Chaudhury, A. et al. Diarrhoea associated with Candida spp: incidence and seasonal variation. *Journal of Diarrhoeal Diseases Research* 14(2): 110–112 (1996).

29. Cook, S.M. et al. Global seasonality of rotavirus infections. *Bulletin of the World Health Organization* 68(2): 171–177 (1990).

30. Checkley, W. et al. Effects of El Niño and ambient temperature on hospital admissions for diarrhoeal diseases in Peruvian children. *Lancet* 355(9202): 442–450 (2000).

31. Singh, R.B.K. et al. The influence of climate variation and change on diarrhoeal disease in the Pacific Islands. *Environmental Health Perspectives* 109(2): 155–159 (2001).

32. Guerrant, R.L. et al. Updating the DALYs for diarrhoeal disease. *Trends in Parasitology* 18(5): 191–193 (2002).

Adaptation and adaptive capacity in the public health context

A. Grambsch,[1,2] B. Menne[3]

Introduction

Understanding of Earth's climate systems has advanced substantially over the past decade (*1*). Research has provided valuable information about the potential health risks associated with climate change (*2*). Despite impressive progress in climate science and health impacts research many unresolved questions remain. While there are considerable uncertainties, as the knowledge base on climate change and health impacts has grown, so has interest in developing response options to reduce adverse health effects. That is, in addition to measures aimed at reducing greenhouse gases and slowing climate change, measures can be aimed at reducing adverse effects (or exploiting beneficial effects) associated with climate change. Even with reductions in greenhouse gas emissions Earth's climate is expected to continue to change so that adaptation strategies are viewed as a necessary complement to mitigation actions. Past experience with climate shows that there is substantial capacity to adapt to a range of conditions through a wide variety of adaptation measures (*2*).

The purpose of this chapter is to describe adaptation in the public health context. The first section defines adaptation and adaptive capacity as used in the climate change community. A discussion of the need to consider adaptation when assessing health impacts and vulnerability also is included. The next section maps adaptation concepts into the more familiar public health concepts of primary, secondary and tertiary prevention. The concept of coping ability is discussed in the third section and adaptive capacity, including the determinants of adaptive capacity, in the next. A brief description of research needs and concluding thoughts complete this chapter. Chapter 12 provides real-world examples of public health adaptations to climate and discusses policy implications.

Adaptation

Several definitions of climate-related adaptation can be found in the literature and continue to evolve (see summary in Smit et al. (*3*)). Many definitions focus on human actions (*4, 5*), some include current climate variability and extreme events (*6*), others are limited to adverse consequences of climate change (*5, 7*). The Intergovernmental Panel on Climate Change (IPCC) (*2*) has developed definitions of adaptation and the closely related concept of adaptive capacity as follows:

---

[1] Global Change Program, US Environmental Protection Agency, Washington DC, USA.
[2] The views expressed are the author's own and do not reflect official USEPA policy.
[3] World Health Organization, Regional Office for Europe, European Centre for Environment and Health, Rome, Italy.

Adaptation: adjustment in natural or human systems in response to actual or expected climatic *stimuli* or their effects, which moderates harm or exploits beneficial opportunities.

Adaptive Capacity: the ability of a system to adjust to climate change (including climate variability and extremes) to moderate potential damages, to take advantage of opportunities, or to cope with the consequences.

These definitions are comprehensive in that they are not limited to either human or natural systems: both current and future changes in climate are encompassed, and beneficial as well as adverse effects of climate change are included. In this chapter, the terms adaptation and adaptive capacity will be used as defined in IPCC (2), although greater emphasis is placed on adaptations that cope with adverse consequences of climate change. That is, the primary objective of adaptation in the public health context is to reduce disease burdens, injuries, disabilities, suffering and deaths (8).

Adaptation, climate impacts, and vulnerability assessment

In order to assess health impacts of, and vulnerability to, climate change and variability it is essential to consider adaptation (2). The ultimate objective of the United Nations Framework Convention on Climate Change (9) (UNFCCC) is "to achieve stabilization of atmospheric concentrations of greenhouse gases at levels that would prevent dangerous anthropogenic (human-induced) interference with the climate system . . .". However, the UNFCCC does not define dangerous levels, although it does refer to levels that "allow ecosystems to adapt, ensure food production is not threatened, and enable economic development to proceed in a sustainable manner" (9). As human population health also depends on these factors, it can serve as an important integrating index of effects of climate change on ecosystems, food supplies, and social-economic development (2). The extent to which the health of human populations is vulnerable or in danger depends on the direct and indirect exposures of human populations (e.g. through disturbances of ecosystems, disruptions in agriculture) to climate change effects; the populations' sensitivity to the exposure; and the affected systems' ability to adapt. To assess the human health risks associated with climate change, impact and vulnerability assessments must address adaptation (see Figure 11.1).

As shown in Figure 11.1, adaptation is considered both in the assessment of impacts and vulnerabilities and as a response option (2). Due to the past accumulation of greenhouse gases (atmospheric concentrations of CO_2 have increased 31% since 1750), the long lifetimes of these gases and the thermal inertia of the climate system, it is likely that global temperatures will increase and other aspects of climate continue to change regardless of the coordinated international mitigation actions undertaken (8, 10). Further, it is unlikely that autonomous actions undertaken by individuals or countries in reaction to climate health impacts will fully ameliorate all impacts (they don't now) (2, 11). As a result, it is prudent to develop planned adaptation strategies that address future changes in climate and impacts. Article 4.1 of the UNFCCC commits parties to formulate and implement national and, where appropriate, regional programmes of "measures to facilitate adequate adaptation to climate change".

Although climate impact and vulnerability studies consider adaptation, they rarely do more than identify potential adaptation options or model them in a simple way, relying on a number of simplifying assumptions. Research is needed

FIGURE 11.1 Climate change and adaptation. *Source: reproduced from reference 2.*

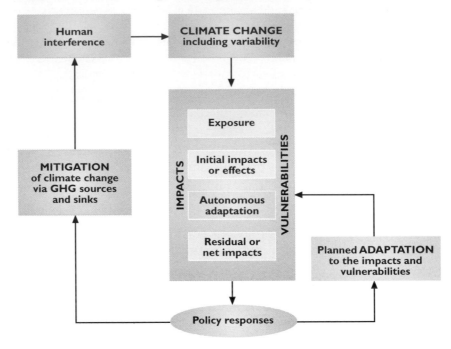

on the dynamics and processes of adaptation decision-making, including the roles and responsibilities of individuals, communities, nations, institutions and the private sector.

Adaptation and prevention

Many of the adaptive measures discussed in health impact and vulnerability assessments are not unique to climate change (*12, 13*). In fact, the IPCC identified rebuilding public health infrastructure as "the most important, cost-effective and urgently needed" adaptation strategy (*2*). Other measures endorsed by the IPCC include public health training programmes; more effective surveillance and emergency response systems; and sustainable prevention and control programmes. These measures are familiar to the public health community and needed regardless of whether or not climate changes: they constitute the basis of a "no-regrets" adaptation strategy.

Adaptive actions to reduce health impacts can be considered in terms of the conventional public health categories of primary, secondary, and tertiary prevention (*8, 13, 14*). Primary prevention refers to an intervention implemented before there is evidence of disease or injury: avoiding hazardous exposure, removing causative risk factors or protecting individuals so that exposure to the hazard is of no consequence. For example, bed nets can be supplied to populations at risk of exposure to malaria and early warning systems (e.g. extreme heat-health warnings, famine early warning) established to provide information on hazards and recommended actions to avoid or reduce risks. Primary prevention largely corresponds to anticipatory adaptation.

Secondary prevention involves intervention implemented after disease has begun, but before it is symptomatic (e.g. early detection or screening), and subsequent treatment that averts full progression to disease. Examples include enhancing monitoring and surveillance; improving disaster response and recovery; and strengthening the public health system's ability to respond quickly to disease outbreaks. Secondary prevention is analogous to reactive adaptation. Finally, tertiary prevention attempts to minimize the adverse effects of an already present disease or injury (e.g. better treatment of heat stroke, improved diagnosis of vector-borne diseases). As the adverse health outcome is not prevented, tertiary prevention is inherently reactive.

Climate-related adaptation strategies should not be considered in isolation of broader public health concerns such as population growth and demographic change; poverty; public health infrastructure; sanitation, availability of health care; nutrition; dangerous personal behaviours; misuse of antibiotics; pesticide resistance; and environmental degradation (12). All of these factors (and others) will influence the vulnerability of populations and the health impacts they experience, as well as possible adaptation strategies.

Coping with climate

Past and current climates have been, and are, variable. This variability is likely to continue with future climate change (1). In the popular literature, global climate change frequently is called 'global warming'" which focuses attention on average global temperature change (a frequently cited IPCC number is average surface air temperature change, projected to warm 1.4 to 5.8 °C over the next century). However, a change in climate actually occurs as changes in particular weather conditions, including extremes, in specific places. In many cases the meteorological variables of interest for public health are not averages and may not be confined to temperature alone. For example, a 2 °C increase in average summer temperature in a specific urban area could result in both higher maximum temperatures and an increase (potentially large) in the number of days over some temperature threshold (say 35 °C), each of which may be important for human health effects associated with heatwaves (15). In addition, changes in humidity and wind speed also may be important in terms of how people experience and are affected by these temperatures (16). Thus, adaptation to climate change necessarily includes adaptation to variability (3, 17, 18).

Given current climate variability, climate change adaptations which enhance a country's coping ability can be expected to yield both near-term benefits, as they enable countries to deal better with current variability, and the longer-term benefits of being able to deal better with future climate (19, 20, 21). Such no-regrets adaptations are likely to be especially important for less developed countries as they result in immediate benefits and are a useful first step in strengthening capacity to deal with future changes (22).

Many social and economic systems have evolved to accommodate the normal climate and some variation around this norm (see Figure 11.2). This evolution takes place in a dynamic social, economic, technological, biophysical and political context, which determines the coping ability of a region or country (2). Coping ability is defined here as the degree to which the public health system and individuals can deal successfully with health effects associated with current climate conditions, including climate variability. It therefore reflects autonomous and planned adaptations that have taken place over time and can be considered the adaptation baseline (i.e. what can be done now given current resources, technology, human capital, institutions, etc.). As shown in Figure 11.2, this ability to

FIGURE 11.2 Climate change and coping ability. The top panel depicts the situation of a country with a well-developed public health system, able to successfully deal with most variations in climate. The second panel depicts a country barely able to deal with the average climate, with little or no capacity to address climate variability. The third panel depicts a situation of increasing coping ability due to investments in adaptation, allowing the country to deal with future mean climate and variability. The last panel shows decreasing coping ability so that even future average climate poses health threats. *Source: adapted from reference 2.*

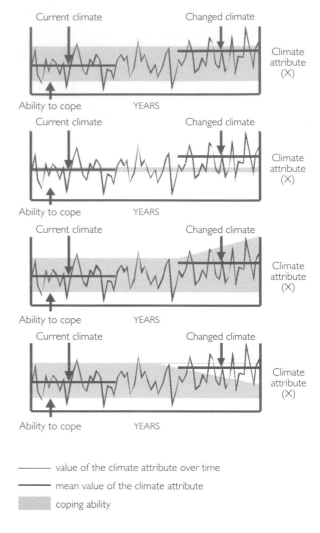

cope with climate varies from place to place and may change over time, both in response to changes in the factors noted above and to investments in specific adaptations to climate (2).

For example, it has been noted that extreme weather events can have "vastly different consequences for those on whom they infringe because of differences in coping ability (19). There is extensive evidence that an extreme climatic event will result in higher losses of life in a developing country than in a developed country because of limited coping ability (23, 24, 25, 26). Cyclones in Bangladesh in 1970 and 1991 are estimated to have caused 300 000 and 139 000 deaths respectively (27). In contrast, Hurricane Andrew caused 55 deaths when it struck the United States of America in 1992 (although estimated to have caused $30 billion in damages) (28). Similarly, 62 514 cases of dengue fever were reported in three Mexican states next to Texas (1980–1999) while only 64 cases were reported across the border in Texas itself (12). This large difference has been attributed in part to differences in living conditions, such as the presence of window screens and air conditioning.

Adaptive capacity

Adaptive capacity encompasses coping ability (i.e. what could be implemented now to deal with current climate and climate variability) and strategies, policies and measures that can expand future coping ability. Adaptive capacity is a theoretical construct because it is not possible to know with certainty whether a country will invest resources to expand its coping ability, how technology and other factors will change, or what adaptations actually will be implemented, until a perturbation or stress occurs. For example, access to clean water and adequate sanitation is part of the coping capacity for developed countries and some economies in transition but part of the adaptive capacity of less developed countries (currently not available but possible with investment in water treatment and sanitation facilities). While not certain, it is hoped that both clean water and sanitation will become part of the adaptation baseline for all countries.

Decisions about public health measures unrelated to climate change, such as sanitation and water treatment, may have a profound influence on health consequences associated with climate change. In fact, adaptation strategies frequently are described as risk management and public health programmes can be characterized as reducing climate change health risks (12, 29). Improved weather warning and preparedness systems, buildings and infrastructure, all can be considered measures to reduce human health risks in the event of a changed frequency of weather disasters. However, there is concern that the adaptive capacity to address changes in the magnitude or frequency of extreme climatic conditions may not be very high even though the adaptive capacity to gradual changes in climate may be relatively high (30).

Highly-managed systems, such as agriculture and water resources in developed countries, are thought to be more adaptable (assuming resources to adapt are available) than less-managed or natural ecosystems (20, 31, 32). Similarly, systems that have coped successfully with historical and/or existing stresses are expected to adapt well to stresses associated with future climate change (33). Both these premises assume that a country's coping ability is maintained or enhanced. Unfortunately, there are numerous examples in public health where this capability is not maintained once the health threat has been brought under control. Thirty years ago the threat of infectious diseases appeared to be decreasing due to advances in antibiotic drugs, vaccines, and chemical pesticides among other developments. Today, most health professionals agree that there has been a general resurgence of infectious diseases throughout the world, partly due to the deterioration of public health infrastructures worldwide (34, 35).

These types of simple assumptions concerning adaptability (i.e. highly managed systems are more adaptable, past successes are likely to continue into the future) have formed the basis for broad assessments of sensitivity and adaptability (36). Based on these factors, usually it is asserted that much can and will be done to reduce the impacts of climate change. However, it is not clear how much adaptation actually will take place given the number of uncertainties surrounding climate change adaptation. These include uncertainties about future climate (especially how extremes may change), potential effects and underlying determinants of adaptive capacity (i.e. how future institutions, technologies, skills, knowledge will evolve). In addition, there are many unknowns with respect to costs, feasibility, unintended consequences and effectiveness of adaptations (2).

Determinants of adaptive capacity

Research on adaptive capacity in climate change is very limited and is a key research need (2). However, substantial literature in other fields (economic development, sustainable development, resource management) can provide insights into the likely key determinants of adaptive capacity. These represent conditions that constrain or enhance adaptive capacity and hence the vulnerability of regions, nations and communities. Consideration of these determinants provides another pathway to the overarching goal of protecting and enhancing human health. The IPCC identified the main features of communities or regions that seem to determine their adaptive capacity: economic wealth, technology, information and skills, infrastructure, institutions and equity. In addition, for public health, the current health status and pre-existing disease burdens must be considered.

Economic resources

The economic status of nations, described in terms of GDP, financial capital, wealth, or some other economic measure, clearly is a determinant of adaptive capacity (37, 38). It is widely accepted that wealthy nations have a greater capacity to adapt because they have the economic resources to invest in adaptive measures and to bear the costs of adaptation (20, 39). It is also recognized that poverty is related directly to vulnerability (19, 40) and that the poorest groups in the poorest countries are the most vulnerable to health impacts of climate change (2).

Approximately one-fifth of the world's population lives on less than US $1 per day. Excluding the three WHO sub-regions with very low child and adult mortality,[1] a strong gradient of increasing child underweight with increasing absolute poverty was found in the remaining eleven (41). Unsafe water and sanitation and indoor air pollution also are associated with absolute poverty in these sub-regions.

The feasibility of adaptation options for many poor countries is constrained by a lack of resources (29). Table 11.1 provides estimates of expenditures on health for the world and by income groups and regions. In 1998 an average of I$523 per person was spent on health services.[2] This average varied significantly across both countries and regions, ranging from only I$82 per person in Africa to I$2078 in the OECD countries (42). Countries with low health expenditures also have poorer health status. The median health-adjusted life expectancy (HALE) in countries that spend less than I$200 per capita on health is 47.1 years (43). Adequate expenditure for health care and public health prevention programmes are fundamental needs for adaptation to climatic change (8).

Income growth, improved educational levels and consequent improvements in nutrition and sanitation have contributed to significant improvements in health and declines in mortality in the twentieth century (44, 45). The link between economic resources and health can be illustrated further by consider-

---

[1] The 191 Member States of the WHO have been divided into five mortality strata on the basis of their level of child and adult male mortality. The matrix defined by the six WHO Regions and the five mortality strata leads to 14 sub-regions, since not every mortality stratum is represented in every region. See WHO 2002 (41) for details.

[2] The "international dollar" (I$) estimates are converted from local currency at purchasing power parity rates.

CLIMATE CHANGE AND HUMAN HEALTH

TABLE 11.1 Health spending in 1998, by income groups and regions.

Income group/region	Total health expenditure (in millions of I$)	Per capita health expenditure (I$)	Share of GDP (%)
World	3072485	523	7.9
Income group			
<1000	9985	24	3.4
1000–2200	156438	88	4.5
2200–7000	518710	206	5.1
>7000	2387353	2042	9.6
Region			
Africa	50170	82	5.1
Americas	176223	438	7.0
Middle East	80932	176	4.8
Eastern Europe and Central Asia	99761	281	5.7
OECD	2317247	2078	9.7
South Asia	141262	95	4.4
Asia and Pacific	206891	142	4.4

Note: I$ are converted from local currency at purchasing power parity rates.
Source: reproduced from reference 42. See Poullier et al. for list of countries in region.

ing episodes of sharp economic downturns that reduce a country's resources to invest in public health. Combined with poor policy decisions and implementation, adverse economic conditions led to reductions in health expenditures in many developing countries and countries of the former Union of Soviet Socialist Republics in the 1980s (46).

Research also has begun to reveal the linkage between health and economic growth. While not definitive, this research consistently finds strong relationships between health, as measured by health indicators such as survival rates and life expectancy, and income levels or economic growth rates (47, 48). Simultaneous impacts of health on wealth and vice versa have been found (49). Improvements in health affect economic growth directly—"healthier people are more productive"—and indirectly, through effects on demography (45). Research on the effects of geography and climate on income suggests that the interaction of tropical climate and diseases, particularly malaria, can significantly affect economic performance (50). Another study found that lagging health and agriculture technology in the tropics opened a substantial income gap between climate zones (51). Other researchers found that the much route by which "tropics, germs, and crops" affect economic development is through institutions, rather than by directly affecting countries' incomes (52).

Technology

Advances in technology, such as new drugs or diagnostic equipment, can increase substantially our ability to solve health problems (53). More generally, the availability and access to technology at the individual, local, and national levels, in key sectors (e.g. agriculture, water resources, health) is an important determinant of adaptive capacity (2). Many of the adaptive strategies that protect human health involve technology (e.g. warning systems, air conditioning, pollution controls, housing, storm shelters, vector control, vaccination, water treatment and sanitation). While much of this technology is well established (water treatment, sanitation), some is relatively new and still being disseminated (well-equipped mobile laboratories, computer information and reporting systems which can support disease surveillance). In other cases there is a need for new technologies (new vaccines and pesticides) to enhance the ability to cope with a changing climate. Countries that are open to the use of technologies and can develop and disseminate new technologies have enhanced adaptive capacity (2).

It is important to assess in advance any risks to health from proposed technological adaptations (11, 13, 22). Increased use of air conditioning would protect against heat stress but could increase emissions of both greenhouse gases and conventional air pollutants. Similarly, if new pesticides are used to control disease vectors their effects on human health, insect predators, and increased insect resistance to pesticides all need to be considered (11, 12). New chemicals or treatments for vector control must be effective but their breakdown products should be non-toxic and non-persistent. The migration of potentially hazardous compounds into air and water should be avoided.

Information and skills

In general, countries with higher levels of "human capital" or knowledge are considered to have greater adaptive capacity (54). The UN reports that more than 850 million people in developing countries are illiterate and about 90 million children worldwide are denied any schooling, raising concerns about their vulnerability to a range of problems (55). Table 11.2 provides estimates of illiteracy rates by income group and region. Illiteracy, as well as poverty, has been listed as a key determinant of low adaptive capacity in north-east Brazil (56). As many adaptive measures involve implementation of effective health education programmes, a high level of illiteracy can seriously compromise their effectiveness. Some of the simple, low-cost, low-technology measures to reduce health effects (e.g. using sari cloth to filter drinking water, removing containers around dwellings that provide habitat for disease vectors) involve educating the public on the feasibility and effectiveness of such measures.

TABLE 11.2 Illiteracy rates in 2000, by income groups and regions.

Income group/region	Illiteracy rate, adult male (% males ages 15 and above)	Illiteracy rate, adult female (% females ages 15 and above)
World	17.02	30.31
Income group		
Low income	28.29	46.85
Lower middle income	9.22	21.47
Middle income	9.17	19.51
Upper middle income	8.95	10.82
High income	. . .	. . .
Region		
East Asia and Pacific	8.09	21.16
Europe and Central Asia	1.38	4.05
Latin America and Caribbean	10.7	12.5
Middle East and North Africa	24.85	45.96
South Asia	33.85	57.3
Sub-Saharan Africa	30.14	46.66

Source: reproduced from reference (57). See the World Development Indicators database of list of countries in region.

Lack of trained and skilled personnel may restrict a nation's ability to implement adaptation measures (11). Health systems in particular are labour intensive and require qualified and experienced staff to function well (53). Health "human capital" can be increased through investment in education and training. Human capital does not deteriorate with use, but can depreciate as old skills become obsolete with the advent of new knowledge, methods, and technologies (53). In addition, individuals' retirement and/or deaths result in the loss of their skills and accumulated knowledge.

Effective adaptation will require individuals skilled at recognizing, reporting and responding to health threats associated with climate change. Researchers trained in epidemiology and laboratory research will be needed to provide a sound basis for surveillance and response. Social scientists can contribute to an understanding of social behaviours and demographics as they relate to causes and control of diseases. Skilled public health managers, who understand surveillance and diagnostic information, will be needed to mobilize the appropriate response. People trained in the operation, quality control and maintenance of public health infrastructure, including laboratory equipment, communications equipment, and sanitation, wastewater, and water supply systems also are required (53).

Infrastructure

Adaptive capacity is likely to vary with the level of a country's infrastructure (2) (see Table 11.3 for estimates of access to water supply and sanitation infrastructure). Adaptive responses to health impacts of climate change are enhanced by infrastructures specifically designed to reduce vulnerability to climate variability (e.g. flood control structures, air conditioning, building insulation, stringent building codes, etc.) and general public health infrastructures (e.g. sanitation facilities, waste water treatment, water supply systems, laboratory buildings).

Infrastructure such as roads, rails and bridges, water systems and drainage, mass transit and buildings can reduce vulnerability to climate change (2). It also has the potential to be adversely impacted (especially if immovable), which can increase vulnerability to climate change. Flooding can overwhelm sanitation infrastructure and lead to water-related illnesses (8, 12). After Hurricane Mitch hit Central America, severe damage to the transportation infrastructure made it more difficult to assist affected populations (2). Similarly, damage to transporta-

TABLE 11.3 Water supply and sanitation coverage in 2000 by region (% of population served).

Region	Water Supply			Sanitation		
	Total	Urban	Rural	Total	Urban	Rural
World	82	94	71	60	86	38
Region						
Africa	62	85	47	60	84	45
Asia	81	93	75	48	78	31
Latin America and the Caribbean	85	93	62	78	87	49
Oceania	88	98	63	93	99	81
Europe	96	100	87	92	99	74
Northern America	100	100	100	100	100	100

Source: reproduced from reference 58. See WHO and UNICEF, 2000 for list of countries in region.

tion infrastructure during the east Africa floods of 1998 and Mozambique floods in 2000 hampered relief efforts (2).

Institutions

Social institutions are considered an important determinant of adaptive capacity. Those countries with less-effective institutional arrangements, commonly developing nations and those in transition, have a lower capacity to adapt than countries with well-established and effective institutions (2). Inadequate institutional support frequently is cited as a hindrance to adaptation. Institutional deficiencies and managerial weaknesses are cited as contributing to Bangladesh's vulnerability to climate change (2). The Democratic People's Republic of Korea experienced strong storms with torrential rain in 1995 and 1996, followed by droughts in 1997 and 1998. Estimates of deaths from famine since 1995 range from 220 000 to 2 million (59). While there were widespread crop failures, the agricultural system's inability to meet the needs of the people is not new. Other features of this society including economic isolation, lack of reserves, highly centralized arrangements for storage and redistribution of foods, lack of variety in agricultural practice and a strictly hierarchical political system exacerbated the situation (60). The IPCC (2) cites "institutional inertia" in the Asia region as limiting investment in environmental protection and increasing climate risks. Inconsistent and unstable agricultural policies have increased the vulnerability of food production in Latin America (2). Political upheavals in African countries, with the accompanying political and economic instability, constrain the implementation of adaptation measures (2). Health systems should offer protection against disease, because of economic and social crises these have (in extreme cases) either collapsed or not been built in many countries (35).

The complex interaction of issues expected with climate change will require new arrangements and collaborations between institutions to address risks effectively, thereby enhancing adaptive capacity. The Environmental Risk Management Authority in New Zealand involves collaboration between the health, forestry, environment and conservation sectors (8). Similarly, nations and international organizations such as WHO can cooperate in coordinating surveillance and response activities to address disease threats more effectively (34).

Finally, increased collaboration between the public and private sectors can enhance adaptive capacity. The Medicines for Malaria venture—a joint initiative by the public and private sectors to develop new antimalarial drugs—is developing new products for use in developing countries (53). Another example is drug donations by industry to help eliminate infectious diseases in developing countries (53). Such collaborations have the potential substantially to reduce health impacts associated with climate change.

Equity

Frequently it is argued that adaptive capacity will be greater if access to resources within a community, nation, or Earth is distributed equitably (2, 61, 62). Universal access to quality services is a bedrock principle of public health. However, while many have broad and advanced access to health care, many have been denied access. WHO estimates that the developing world carries 90% of the disease burden, yet poorer countries have access to only 10% of the resources for health (53).

Demographic variables such as age, gender, ethnicity, educational attainment and health often are cited in the literature as related to the ability to cope with risk (11, 37, 63). WHO notes that for a large group of people long-term unemployment results in exclusion from the mainstream of development and society (64). The combination of homelessness and lack of access to financial resources and infrastructure restricts adaptation options (2). Similar conclusions about the marginalization of minority groups have been drawn (65).

Health status and pre-existing disease burdens

Population well-being is an important ingredient and determinant of adaptive capacity. Table 11.4 provides estimates of deaths and disability-adjusted life years for 2001 (41). Great progress has been achieved in public health, particularly through the improvement of drinking water and sanitation; development of national health systems; introduction of antibiotics and mass immunization; and the improvement of nutrition. Yet 170 million children in poor countries are underweight: over 3 million of them die each year as a result. The African region remains the region most affected by infectious and parasitic diseases. Malaria, HIV/AIDS, childhood vaccine preventable diseases and diarrhoea represent the highest estimated deaths in Africa. Malaria is estimated to have caused 963000 deaths and the loss of around 36 million years of "healthy" life in this region in 2001.

Non-communicable diseases, in particular cardiovascular diseases, represent the highest mortality in countries with very low child and adult mortality. However, they are now becoming more prevalent in developing nations, where they create a double burden of disease—a combination of long-established infectious diseases and increasing chronic, noncommunicable diseases. For example, countries in south-east Asia with high child and adult mortality show an estimated 2.7 million deaths from infectious and parasitic diseases, 3.2 million from cardiovascular diseases.

Research needs

Few studies of climate change and health go beyond identifying adaptation options that might be possible and/or incorporating simplified representations of adaptive responses. It is important to improve understanding of the process of adaptation. This includes gaining better knowledge of the processes of adaptation decision-making; roles and responsibilities in adaptation of individuals, communities, nations, institutions and the private sector; conditions that stimulate or act as a barrier to adaptation; and what level of certainty is needed for public health decision-makers to act.

Research on barriers and opportunities for enhancing adaptive capacity in order to protect human health, as well as potential interactions with ongoing development projects and programmes, also is a key research need.

Determining the benefits of health adaptations—reduction in the effects of climate change on health—is likely to be complex and controversial. In general, there is little scientific literature on the population burden of disease attributable to current or future climate change (2) (but see chapter seven), and much less on the economic and non-economic valuation of those effects. There is therefore little basis for making aggregate estimates of the costs avoided (or benefits gained) by successful adaptation measures to compare with the costs of such

TABLE 11.4 Deaths and burden of disease in DALYs in 2001, by mortality stratum in WHO Regions (in thousands).

Region and mortality stratum	Deaths				Disability-adjusted Life Years (DALYs)			
	Total	Communicable	Noncommunicable	Injuries	Total	Communicable	Noncommunicable	Injuries
World	56554	18374	33077	5103	1467257	615737	672865	178656
Region								
Africa								
High child and adult	4365	2968	1098	298	147899	105097	30030	12771
High child, very high adult	6316	4615	1264	437	209985	156359	36075	17551
The Americas								
Very low child and adult	2747	172	2400	176	46520	3250	38642	4628
Low child and adult	2619	485	1810	324	81270	17105	50328	13837
High child and adult	544	198	288	57	17427	6761	8432	2235
Eastern Mediterranean								
Low child and adult	707	126	475	106	23007	5691	13282	4034
High child and adult	3449	1700	1454	295	113214	61446	39329	12439
Europe								
Very low child and adult	4076	236	3643	197	53075	2579	46259	4237
Low child and adult	1969	193	1664	113	38936	7029	27473	4434
Low child, high adult	3658	157	3048	453	59212	4999	42170	12042
South-east Asia								
Low child and adult	2194	644	1275	275	61290	20403	31866	9021
High child and adult	12273	5171	5913	1188	357554	167749	144703	45102
Western Pacific								
Very low child and adult	1161	138	939	84	16430	1064	13720	1646
Low child and adult	10475	1572	7805	1098	241438	56205	150556	34677

Source: reproduced from reference 41. See WHO, 2002 for definitions of mortality stratum and list of countries in sub-region.

measures. Information on costs and benefits of adaptation measures is likely to be important for decision-makers.

Finally, integrated frameworks will be needed to evaluate adaptation options. Designing such frameworks is challenging because they must organize information from a wide array of sources, including: epidemiological studies; models that incorporate identified exposure-response relationships to project future health outcomes; adaptation options; cost and effectiveness analyses; monitoring and evaluation systems; and feedback from decision-makers.

Conclusions

Building capacity is an essential step in preparing adaptation strategies. Education, awareness raising and the creation of legal frameworks, institutions and an environment that enables people to take well-informed, long-term, sustainable decisions, all are needed. Building adaptive capacity in public health will require a forward-looking, strong and unifying vision of health care in the twenty-first century as well as an understanding of the problems posed by climate change. However, it must be stressed that adapting to climate change will require more than financial resources, technology, and public health infrastructure. Human resources and knowledge are critical; institutions that are committed to and support the goals articulated in "Health for All by the Year 2000" are essential. The public health community is in a key position to help nations face the health challenges associated with climate change.

References

1. IPCC. 2001a. *Climate change 2001: the scientific basis.* Contribution of Working Group I to the Third Assessment Report of the Intergovernmental Panel on Climate Change. Cambridge, UK, Cambridge University Press, 2001.
2. IPCC. 2001b. *Climate change 2001: impacts, adaptation, and vulnerability.* Contribution of Working Group II to the Third Assessment Report of the Intergovernmental Panel on Climate Change. Cambridge, UK, Cambridge University Press, 2001.
3. Smit, B. et al. An anatomy of adaptation to climate change and variability. *Climatic Change* 45: 223–251 (2000).
4. Burton, I. *Adapt and thrive.* Unpublished manuscript, Downsview, Ontario, Canadian Climate Center, 1992.
5. Smith, J.B. et al. eds. *Adapting to climate change: an international perspective.* New York, USA, Springer-Verlag, 1996.
6. Smit, B. *Adaptation to climate variability and change.* Report of the Task Force on Climate Adaptation. Canadian Climate Program Board Department of Geography, University of Guelph, Occasional Paper No. 19. Downsview, Ontario, Environment Canada, 1993.
7. Stakhiv, E. *Evaluation of IPCC adaptation strategies.* Draft Report. Fort Belvoir, VA, USA, Institute for Water Resources, US Army Corps of Engineers, 1993.
8. Kovats, R.S. et al. *Climate change and human health: impact and adaptation.* Geneva, Switzerland & Rome, Italy, WHO European Centre for Environment and Health, 2000 (document WHO/SDE/OEH/00.4).
9. UNFCCC. Text of the United Nations Framework Convention on Climate Change. 1992. http://unfccc.int/resource/docs/convkp/conveng.pdf.
10. Wigley, T.M.L. *The science of climate change: global and US perspectives.* Arlington VA, USA, Pew Center on Global Climate Change, 1999. http://www.pewclimate.org/projects/env_science.cfm.

11. Scheraga, J. & Grambsch, A. Risks, opportunities, and adaptation to climate change. *Climate Research* 10: 85–95 (1998).
12. Patz, J.A. et al. The potential health impacts of climate variability and change for the United States: executive summary of the report of the health sector of the US national assessment. *Environmental Health Perspectives* 108: 4 (2000).
13. Patz, J.A. Health adaptation to climate change: need for far-sighted, integrated approaches. In: *Adapting to climate change: an international perspective.* Smith, J. et al. eds. New York, USA, Springer-Verlag, pp. 450–464, 1996.
14. McMichael, A.J. & Kovats, R.S. Climate change and climate variability: adaptations to reduce adverse health impacts. *Environmental Monitoring and Assessment* 61: 49–64 (2000).
15. National Climatic Data Center (NCDC). *Probabilities of temperature extremes in the USA:* Version 1. Asheville, NC, USA, National Climatic Data Center, 1999.
16. Kalkstein, L.S. & Greene, J.S. An evaluation of climate/mortality relationships in large US cities and the possible impacts of a climate change. *Environmental Health Perspectives* 105: 84–93 (1997).
17. Downing, T.E. et al. *Climate change and extreme events: altered risks, socio-economic impacts and policy responses.* Amsterdam, The Netherlands, Vrije Universiteit, 1996.
18. Smithers, J. & Smit, B. Human adaptation to climatic variability and change. *Global Environmental Change* 7(2): 129–146 (1997).
19. Rayner, S. & Malone, E.L. eds. *Human choice and climate change volume 3: the tools for policy analysis.* Columbus, OH, USA, Battelle Press, 1998.
20. Burton, I. The growth of adaptation capacity: practice and policy. In: *Adapting to climate change: an international perspective.* Smith, J. et al. eds. New York, USA, Springer-Verlag, pp. 55–67, 1996.
21. Ominde, S.H. & Juma, C. Stemming the tide: an action agenda. In: *A change in the weather: African perspectives on climate change.* Ominde, S.H. & Juma, C. eds. Nairobi, Kenya, ACTS Press, pp. 125–153, 1991.
22. Klein, R.J.T. & Tol, R.S.J. *Adaptation to climate change: options and technologies, an overview paper.* Bonn, Germany. United Nations Framework Convention on Climate Change Secretariat, 1997 (Technical Paper FCCC/TP/1997/3 p. 33). http://www.unfccc.int/resource/docs/tp/tp3.pdf.
23. Burton, I. et al. *The environment as hazard.* New York, USA, The Guilford Press, 1993.
24. Blaikie, P. et al. *At risk: natural hazards, people's vulnerability, and disasters.* New York, USA, Routledge, 1994.
25. Kundzewicz, Z. & Takeuchi, K. Flood protection and management: quo vadimus? *Hydrological Sciences* 44(3): 417–432 (1999).
26. Noji, E.K. ed. *The public health consequences of disasters.* New York, USA, Oxford University Press, 1997.
27. NOAA. NOAA releases century's top weather, water, and climate events. 1999. http://www.noaanews.noaa.gov/stories/s334b.htm.
28. US Centers for Disease Control (CDC). Rapid health needs assessment following Hurricane Andrew—Florida and Louisiana, 1992. *Morbidity and Mortality Weekly Report* 41(37): 685 (1992).
29. McMichael, A.J. et al. eds. *Climate change and human health.* An assessment prepared by a task group on behalf of the World Health Organization, the World Meteorological Organization, and the United Nations Environment Programme, Geneva, Switzerland, 1996 (document WHO/EHG/96.7).
30. Appendi, K. & Liverman, D. Agricultural policy and climate change in Mexico. In: *Climate change and world food security.* Downing, T.E. ed. Berlin, Germany, Springer-Verlag, 1996.
31. Strzepek, K.M. & Smith, J.B. eds. *As climate changes, international impacts and implications.* Cambridge, UK & New York, USA, Cambridge University Press, 1995.

32. Toman, M. & Bierbaum, R. An overview of adaptation to climate change. In: *Adapting to climate change: an international perspective*. Smith, J. et al. eds. New York, USA, Springer-Verlag, 1996.

33. Ausubel, J. A second look at the impacts of climate change. *American Scientist* 79: 211–221 (1991).

34. Committee on International Science, Engineering, and Technology (CISET). *Infectious disease—a global health threat*. Report of the NSTC CISET working group on emerging and re-emerging infectious diseases, 1995. http://www.ostp.gov/CISET/html/ciset.html.

35. World Health Organization (WHO). *World health report 1996: fighting disease, fostering development*. Geneva, Switzerland, World Health Organization, 1996.

36. US National Academies of Science and Engineering (USNAS). *Policy implications of greenhouse warming: mitigation, adaptation, and the science base*. Institute of Medicine, Washington, DC, National Academy Press, 1992.

37. Burton, I. et al. Adaptation to climate change: theory and assessment. In: *Handbook on methods for climate change impact assessment and adaptation strategies*. Feenstra, J.F. et al. eds. Free University of Amsterdam, United Nations Environment Programme and Institute for Environmental Studies, pp. 5.1–5.20 1998.

38. Kates, R.W. Cautionary tales: adaptation and the global poor. *Climatic Change* 45(1): 5–17 (2000).

39. Goklany, I.M. Strategies to enhance adaptability: technological change, sustainable growth and free trade. *Climatic Change* 30: 427–449 (1995).

40. World Health Organization (WHO). *World health report 1997: conquering suffering, enriching humanity*. Geneva, Switzerland, World Health Organization, 1997.

41. World Health Organization (WHO). *World health report 2002: reducing risks, promoting healthy life*. Geneva, Switzerland, World Health Organization, 2002.

42. Poullier, J-P. et al. *Patterns of global health expenditures: results for 191 countries*. Geneva, Switzerland, World Health Organization, 2002. (EIP/HFS/FAR Discussion Paper No. 51). http://www3.who.int/whosis/discussion_papers/pdf/paper51.pdf.

43 World Health Organization (WHO). World Health Report 2001: Mental Health – New Understanding New Hope. Geneva, Switzerland, World Health Organization, 2001.

44. World Health Organization (WHO). *World health report 1998: life in the 21st century: a vision for all*. Geneva, Switzerland, World Health Organization, 1998.

45. World Health Organization (WHO). *World health report 1999: making a difference*. Geneva, Switzerland, World Health Organization, 1999.

46. Evlo, K. & Carrin, G. Finance for health care: part of a broad canvas. *World Health Forum* 13: 165–170 (1992).

47. Fogel, R.W. New findings on secular trends in nutrition and mortality: some implications for population theory. In: *Handbook of population and family economics*, Rosenzweig, M.R. & Stark, O. eds. Vol.1A: 433–481. Amsterdam, The Netherlands, Elsevier Science, 1997.

48. Jamison, D.T. et al. Health's contribution to economic growth, 1965–90. In: *Health, health policy and economic outcomes*. Geneva, WHO Director-General's Transition Team, (Health and Development Satellite, Final Report): 61–80, 1998.

49. Hamoudi, A. & Sachs, J.D. *Economic consequences of health status: a review of the evidence*. Cambridge, MA, USA, Harvard Institute for International Development, 1999.

50. Gallup, J.L. & Sachs, J.D. *The economic burden of malaria*. Cambridge, MA, USA, Harvard Institute for International Development, 2000.

51. Sachs, J.D. *Tropical underdevelopment*. Cambridge, MA, USA, National Bureau of Economic Research, 2001 (Working Paper No w8119).

52. Easterly, W. & Levine, R. *Tropics, germs, and crops: how endowments influence economic development*. Cambridge, MA, USA, National Bureau of Economic Research, 2002 (Working Paper No 9106).

53. World Health Organization (WHO). *World health report 2000: health systems: improving performance.* Geneva, Switzerland, World Health Organization, 2000.

54. Smith, J.B. & Lenhart, S.S. Climate change adaptation policy options. *Climate Research* 6(2): 193–201 (1996).

55. United Nations Development Program (UNDP). *Human development report 2000: human rights and human development.* New York, USA, Oxford University Press, 2000.

56. Magalhães, A.R. Adapting to climate variations in developing regions: a planning framework. In: *Adapting to climate change: an international perspective.* Smith, J. et al. eds. New York, USA, Springer-Verlag, 1996.

57. World Bank. *2002 World development indicators* database. Available online at www.worldbank.org/data/onlinedatabases/onlinedatabases.html.

58. WHO and UNICEF. *Global water supply and sanitation assessment 2000 report.* Geneva, Switzerland, World Health Organization, 2000.

59. UNICEF *Alert: North Korea 2002.* http://www.unicefusa.org/alert/emergency/nk/nk.html.

60. International Federation of Red Cross and Red Crescent Societies (IFRC) *World disaster report 1996.* Oxford, UK, Oxford University Press, 1996.

61. Toth, F. Development, equity and sustainability concerns in climate change decisions. In: *Climate change and its linkages with development, equity and sustainability.* Proceedings of the IPCC Expert Meeting held in Colombo, Sri Lanka, 27–29 April, 1999. Munasinghe, M. & Swart, R. eds. Colombo, Sri Lanka, LIFE; Bilthoven, The Netherlands, RIVM; and Washington D.C., USA, World Bank, pp. 263–288, 1999.

62. Rayner, S. & Malone, E.L. Climate change, poverty and intragenerational equity: the national level. In: *Climate change and its linkages with development, equity and sustainability.* Proceedings of the IPCC Expert Meeting held in Colombo, Sri Lanka, 27–29 April, 1999. Munasinghe, M. & Swart, R. eds. Colombo, Sri Lanka, LIFE; Bilthoven, The Netherlands, RIVM & Washington DC, USA, World Bank, pp. 215–242, 1999.

63. Chan, N. & Parker, D. Response to dynamic flood hazard factors in peninsular Malaysia. *The Geographic Journal* 162(3): 313–325 (1996).

64. World Health Organization (WHO). *World health report 1995: bridging the gaps.* Geneva, Switzerland, World Health Organization, 1995.

65. Bolin, R. & Stanford, L. Shelter, housing and recovery: a comparison of US disasters. *Disasters* 15(1): 24–34 (1991).

CHAPTER 12

From science to policy: developing responses to climate change

J.D. Scheraga,[1,2] K.L. Ebi,[3] J. Furlow,[1,2] A.R. Moreno[4]

Introduction

Climate change poses risks to human health, ecosystems, social and cultural systems, and economic development. It also provides opportunities. The goals of climate policy should be to reduce the risks and take advantage of the opportunities.

Adapting to the potential effects of climate change is a complex and ongoing process requiring actions by individuals, communities, governments and international agencies. In order to make informed decisions, policymakers will need timely and useful information about the possible consequences of climate change, people's perceptions of whether the consequences are positive or negative,[4] available adaptation options, and the benefits of slowing the rate of climate change. The challenge for the assessment community is to provide this information. Assessments must be made of the potential consequences of climate change, as well as of opportunities to adapt in order to reduce the risks, or take advantage of the opportunities, presented by change.

A policy-focused assessment[5] is an ongoing process designed to provide timely and useful information to decision-makers (1). Assessment products that address decision-makers' specific questions must be produced so that they coincide with decision points. Decisions will be made, including decisions to do nothing, whether or not the science is complete and the scientific community is prepared to provide input (2). Decisions made today might influence future opportunities to mitigate greenhouse gas emissions and to adapt to a changing climate.

Once policymakers have received information from the assessment community, their challenge is to integrate this information (where appropriate) to

---

[1] Global Change Research Program, US Environmental Protection Agency, Office of Research and Development, Washington, DC, USA.

[2] The views repressed are the author's own and do not reflect official USEPA policy.

[3] World Health Organization Regional Office for Europe, European Centre for Environment and Health, Rome, Italy.

[4] The United States-Mexico Foundation for Science, Mexico D.F., Mexico.

[5] Positive impacts from one person's perspective may be negative from another's.

[6] Scheraga and Furlow note that the term "policy-relevant" is more commonly used. "Policy-relevant" is defined as an iterative process that engages both analysts and end-users to evaluate and interpret the interactions of dynamic physical, biological, and social systems and communicate useful insights in a timely fashion. They choose to refer to this type of assessment as "policy-focused" because many types of research and assessment activities may, at some time, be relevant for policy, but are not focused on answering specific questions being asked by policy makers by a specific point in time. Policy-focused assessments are intended to inform in a timely fashion decision-makers asking specific questions. The terminology is intended to distinguish between these different types of assessment activities.

develop a policy portfolio consisting of a sensible mix of approaches for reducing the risks and taking advantage of the opportunities presented by climate change. An array of response options exists, including actions to mitigate greenhouse gas emissions in order to slow the rate of climate change; actions to adapt to a changing climate in order to increase society's resilience to change; communication strategies to increase public awareness of climate change; investments in monitoring and surveillance systems; and investments in research to reduce key policy-relevant uncertainties and increase understanding of the potential effects of climate change.

The challenge for decision-makers is complicated by the fact that climate change is only one of many factors that influence human health, ecosystems and social well-being. Climate change cannot be considered in isolation from other important stresses. Earth's ecological, socioeconomic and climate systems are closely linked (3). Human health is affected by a variety of social, political, economic, environmental, technological and demographic factors, including:

- urbanization
- level of economic development and availability of wealth (e.g. funds available for research, sanitation, surveillance and monitoring)
- technological innovation (can pose its own environmental and health risks)
- scientific breakthroughs
- individual behaviour
- environmental conditions (e.g. air or water quality).

Any public health programme should consider these factors holistically to identify the important stressors and the most effective remedial mechanisms. The resources available to the public health community need to be shared among a variety of public health problems (and other problems of concern to society), therefore the ideal situation is to direct resources to their highest valued uses for the greatest public good. Often it is said that public health resources are scarce, in the sense that a resource is scarce if it is desired but limited.

"Highest valued" is a value-laden concept and therefore a social choice, not a scientific decision. It may include equity considerations (e.g. a decision that leads to differential health impacts among different demographic groups), efficiency considerations (e.g. targeting those programmes that will save the greatest number of lives) and political feasibility. It may also include ethics considerations, which are particularly important in public health. The practice of public health is guided by four ethical principles (4): respect for autonomy, nonmaleficence, justice and beneficence. Beneficence, the principle of doing good, is the dominant ethical principle of public health—but may not be the dominant value for decision-makers faced with competing demands. Respect for autonomy means concern about human dignity and freedom, and the rights of individuals to make choices and decisions for themselves, rather than others deciding for them. Nonmaleficence is the principle of not harming, derived from the ancient medical maxim: first, do no harm. Justice in this sense means social justice—fairness, equity and impartiality.

Public health frequently is faced with situations where it is not possible to uphold each of these principles. For example, a decision to quarantine a group of individuals to prevent the spread of a communicable disease emphasizes beneficence over the respect for autonomy and justice. Public health places the need to protect society as a higher imperative than the rights of an individual. Public health practitioners believe that as long as the community perspective (driven

CLIMATE CHANGE AND HUMAN HEALTH

by shared customs, institutions and values) and the perspective of a common good (including the desire to reduce disease, save lives and promote good health) remain intact, actions can be taken that limit individual autonomy and justice. Nevertheless, there isn't always consensus—even among public health officials—on the best way to promote the public's health and how to weigh individual liberties against the welfare of the whole.

Decision-makers outside the public health community are faced with similar challenges. Climate change affects multiple sectors and resources including (but not limited to) agriculture, forestry, water resources, air quality, ecosystems and biodiversity, and cultural resources. Stakeholders with an interest in each of these may have conflicting desires and conflict resolution is likely to be required. Policy-makers dealing with multiple social objectives (e.g. elimination of poverty, support for agriculture, promotion of economic growth, protection of cultural resources) and competing stakeholder desires must make difficult choices as they allocate scarce human and financial resources. For this reason, the Intergovernmental Panel on Climate Change (IPCC) suggested that it is helpful to view climate change as part of the larger challenge of sustainable development (5). Climate policies, including those intended to protect public health, can be more effective when consistently embedded within broader strategies designed to make national and regional development paths more sustainable. The impact of climate variability and change, climate policy responses, and associated socioeconomic development will affect the ability of communities to achieve sustainable development goals. Conversely, the pursuit of sustainable development goals will affect the opportunities for, and success of, climate policies.

This chapter addresses the health impacts assessment community's challenge to provide timely and useful information to decision-makers. We discuss the notion of policy-focused assessment and demonstrate how it can be a bridge between the research community and decision-makers. The key characteristics of a successful policy-focused assessment are identified. Case studies are presented to illustrate how actions and policy decisions already have been informed in a timely and useful way by policy-focused assessments.

A theme that pervades the entire chapter is that the existence of scientific uncertainty about climate change and its potential health impacts does not preclude policy-makers from taking actions now in anticipation of the health impacts of climate change. In fact, decisions are made every day where uncertainties about factors unrelated to climate are as high or higher than those associated with climate change (e.g. the financial markets). While there is uncertainty about some future climatic changes, also there are some for which there is relative certainty (5). The process of policy-focused assessment receives increased attention because of the need, the imperative, for scientists to provide timely insights to risk managers who must make decisions every day despite the existence of scientific uncertainties. A community building a drinking water plant in a coastal zone today cannot wait for projections of future sea level rise to be perfected. Decisions about plant location and investments in water purification equipment that may have a long lifetime will be made despite the existence of uncertainties. Policy-focused assessments analyse the best available scientific and socioeconomic information to answer questions posed by risk managers. They characterize and, if possible, quantify remaining scientific uncertainties and explain the potential implications of the uncertainties for the outcomes of concern to the decision-makers.

Ultimately, it is society's choice to decide whether a perceived risk (however large or small) warrants action. The existence of scientific uncertainties, by itself, is not an excuse for delay or inaction on the part of decision-makers.

Boundaries between assessment and policy formation

Only society can decide whether or not particular risks and opportunities are of concern. The extent to which it is willing to expend resources to avoid the effects of climate change will depend in part on its perceptions of the risks posed, perceived costs of the effort and how much it is willing to risk possible negative consequences (6, 7). Strategies to cope with or take advantage of climate change or to mitigate greenhouse gas emissions compete for scarce resources that could be used to address other societal problems (e.g. health care and poverty) that will be of varying degrees of concern to different individuals and groups.

Society is composed of individuals and groups that often have very different interests, motivations and levels of access to levers of power. The process by which society makes decisions to form and implement policy is complex. For this reason, care must be taken to respect the boundary between assessment and policy formation. Policy-focused assessment is an ongoing process, the goal of which is to inform policy decisions in a timely fashion. However an assessment should not make specific policy recommendations, unless commissioned by a deliberative, decision-making body that asks explicitly for these. In order to maintain the credibility, usefulness and effectiveness of the assessment process, a policy-focused assessment must be divorced from the actual process by which policy decisions are made. The process must be not only apolitical and unbiased, but also perceived as such by all parties participating in the assessment process, including decision-makers relying upon the results of the assessment.

It is the job of policy-makers, not researchers, to decide which criteria should be used. Such decisions are informed by the science but include other considerations that reflect societal values (e.g. equity considerations) and other factors affecting decision-making (e.g. political feasibility).

Decision-making criteria

Many criteria and approaches exist for making decisions under uncertainty. One criterion that increasingly has been included in national legislations and international treaties is the precautionary principle. This is a risk management policy that is applied when a potentially serious risk exists but with significant scientific uncertainty (8). Like other approaches, it requires that decision-makers take into account the likelihood of a particular health hazard and the nature and scale of the consequences should it occur. The precautionary principle allows some risks to be deemed unacceptable not because they have a high probability of occurring but because there may be severe or irreversible consequences if they do. This was featured in the 1992 Rio Declaration on Environment and Development as principle 15 (9):

> "In order to protect the environment, the precautionary approach shall be widely applied by States according to their capabilities. Where there are threats of serious or irreversible damage, lack of full scientific certainty shall not be used as a reason for postponing cost-effective measures to prevent environmental degradation."

The precautionary principle is one approach to risk management. There are other criteria and approaches for making decisions under uncertainty. One example is the "benefit-cost" criterion that weights the expected benefits and expected costs of an action taken to address an uncertain health risk. The benefit-cost criterion is straightforward: if the winners under a proposed policy value the change more highly than do the losers, it is said that the policy has positive net benefits and enhances efficiency (10). In calculating net benefits, allowance must be made for the possibility that benefits and costs may not accrue instantaneously, but over time. Discounting is the process by which measured future economic effects are transformed into their present values and compared. The weight given to benefits and costs at different points in time is critical and determined by the discount rate that is used. The evaluation of projects and their relative rankings according to a benefit-cost criterion is highly sensitive to the discount rate, particularly when the time-paths of benefits and costs are markedly different.

There are questions about how benefits and costs should be measured and how they should be compared among different societies. Often, benefits and costs must be measured using different metrics. Benefits frequently are measured in non-monetary terms (e.g. fewer hospital visits, fewer days of work/school missed, QALYs, DALYs) while costs are measured in dollar terms (e.g. the price of a smoke-stack or catalytic converter). These differences in metrics can lead to debate.

The benefit-cost criterion emphasizes the efficient use of scarce resources. It is not, however, a criterion for dealing with equity. As environmental issues (like climate change) become more complex, global and long-term in nature, policy analysts increasingly are questioning whether the process of discounting in benefit-cost analysis biases decisions against the interests of future generations. Such issues have the potential for catastrophic outcomes in the distant future that would be trivialized by the process of discounting. Despite these concerns benefit-cost analysis should not be dismissed completely. This would only deprive decision-makers of one set of insightful information. Further, advances have been made in the development of discounting approaches that account for situations in which the length of the stream of benefits from an environmental policy increases relative to the cost stream, and in which lags may exist between the time that costs are incurred and benefits are realized (e.g. health effects with latency periods) (11, 12).

Decision-support tools

Once society has made a decision to act, the choice of a "best" policy for coping with climate change is a decision inherently dependent upon social values and selection criteria that must be identified by decision-makers and stakeholders (not by researchers or assessors). Policy decisions often are complex because of the need to consider multiple social objectives and to assess the importance and relevance of these in some consistent way—which requires their own set of tools (13). Another important role for the assessment community is to develop decision-support tools that will help risk managers organize and visualize information, analyse potential trade-offs between social objectives as they make decisions under uncertainty and understand the implications of remaining scientific uncertainties for the outcomes they care about.

Response options

An array of response options exist should society decide to expend resources to protect public health, ecosystems and social well-being from climate change. The

CASE STUDY #1 Hantavirus pulmonary syndrome in the southwestern United States of America

Insights from this case study

The following case study illustrates how a policy-focused health-impacts assessment ultimately can lead to on-the-ground interventions to prevent disease and protect the public's health. Using information derived from an assessment of the environmental conditions associated with outbreaks of hantavirus pulmonary syndrome, decision-support tools in the form of risk maps have been developed that support public health officials trying to prevent outbreaks in the southwestern United States of America. This case study is based on the work of Glass et al (*14*).

Motivation for the assessment

In 1993, a disease characterized by acute respiratory distress with a high death rate (>50%) among previously healthy persons was identified in the southwestern United States. Hantavirus pulmonary syndrome (HPS) was traced to a virus maintained and transmitted primarily within populations of a common native rodent, the deer mouse.

After the outbreak, researchers hypothesized that it was due to environmental conditions and increased rodent populations caused by unusual weather associated with the El Niño Southern Oscillation (ENSO) in 1991–92. It was suggested that a cascading series of events from weather (unseasonable rains in 1991 and 1992, mild winter of 1992) through changes in vegetation, to virus maintenance and transmission within rodent populations, culminated in changes in human disease risk from HPS. Public health officials wanted to understand the cause of the outbreak in order to develop effective techniques for intervention and prevention of the disease.

Results of the assessment

A study at The Johns Hopkins School of Hygiene and Public Health, sponsored by the United States Environmental Protection Agency (EPA), explored this hypothesis by comparing the environmental characteristics of sites where people were infected with those sites where people were not. This research found that high-risk areas for hantavirus pulmonary syndrome can be predicted over six months in advance based on satellite generated risk maps of climate-dependent land cover. Predicted risk paralleled vegetative growth, supporting the hypothesis that heavy rainfall from El Niño in 1992 was associated with higher rodent populations that triggered the hantavirus outbreak in 1993. Landsat satellite remote sensing images from 1995, a non El Niño "control" year, showed low risk in the region: images from the 1998 strong El Niño again showed high risk areas as in 1992–93. Trapping mice in the field (collectors blinded to risk category) validated these satellite generated risk maps with mouse populations directly related to risk level, with a correlation factor of over 0.90. Risk classification also was consistent with the numbers of HPS cases in 1994, 1996, 1998, and 1999. This information was used to develop an early warning system, with intervention strategies designed to avoid exposure.

These strategies, developed in partnership with the centers for Disease Control and Prevention (CDC) and the Indian Health Service, already are being implemented for disease prevention in the south-west by the United States Department of Health and Human Services.

---

mitigation of greenhouse gases provides a mechanism for slowing, perhaps eventually halting, a build-up of greenhouse gases in the atmosphere. There is growing concern over the need to slow the rate of growth of atmospheric concentrations of greenhouse gases, particularly from anthropogenic sources. As noted in the 1995 Second Assessment of the Intergovernmental Panel on Climate Change: "with the growth in atmospheric concentrations of greenhouse gases, interference with the climate system will grow in magnitude, and the likelihood of adverse impacts from climate change that could be judged dangerous will become greater" (*15*). More recently, the IPCC 2001 Third Assessment concluded: "there is new and stronger evidence that most of the warming observed over the last 50 years is attributable to human activities". The potential for unexpected, large and rapid climate changes (surprises) may increase due to the non-linear nature of the climate system. A slowing of the rate of warming could yield important benefits in the form of reduced impacts to human health and other systems.

Adaptation is another important response option (*6*). As discussed in chapter 11, adaptive responses are those actions taken to enhance the resilience of vulnerable systems, thereby reducing damages from climate change and climate variability. Adaptation in the form of adjustments in behaviours, practices, processes, or structures of systems is important for protecting human health from the risks posed by climate change and for exploiting beneficial opportunities provided by a changing climate. Actions may be taken in reaction to climate change as it occurs (reactive adaptation) or in anticipation of future climate change (anticipatory adaptation).

Adaptation strategies need to be considered as part of any policy portfolio. Some level of climate change is inevitable since the climate is already changing and mitigation efforts over the near future cannot stop climate change, only slow the rate of change. Some of this change will occur as the result of natural climatic variation, some as the result of human activities that already have altered the atmosphere and committed us to future climate change. Regardless of the source of change, systems that are sensitive to changes in climatic conditions will be affected, including human health. Failure to invest in adaptation may leave a nation poorly prepared to cope with adverse changes and take advantage of opportunities, thereby increasing the probability of severe consequences (*16*).

Communication of information about climate change, potential health impacts and response strategies is itself a public policy response to climate change. The development and implementation of monitoring and surveillance systems and investments in research, also are adaptive responses. All of these should be considered when developing portfolios of policies for responding to climate change.

Expenditures on adaptation and mitigation measures will likely not be politically feasible unless an informed public understands the potential consequences of climate change and decides that there is a need to respond. An informed public also is required to ensure the continued effectiveness of any adaptation and mitigation policies that are implemented.

Monitoring and surveillance systems are integral and essential to providing the necessary information and continuity in the data to support decisions by public health officials (*17*). Sustained investments in monitoring systems and their evolution are critically important for tracking changes in the environment, particularly those factors that affect human health. Monitoring systems also are required to track the effectiveness of adaptation and mitigation policies on an ongoing basis. Investments in public health surveillance systems are important for detecting changes in health status and the potential effects of climate change (*18*). Vital elements of such a system are:

- a rapid and comprehensive communications network
- accurate, reliable, laboratory-based diagnosis capabilities in host countries or regional centres
- a mechanism for rapid response.

The functioning of such systems would be aided by heightened cooperation among local, regional, national and international health organizations (*19*).

Ongoing investments in research are necessary to provide the scientific information required for the conduct of policy-focused assessments. Research and assessment activities are complementary. Regular interactions between assessment processes and research activities can help ensure that the ongoing production of scientific information will help to answer new stakeholder questions as they emerge. When successfully implemented, such interactions permit the research community (health, physical, biological and social sciences) to identify and communicate new information and data to the assessment and policy communities on a regular basis. At the same time assessments identify and prioritize research needs in order to better answer questions being asked by the stakeholder community.

Building the bridge from science to policy: policy-focused assessment

Policy-focused assessment is a process that can help resource managers and other decision-makers to meet the challenge of assembling an effective policy portfolio. As noted earlier, it is a process by which the best-available scientific information can be translated into terms that are meaningful to policy-makers. A policy-focused assessment is more than just a synthesis of scientific information or an evaluation of the state of science. Rather, it involves the analysis of information from multiple disciplines—including the social and economic sciences—to answer the specific questions being asked by stakeholders. It includes an analysis of adaptation options to improve society's ability to respond effectively to risks and opportunities as they emerge.

Policy-focused health assessments should be designed to identify key stresses on human health within a particular region or community under current climatic conditions, and the extent to which society is vulnerable to these stresses. Formulation of good policy requires understanding of the variability in vulnerability across population sub-groups and the reasons for that variability. The vulnerability of a population is a function of the:

- extent to which health, or the natural or social systems on which health outcomes depend, are sensitive to changes in weather and climate (exposure-response relationships);
- exposures to the weather or climate-related hazards;
- population's ability to adapt to the effects of the weather or climate-related hazards.

Understanding a population's vulnerability to climate variability and change thus depends on knowing the baseline associations between specific health outcomes and specific weather and climate conditions. It also depends on the capacity of individuals and the population to adapt to changes in weather and climate conditions. The consequences are not just severe in countries with less adaptive capacity but also within specific communities and regions with less adaptive capacity. Individuals, populations and systems that cannot or will not adapt are more vulnerable, as are those who are more susceptible to weather and climate changes (e.g. consequences of storms and floods are more severe in countries with less adaptive capacity). Once current conditions are understood, the extent to which climate change may exacerbate or ameliorate health outcomes can be considered. Climate change should be considered in the context of the changes that may occur in other stresses: changes in these and in adaptive capacity can be expected to influence strongly the anticipated future benefits and costs of adaptation policies. Such a multiple stressor approach permits identification of the priority risks in a region or nation.

A complete health impact assessment goes beyond risk assessment. It also includes an adaptation assessment: an evaluation of society's capacity to adapt to change and of alternative risk management options. Once the risks have been identified, an assessment should identify and analyse appropriate adaptive responses to improve society's ability to respond effectively to risks and opportunities as they emerge. Barriers to successful adaptation and the means of overcoming such barriers need to be identified. Adaptive responses may have multiple benefits: reducing risks from climate change while, at the same time, addressing other risks to public health (also known as win-win strategies). Elimination of poverty in a community will not only reduce hunger but also facilitate improvements in sanitation, medical care and other public health systems, and thus increase resilience to climate change.

Earlier it was noted that stakeholders—both inside and outside the public health community—may have conflicting desires and conflict resolution often is needed to formulate and implement an effective policy response. Assessments of the potential health consequences of climate change can facilitate the process of conflict resolution and the allocation of scarce resources to their highest valued uses by considering multiple stresses on multiple systems and across multiple species that share interactive and interdependent relationships. Communities and sectors that share interactive and interdependent relationships also should be considered. Assessments that do not account for multiple interactions and feedbacks between systems and sectors may provide inadequate or inaccurate information for developing adaptive responses and may increase the likelihood of implementing ineffective or maladaptive strategies.

There are several key characteristics of a successful assessment of the health consequences of climate change (1):

- draws upon expertise from multiple disciplines;
- engages stakeholders throughout the assessment process to ensure timeliness and relevance of assessment results;
- analyses potential adaptive responses and options presented to decision-makers;
- identifies uncertainties and characterizes their implications for the specific decisions being made by policy-makers;
- produces a research agenda that identifies and prioritizes key knowledge gaps.

Assessment as a multidisciplinary activity

Ideally, assessment teams are composed of researchers from a variety of disciplines working together to address complex research and assessment questions. By its very nature, assessment of the potential health consequences of climate change is a multidisciplinary activity. It includes an evaluation that engages the human health, biological, physical and socioeconomic sciences in order to understand health outcomes' sensitivity to changes in weather and climate, and the possible risks under different scenarios of climate change, economic growth and technological change. Assessment also entails consideration of how human behaviour might contribute to, or ameliorate, the risks.

The extent to which climate change may damage or harm human health will depend on the magnitude of the exposure and the ability of people to adapt successfully to new climatic conditions. The ability to adapt effectively may depend upon the social, political, economic, environmental, technological and demographic factors that may affect human health. Hence, to capture the full array of factors that may affect human health directly or indirectly, assessments also should include expertise from the social, political, environmental and engineering sciences. Researchers and modellers in all of these disciplines should be considered for inclusion in a health impact assessment team.

Stakeholder engagement

Assessment of the potential health effects of climate variability and change should be guided by the priority vulnerabilities in a region or nation. The assessment should consider new or potentially evolving vulnerabilities that may or may not be climate driven. One way to determine priorities is through a stakeholder-oriented assessment process with involvement of stakeholders throughout the process.[1]

Sometimes it is difficult to identify all constituencies that might have an interest in a particular health impacts assessment. New stakeholders often are identified during the course of an assessment process, as understanding of the potential health consequences of climate change evolves. For this reason, the process of identifying and involving stakeholders is ongoing.

For an assessment to be informative, assessors must know the particular issues and questions of interest that the stakeholders and decision-makers want answered. This includes consideration of relevant questions suggested by scientists. At the outset of an assessment, the questions and outcomes of greatest concern (which may not be limited to health-related effects) should be elicited. The decision-makers who are clients for the assessment results often have multiple objectives and concerns, not necessarily limited to concerns about climate change—or even public health. Also, decision-makers may need very specific types of information in order to incorporate climate change into their decision-making and to formulate and implement new and effective adaptive responses. Stakeholders can be a source of this information.

An assessment begins with a framing exercise to determine process and goals. A multidisciplinary steering group established to agree upon the terms of reference for the assessment should conduct this exercise. This group also should be available to provide advice and support during the assessment process. The terms of reference will be specific to each assessment. Common elements should include detailed descriptions of the conduct and evaluation of the assessment, including such questions as the scope; timetable; budget and funding sources; methods used for identifying members of the team; methods used in the assessment and its review/evaluation; form and content; and the nature and frequency of feedback from the steering group and from stakeholders.

Stakeholders should be involved in the analytical process on an ongoing basis. Assessors and stakeholders are not necessarily distinct communities. In many

---

[1] Stakeholders in a health impacts assessment process might, for example, include public health officials, doctors, health insurance companies, drug companies, managers of public/private research programmes, disaster preparedness officials, concerned members of society and the decision and policymakers. They might also include agencies that fund priority research and the researchers themselves.

cases, stakeholder communities can offer data, analytical capabilities, insights and understanding of relevant problems that can contribute to the assessment. For example, stakeholders often can share traditional knowledge from their communities about the effectiveness of previous attempts to adapt to change.

In order to ensure that an assessment is timely and useful, assessors need to understand how stakeholders will use results of the assessment and the time frame within which the assessment is needed. When scientific uncertainties still exist research scientists often are reluctant to make statements that might be used by policymakers. Yet, policymakers frequently make decisions under uncertainty, whether or not scientists are prepared to inform those decisions. A decision to wait until the science is more certain is still a decision and opportunities may change during the waiting period. Assessors of health impacts strive to answer decision-makers' questions to the extent possible given the level of uncertainties of this science, in the belief that informed decisions are better than uninformed decisions. They also characterize the uncertainties and explore their implications for different health policy or resource management decisions in the belief that a better understanding of the quality and implications of scientific information leads to more informed decisions.

As an assessment process evolves and better understanding is gained of the potential consequences of climate change for human health, it is important to re-evaluate stakeholder concerns about issues previously identified and to identify new or evolving issues of concern as stakeholders are informed of assessment results.

Evaluation of adaptation options

Chapter 11 provided a thorough discussion of the topic of adaptation. In this section, we discuss the challenges faced by assessors as they evaluate alternative adaptation options for responding to the risks and opportunities of climate change, as well as potential barriers to implementation.

The evaluation, design and implementation of an effective adaptation strategy are complex undertakings. Assessors and policymakers should not be cavalier about the ease with which adaptation can be achieved, or the expected effectiveness of any policies they implement. Not only must the potential health impacts of climate change and options for responding to these impacts be identified, but also barriers to successful adaptation and the means of overcoming them need to be evaluated. Consideration of barriers to successful adaptation, as well as potential for unintended negative consequences of adaptation measures, is critical to the translation of assessment results into effective public health policies. Also, sustaining adaptation needs to be considered, as well as the continued evaluation of effectiveness recognizing that climate, socioeconomic factors and other drivers continue to change. The effectiveness of an introduced adaptive response may need to be reconsidered over time.

The difficulties involved in ensuring the effectiveness of future adaptive responses are illustrated by existing efforts to cope with the effects of climate variability under current climatic conditions. Historical evidence demonstrates that societies have not always adapted effectively to existing risks. For example, exposure to extreme heat causes deaths in urban areas throughout the world, even during years with no heatwaves. During heatwaves these numbers can increase dramatically. Many of these deaths are preventable, yet they continue.

CASE STUDY #2 Preparing for a changing climate in the Great Lakes region

Insights from this case study

This is another illustration of how a policy-focused assessment can be used to inform policy decisions, despite the existence of scientific uncertainties. It highlights the recent efforts by the Water Quality Board of the United States-Canada International Joint Commission to begin adapting to a changing climate in order to protect the beneficial uses provided by the Great Lakes system. These beneficial uses include agriculture, fish and wildlife consumption and other factors that can affect human health.

Great Lakes Water Quality Agreement of 1978

The Governments of Canada and the United States entered into a Great Lakes Water Quality Agreement (GLWQA) in 1978. This committed both countries to restore and enhance water quality in the Great Lakes system (20). The GLWQA has been amended several times, most recently in 1987. Beneficial uses provided by the Great Lakes system are defined in the agreement. The Water Quality Board of the International Joint Commission (IJC) is responsible for fulfilling this goal. Among its responsibilities is the development of remedial action plans and lakewide management plans that embody a systematic and comprehensive ecosystem approach to restoring and protecting beneficial uses in geographical areas that fail to meet the objectives of the agreement, or in open lake waters.

IJC focus on climate change

Each of the Governments of Canada and the United States recently completed policy-focused assessments of the potential effects of climate change on the Great Lakes system (21, 22). These assessments demonstrated that climate variability and change would likely have important consequences for water supplies (both quantity and quality) and therefore human health in the Great Lakes region. Examples of the implications for human health include:

- changes in the availability and quality of drinking water
- potential spread of water-borne diseases
- impacts on habitat, leading to changes in fish and wildlife consumption.

Given the findings of these and other assessments, the IJC Water Quality Board is beginning to explore and assess opportunities to adapt to a changing climate in order to protect the beneficial uses derived from the Great Lakes system. Specific questions being addressed by the Water Quality Board as it develops recommended adaptation strategies include:

- What are the Great Lakes water quality issues associated with climate change?
- What are the potential impacts of climate change on the "beneficial uses" identified in Annex 2 of the Great Lakes Water Quality Agreement of 1978?
- How might these impacts vary across the region and across demographic groups?
- What are the implications for decision-making?
- What specific advice can the Board provide within the context of the agreement, that is, how can we adapt in order to mitigate?

It is anticipated that the Board will finalize a set of recommended adaptation options sometime in 2003.

There is a wide array of possible explanations for society's failure to adapt effectively to existing risks:

- failure to identify and understand factors that affect the risk and the ability of society and individuals to respond;
- limited resources available for adaptation;
- conscious decision by society not to invest scarce resources in adaptive responses;
- perceived lack of vulnerability or perceived elimination of the threat.

Regardless of the reasons for the limited effectiveness of existing adaptive responses, historical evidence suggests that one should not be over-confident about the effectiveness of adaptive strategies when making projections of future vulnerabilities to climate change.

In the assessment of adaptation options, a number of factors related to the design and implementation of strategies need to be considered, including:

- appropriateness and effectiveness of adaptation options will vary by region, across demographic groups, and with time;
- adaptation comes at a cost;
- some strategies exist that would reduce risks posed by climate variability, whether or not the effects of climate change are realized;
- the systemic nature of climate impacts complicates the development of adaptation policy;
- maladaptation can result in negative effects that are as serious as the climate-induced effects being avoided.

Adaptation varies by region and demographic group

Policy-focused assessments must account for the fact that the potential risks, and human capacity to respond to these, vary by location and across time in scope and severity. In addition, the vulnerability of populations within and across regions will vary depending upon the health effect being considered. Consequently, appropriate adaptive responses will vary across geographical regions, demographic groups and with time.

Consider, for example, that climate change will likely increase the frequency and severity of very hot days and heatwaves during the summer. Studies in urban areas, mostly in temperate regions, show an association between increases in heat and increases in mortality (23). The risk of heat stress may rise as a result of climate change (24). The most vulnerable populations within heat-sensitive regions are urban populations. Within these vulnerable populations, the elderly, young children, the poor and people who are bedridden or on certain medications are at particular risk.

To be effective, adaptive responses must target these vulnerable regions and demographic groups, some of which may be difficult to reach (25). For example, the elderly are less likely to perceive excess heat (26); they may be socially isolated and physically frail (27, 28). This may make it difficult to convince them to use air conditioning (because they do not feel the heat) or to travel to air-conditioned environments (e.g. they may have no one to take them and may be unable to travel on their own). The poor may not be able to afford air conditioning, and if they live in high crime areas may be afraid to visit cooling shelters. Finally, for infants and young children, decisions about how warmly to dress and how much time to spend

CASE STUDY #3 Hot weather watch/warning systems

Insights from this case study

This describes the development of hot weather watch/warning systems intended to reduce the number of deaths due to extreme heat in urban areas. It illustrates several important aspects of an assessment process and the challenges faced in developing and implementing effective adaptation strategies.

The assessors drew together information from climatology, meteorology, medicine and public health to develop a city-specific system to alert the public to changing day-to-day health risks. The system is an example of a no-regrets adaptation that addresses a current problem and increases resilience in the future. Many strategies that would reduce risks posed by climate change or exploit opportunities make sense whether or not the effects of climate change are realized. Enhanced responses to urban heatwaves can save lives now.[1]

This case study also highlights the importance of assessing potential barriers to effective adaptation and the need to monitor performance over time. In particular, the critical importance of combining the watch/warning system with an effective communication strategy to provide advice on adaptive responses to limit the risk of heat stress is discussed. In some cases the watch-warning systems have been combined with communications strategies that may result in maladaptation.

Heatwaves and human mortality

The watch/warning systems originated with a fundamental assessment question posed by stakeholders: what might be the impacts of climate change on deaths due to heat stress? Assessors analysed the relationship between various climatological variables (including temperature) and human mortality in order to answer this. They turned to the climatological research community to understand expected changes in the frequency and intensity of heatwaves as the climate changes, and evaluated the potential changes in numbers of deaths in particular urban areas as climate changes.

Heat and heatwaves are projected to increase in severity and frequency with increasing global mean temperatures. Studies in urban areas show an association between increases in mortality and increases in heat, measured by maximum or minimum temperature, heat index and, sometimes, other weather conditions (23). The elderly (65 years or older), infants and young children, the poor, mentally and chronically ill people, and those who are socially isolated, are at increased risk.

Several different methods for characterizing climate have been used to assess daily mortality associated with heat episodes. One such approach hypothesizes that humans respond to more than just temperature; rather, they respond to the entire blanket of air (or air mass) around them (29). Air masses were characterized for specific localities using statistical methods to separate the air masses into area-specific categories based on a number of meteorological variables (30). This synoptic approach has led to important findings about the effect of different types of air mass on mortality. For example, daily mortality in 44 United States' cities with populations greater than one million was analysed in relation to the frequency of particular air masses. Two types of air masses associated with particularly high mortality were identified (31).

After this initial research, assessments were conducted to identify potential adaptation strategies. Since people die of heat stress every year and many of these deaths are avoidable, strategies were

---

[1] In some cases, existing institutions and public policies may result in systems that are more rigid and unable to respond to changing conditions. Elimination of these institutions and policies—if it makes sense to do so under current climatic conditions—may increase resilience to change. For example, the existence of federal flood insurance in the United States of America provides an incentive for development in high-risk coastal areas. Elimination of federal flood insurance would reduce the inventory of private property that is at risk today, and in the future when sea level rises further.

considered that would reduce avoidable deaths today while increasing resilience to future climate change. Information from the public health community was used to identify the most vulnerable populations. From this emerged the concept of watch-warning systems that have been implemented in the United States, Italy and China to protect these populations (32). These systems are used to alert the public about the timing and severity of a heatwave and, more specifically, to the presence of air masses that are known to pose the highest risks to human health. This contrasts with other systems that issue alerts and warnings using arbitrary temperature thresholds. The systems are city-specific, using information about the population and air masses for a particular city. Once the alerts have been issued, the media provide advice about risk factors and beneficial response measures that can be taken by individuals.

Prototype systems were first established by the University of Delaware, in partnership with the U.S. Environmental Protection Agency (USEPA), in Philadelphia and Washington, DC, in the United States. Systems were then implemented in Rome and Shanghai by the United Nations Environment Programme (UNEP) in partnership with the University of Delaware, World Meteorological Organization (WMO), USEPA and World Health Organization (WHO).

The effectiveness of these systems is now being monitored and evaluated. Bernard (33) notes that when alerts and warnings are issued, media sources provide advice on adaptive responses to limit the risk of heat stress (e.g. limitation of outdoor physical activity; use of sun protection cream, hats and light-coloured clothing; increased hydration with non-alcoholic and non-caffeinated beverages). However, the effectiveness of this advice is questionable. In some cases, the most vulnerable populations are identified but limited information targeted at these specific groups is provided. For example, media advice rarely specifies that elderly people and others who might be confined to the home by illness should be visited in person, rather than simply contacted by telephone, because they might be unaware of either the elevated heat in their residence or of symptomatic heat-related health impairment. Even worse, advice that could lead to maladaptation sometimes has been provided. Some sources have provided potentially dangerous and confusing information by combining advice about poor air quality with advice about heatwaves, recommending that the elderly remain indoors.

These findings suggest that advisory messages should be revised in light of established information about risk factors. There must be further evaluation of the effectiveness of the weather watch/warning systems combined with media advice.

in hot environments often are made by adults, with the children and infants unable effectively to communicate their discomfort (26).

Adaptation comes at a cost

The resources used to adapt to a changing climate may be diverted from other productive activities, such as reducing other stresses on human health, ecosystems and economic systems. This could mean reducing the resources available for remediation of public health problems unrelated to climate. Similarly, society may have to divert natural and financial resources from addressing social problems in other sectors. In the vernacular of economics, there are opportunity costs to using scarce resources for adaptation.

Once society has decided to invest in adaptation, having identified what and how it wants to adapt, it has the option of incurring the costs of adaptation at different times: either to invest immediately or to delay until a future time (assuming effectiveness of the adaptation is the same). In either case, there are costs associated with adaptation. It is a question of when the costs are incurred and what they buy. The decision of whether to adapt now or later should be based on a comparison of the present value of expected net benefits associated with acting sooner versus later.

Whenever incurred, adaptation costs must be weighed carefully when decision-makers consider the tradeoffs between alternative adaptation strategies, reducing the cause of the change and living with the residual impacts (34). It is therefore important that a health impact assessment evaluates the availability of the resources required to implement alternative adaptive strategies. A nation's ability to implement adaptation measures may be restricted by a lack of appropriate technology and trained personnel; financial limitations; cultural and social values; and political and legal institutions. Society's willingness to divert the required resources from other desired uses also must be evaluated.

Opportunities exist to adapt to multiple factors

Assessments of adaptation options should consider no-regrets strategies that would reduce risks posed by climate change and are sensible whether or not the effects of climate change are realized. These measures result in human systems that are more resilient to current climate variability and hopefully to future climate change (depending on what happens with climate change, successful adaptation to climate variability may or may not prepare a community for the future). The public health community recognizes adaptation strategies such as heatwave early warning and vector-borne disease surveillance systems to be important to the protection of lives and health regardless of future climate change (35).

In some cases existing institutions and public policies result in systems that are rigid and have limited ability to respond to changing conditions. For example, as noted earlier, the existence of federal flood insurance in the United States provides an incentive for development in high-risk coastal areas, increasing the risk of injury and death to coastal populations. Elimination of the federal flood insurance today would reduce the size of coastal communities currently at risk (at a financial cost to those individuals already living in coastal communities), as well as in the future when sea level rises further.

The systemic nature of climate change complicates efforts to adapt

Climate change will have wide-ranging effects that may occur simultaneously. Many of the effects are likely to be interdependent. The systemic nature of climate change and its effects poses unique challenges to resource managers developing adaptive responses (34). An adaptation strategy that may protect human health may, inadvertently, increase risks to other systems. In some cases it may be impossible to avoid all risks and exploit all opportunities. Society may have to choose between alternative outcomes.

Consider, for example, the increased risk of injury and death to populations in coastal areas due to more severe storm surges and flooding that may result from a rapid rate of sea level rise. If the only concern was protection of these coastal populations, one adaptive response might be to build sea walls: but the rise in sea level also threatens wetlands and the building of sea walls prevents new wetlands from forming. Destruction of wetlands can affect water quality, which in turn could have implications for the public's health. Destruction of wetlands can affect bird migration patterns and biodiversity too. Yet if the sea walls are not built, the rise in sea level could lead to salt water intrusion: threatening freshwater aquifers and drinking water, as well as fresh water required for other uses (e.g. irrigation).

A policy-focused assessment must identify for decision-makers the tradeoffs that society may have to make between future outcomes, effective risk reduction, exploitation of new opportunities presented by climate change and maximization of social well-being.

Maladaptation is possible

Adaptive responses may have unintended secondary consequences that outweigh the benefits of undertaking the strategy. For example, one possible adaptive response is the use of pesticides for vector control (36). Pesticides' effects on human health and insect predators and on increased pesticide resistance need to be considered when evaluating new pesticides. Pesticides used to eradicate mosquitoes that may carry infectious diseases (e.g. dengue fever) may have their own adverse impacts on human health. These offsetting effects must be considered prior to implementation of an eradication programme. Programmes like mosquito eradication require long-term commitment; failure to keep that commitment may result in adverse consequences. In the 1940s and 1950s for example, the Pan American Health Organization (PAHO) undertook an *Aedes aegypti* eradication programme to prevent urban epidemics of yellow fever in south and central America. The programme was successful in most countries and was discontinued in the early 1970s. Failure to eradicate *Aedes aegypti* from the whole region resulted in repeated invasions by this mosquito into those countries that had achieved eradication. By the end of the decade, many countries had been reinfested. The reinfestation continued during the 1980s and 1990s (37). This is also a good example of how a perceived lack of vulnerability can have an impact on the perceived need for adaptation (or in this case, continued investment in an adaptive strategy).

It is also important to assess in advance the risks to health from proposed technological adaptations to climate change. Increased use of air conditioning would protect against heat stress but also could increase emissions of greenhouse gases and conventional air pollutants, assuming the current proportion of coal-fired power plants (15, 38).

A well-informed decision-maker, weighing the risks and adaptation options, may decide that the adverse effects of the adaptive measures are of greater concern than the risks posed by climate change itself.

Characterization of uncertainties

Significant scientific and socioeconomic uncertainties related to climate change and the potential consequences for human health complicate the assessment process. Uncertainties exist about the potential magnitude, timing and effects of climate change; the sensitivity of particular health outcomes to current climatic conditions (i.e. to weather, climate and climate-induced changes in ecosystems); the future health status of potentially affected populations (in the absence of climate change); the effectiveness of different courses of action to address adequately the potential impacts; and the shape of future society (e.g. changes in socioeconomic and technological factors). A challenge for assessors is to characterize the uncertainties and explain their implications for the questions of concern to the decision-makers and stakeholders. If uncertainty is not directly

CASE STUDY #4 Human dependence on food from coral reef fisheries

Insights from this case study

This illustrates how research and policy-focused assessments have evaluated the linkages between global change (multiple stressors that include climate variability and change, UV radiation and other human-induced stresses on the environment), ecosystem change and human health. In particular it examines the effects of global change on the health of coral reefs, which among other uses supply a wide variety of valuable fisheries: an important source of food and animal protein in the human diet. Insights from policy-focused assessments are being used already to develop management options for protecting coral reefs around the world.

Coral reefs and food supplies

Coral reefs are one of the most threatened global ecosystems and also one of the most vital. They offer critical support to human survival, especially in developing countries, serving as barriers for coastal protection; major tourist attractions; and especially as a productive source of food and trade for a large portion of the population (*39, 40*). Coral reefs supply a wide variety of valuable fisheries, including both fish and invertebrate species (*41*). Some fisheries are harvested for food, others are collected for the curio and aquarium trades.

Reefs have been an abundant and productive source of food for millennia. In many nations, particularly those of the Pacific islands, reefs provide one of the major sources of animal protein in the human diet, with over 100 kilos of fish consumed per person per year. In the case of many small island developing nations, the majority of fisheries' harvest is small-scale and subsistence in nature; however, commercial fisheries have developed rapidly, for export markets as well as local sales. Target species include grouper, lobster, parrotfish, rabbitfish, emperors and snappers; and in areas with tourism, tourist preferences result in concentration on conch fisheries as well as grouper, snapper and lobster (*42*).

Other uses of coral reefs

The aquarium trade serves approximately two million hobbyists worldwide who keep marine aquaria, most of which are stocked with wild-caught coral reef species. This industry has attracted controversy recently as opponents point out the damaging collection techniques that often are used and high mortality rates that result. Supporters point out that proper collection techniques can avoid major impacts to the reef and the industry involves low-volume use with very high value. Indeed, in developing countries, collectors can attain incomes that are many times the national average; recently, a kilo of aquarium fish from one island country was valued at almost US $500 in 2000, compared to reef fish harvested for food that were worth only US $6. Aquarium trade target species include not only fish but also a variety of hard and soft corals, clams and snails.

Impacts of coral bleaching

Reef-building corals live in symbiosis with tiny single-celled algae (zooxanthellae) that reside in the corals' tissues and provide them with most of their colour and much of their energy (*43*). Coral bleaching occurs when, alone or in combination, stressors in the environment cause the degeneration and expulsion of zooxanthellae from the coral host, such that the white skeleton becomes visible through the transparent coral tissues. Depending on the intensity and duration, once the stress is removed corals often recover and regain their zooxanthellae (*44*). Prolonged exposure can result in partial or complete death of not only individual coral colonies, but also large tracts of reef. Bleached corals, whether they die totally or partially, are more vulnerable to algal overgrowth, disease and reef organisms that bore into the skeleton and weaken the reef structure (*41*). As reefs disintegrate, patterns of coral species diversity can alter dramatically and the reef community may be restructured (*45, 46*), with consequent impacts on the diversity of fish and other organisms within the reef ecosystem.

Stressors that trigger bleaching include freshwater flooding (47, 48), pollution (49, 50), sedimentation (51), disease (52, 53), increased or decreased light (54, 55) and especially elevated or decreased sea surface temperatures (SSTs) (44, 56, 57). Elevated SSTs during the 1997/1998 El Niño Southern Oscillation (ENSO) triggered mass coral bleaching that resulted in extensive reef damage in many regions of the world. In severely impacted regions such as the Indian Ocean (where mortality of reef-building corals reached over 90% in some areas), some countries are now at serious risk of losing this valuable ecosystem and the associated economic benefits of fisheries and tourism (58, 59). Furthermore, if average baseline temperatures continue to increase due to global climate change, then corals will be subjected to more frequent and extreme bleaching events.

Designing effective management options

Small-scale, localized bleaching events that are due to direct anthropogenic stressors (e.g. pollution or freshwater runoff) can be addressed directly to minimize the threat at its origin. In contrast, coral reef managers cannot readily address large-scale bleaching events linked to global warming and ENSO events. Climate-related threats therefore must be tackled indirectly, through thoughtful planning and strategic care of reefs within existing and future marine protected areas (MPAs), to take advantage of natural properties of coral reef ecosystems and mitigate the impact of bleaching and related mortality.

The Nature Conservancy (TNC) and World Wildlife Fund (WWF) launched a joint initiative in 2001 to develop strategies for mitigating the impacts of coral bleaching through MPA design. They have since been joined by Conservation International (CI) and participating scientists from the Australian Institute of Marine Science (AIMS) and ReefBase, among others. The new MPA initiative seeks to identify for strict protection specific patches of reef where environmental conditions favour low or negligible temperature-related coral bleaching and mortality and to enhance reef recovery by ensuring optimal conditions for larval dispersal and recruitment among sites within a strategically designed network.

The approach is to develop a set of science-based, empirically testable principles to help managers identify, design and manage such networks, in order to maximize overall survival of the world's coral reefs in the face of global climate change. Though the intent may be to save the reefs for tourism, biodiversity or other reasons, the effort will also benefit human health by preserving an important source of protein.

addressed as part of the analysis, a health impacts assessment can produce misleading results and possibly contribute to ill-informed decisions.

A variety of methods is available to deal with the existence of scientific uncertainties, while still illustrating, analysing and providing useful insights into how climate change may influence human health. These methods also can be used to inform the design and implementation of effective adaptation options intended to increase resilience to change. The method chosen depends upon a variety of factors, including the type of question being asked by a policy-maker, public health official, or resource manager and the types of scientific uncertainties that exist.

Historical records

Data and records from the past provide an essential perspective on how changes in climate affect human and natural systems. Gaining an understanding of present vulnerabilities and adaptive capacity of human populations, how those vulnerabilities are affected by variations in climate and other stressors, and which strategies have and have not worked to ameliorate the vulnerabilities, can be illuminating for understanding possible future vulnerabilities and adaptations to climate change (*60*).

Adaptive capacity, which will change over time, is determined by socioeconomic characteristics (i.e. non-climate variables). Different communities, regions and systems have different capacities to adapt. Understanding of the factors that presently contribute to health vulnerabilities that are sensitive to climate, and a sense of how these factors may change in the future, may enable effective actions to be devised even if there is uncertainty about how the climate drivers will change in the future. Also, one can ask the question: how bad would things have to be in order for the community to find itself outside the range of its ability to cope?

A word of caution is in order. It cannot simply be asserted that by increasing the capacity to cope/adapt to climate variability under current climatic conditions, future vulnerabilities to climate change will be reduced (i.e. increase resilience to change). Coping now does not always reduce future vulnerabilities to climate change. In fact, investments to increase current coping capacity may exacerbate the effects of climate change. A resource manager who knew with certainty what future climate would be might discover that planned investments would actually increase vulnerability to future climate. A completely different investment would increase resilience to change. Since the future cannot be known with certainty, scenario analysis (or some other approach) can help to reveal cases in which this might occur.

Two of the case studies presented in this chapter rely primarily on such assessments, not on the use of scenario analysis. In the case study that follows, Focks et al. (*61, 62, 63*) establish critical thresholds for dengue based upon past/present disease incidence. Adaptation responses that would reduce vulnerability to dengue can be evaluated from this, without relying upon scenario analyses. Similarly, the hot weather watch/warning system described in Case Study #3 is based largely upon analysis of health effects of weather events of the recent past and not on scenario projections of the future climate.

Scenario analyses

Given the current state of climate science it is not yet possible to make predictions of the impacts of climate change, with the exception of future sea level rise. Nevertheless, useful insights that inform risk management decisions can be provided to decision-makers. One such approach is scenario analysis.

Scenarios are plausible alternative futures that paint a picture of what might happen under particular assumed conditions. Scenarios are neither specific predictions nor forecasts. Rather, they provide a starting point for investigating questions about an uncertain future and for visualizing alternative futures in concrete and human terms. The use of scenarios helps to identify vulnerabilities and explore potential response strategies.

Scenarios can be derived in a number of ways. Projections from sophisticated climate models (e.g. General Circulation Models [GCMs]) are one tool for understanding what future climate might be like under particular assumptions. Scenarios also can take the form of "what if . . ." or "if . . . then . . ." questions. These sorts of questions, together with sensitivity analyses, are used to determine under what conditions and to what degree a system is sensitive to change. Sensitivity analyses help to identify the degree of climate change that would cause significant impacts to natural and human systems, i.e. how vulnerable and adaptable these systems are.

There is a pressing need for the research and assessment communities to develop meaningful and credible scenarios of the potential health effects of global climate change in the context of other major risk factors for adverse health outcomes. This requires analysis of the relationships between health status and socioeconomic variables in order to develop a model that relates regional health status to potential changes in socioeconomic status over the next century. Once a health model has been developed successfully, the health risk model can be used to develop health scenarios based on future climate and technological change (e.g. scenarios derived from the IPCC's Special Report on Emissions Scenarios) (64).

For health outcomes where knowledge of the potential consequences of climate change is not sufficient to support modelling, assessors might rely upon expert judgment and existing peer-reviewed studies to provide qualitative insights to stakeholders.

Describing uncertainties

Uncertainty can be expressed in a variety of ways. Because the types of uncertainty in a health impacts assessment are diverse, a multifaceted approach for characterizing uncertainties often is desirable.

Expressions of uncertainty can be categorized by the degree to which they are based on quantitative techniques versus qualitative techniques. Quantitative approaches have advantages because they enable uncertainty to be estimated when they propagate through a set of linked models, thus specific bounds on the outputs of an assessment model can be derived. The combined effect of many sources of uncertainty can be assessed formally. Uncertainties due to inherent randomness in the information base, and model formulations that use simplifying assumptions about that variability, tend to be easier to express quantitatively than uncertainties due to lack of knowledge. Uncertainty can be expressed quantitatively by probability density functions and summary statistics. Numerous sources of uncertainty can then be combined using either probability trees or Monte Carlo analysis to assess their overall uncertainty in a health impacts assessment. However, a shortcoming of this form of quantitative expression is that the propagation of uncertainty can become analytically impossible when many uncertainties are combined.

Qualitative approaches permit additional screening for potential biases and weak elements in the analysis that cannot be captured quantitatively. Either approach alone is insufficient in a complex analysis but overlaying the two provides more complete characterization of the issues. (It is noteworthy, however, that formal methods for combining qualitative information with quantitative estimates to aid decision-making have not yet been developed.)

The benefits of uncertainty analysis

Arguably, the inclusion of uncertainties might create so much fuzziness in the results that distinguishing among policy alternatives is not possible. It should be noted, however, that many of the uncertainties might have similar effects on each of the health-related policy options being analysed. These dependencies should be taken into account in the uncertainty analysis by assessing the uncertainties on policy differences (under different climate scenarios) rather than the uncertainties of each individual policy outcome and subsequently examining the differences among them. If the uncertainty in the policy differences is built into the uncertainty analysis process, there is considerably less chance that the uncertainties will overwhelm any distinctions among types of policies.[1]

Careful and explicit assessment of uncertainties can lead to:

- better understanding of the limitations of the health impacts assessment itself;
- better understanding of the implications of the identified imprecision and biases for the makers of health policy;
- characterization of the robust properties of comparisons among types of health policies;
- identification of the most pressing areas for further research as selected health policies are implemented.

CASE STUDY #5 Dengue simulation modelling and risk reduction

Insights from this case study

This illustrates how decisions to make public health interventions can be taken despite the existence of uncertainties. This case study is based on the work of Focks et al. (*61, 62, 63*).

Risk posed by dengue fever

It is estimated that dengue fever, and associated dengue haemorraghic fever, are responsible for the loss of over 20,000 lives and 653 000 disability adjusted life years (DALYs) annually (*17*). The dengue virus needs a mosquito vector for disease transmission to occur, the principal one being *Aedes aegypti*. A secondary vector is *Aedes albopictus*.

Temperature, climate change and dengue transmission

Temperature affects the rate at which the virus develops inside the mosquito. The warmer the temperature, the faster the incubation and the faster the mosquito becomes infectious. Also, in higher temperatures the mosquito reproduces more quickly and bites more frequently. All of these factors increase transmission of dengue (*ceteris paribus*). It is expected that as global average temperatures increase, the risk of dengue epidemics will rise (*ceteris paribus*).

Challenge faced by public health officials

Public health officials in developing countries usually have limited resources with which to protect the public's health. These must be targeted at the most serious health risks and needs. However, public health officials often are faced with significant uncertainties about the severity of a particular risk. An example of such a risk for many developing countries is dengue fever.

---

[1] Alternatively, as suggested earlier, in some cases vulnerability assessment can help to circumvent uncertainty about future climate for the purposes of developing adaptation strategies that are robust across climate scenarios. For example, whether or not climate change will increase breeding habitat for a disease vector, an adaptation strategy that succeeds in getting households to reduce breeding opportunities (e.g. eliminating standing water in and around their homes) will reduce health risks.

Decision-support tool to aid public health interventions

Focks et al. have developed an effective computer-based decision-support tool that enables public health officials to make decisions about when and where to target public health interventions. Communities that are resource-constrained realistically can employ this tool (65).

Several factors must be present for a dengue epidemic to occur:

- the dengue virus;
- a sufficient number of *Aedes aegypti* mosquitoes to spread the virus effectively (entomologic factors);
- people who have not had that type of dengue virus before and therefore are not immune (seroprevalence).

Focks' computer simulation model uses these factors to determine a "threshold" number (pupae/person/area). Risk of dengue epidemic is low below the threshold; risk increases above the threshold.

Calculation of the threshold number requires an accurate survey to determine the number of pupae/person/area. To obtain this number, Focks has designed an inexpensive survey of households that uses the following steps:

- map out area to be surveyed;
- enter every fifth house to look for suspicious containers in which water accumulates and becomes a potential breeding ground for mosquitoes (e.g. tyres or buckets);
- record the number of people living in the house;
- empty the containers, pouring the water through a screen;
- rinse the screen with clean water, allowing the water to pour into a white basin so that any pupae can be seen;
- use a dropper to capture the pupae and place in a labelled vial;
- use a microscope to determine the type of mosquitoes present;
- count the number of pupae per person per area.

Once the household survey has been completed, Focks' computer model can be used to compute the threshold number using the survey data, current or anticipated temperature, and seroprevalence rate. The threshold number also is a function of the prevalence of containers of different types (e.g. tyres, drums, plant dishes, pools, vases) in a given area. Thus, the model can be used to conduct "what if" analyses to inform public health officials about how the threshold number changes as the number of containers of different types is varied. Officials can use this information to target specific types of containers to empty or scrub in order to reduce the number of pupae/person below the threshold number and avert an epidemic.

The Focks decision-support tool is attractive for several reasons. First, it uses as input household survey data that can be obtained at reasonable cost. Second, it enables public health officials to minimize intervention costs by targeting those breeding ground containers that are most important. Certain types of containers have been found regularly to harbour a greater number of pupae in a particular area. In fact, in some areas, less than 1% of the containers have been found to produce more than 95% of the adult mosquitoes. The Focks method of targeting especially productive breeding containers has been found to be more cost effective than mosquito eradication programmes that use insecticides that also can have undesirable health and ecosystem side-effects (i.e. maladaptation).

Formulating a research agenda

Assessment is an ongoing, iterative *process* that yields specific assessment *products* (e.g. reports) at various points in time. Given the extensive scientific and socioeconomic uncertainties surrounding the issue, it is unlikely that any particular assessment report will answer all of the questions posed by decision-makers. It is therefore important that each assessment report identifies and prioritizes remaining key research gaps. Also, the assessment should include science/policy linkage gaps, gaps in those things needed to inform the process (e.g. implementation and adaptation evaluation gaps such as monitoring and surveillance) and communication gaps. Key gaps are those that must be filled to answer current stakeholder questions.

It is not a simple matter to identify and prioritize key research gaps because stakeholder needs change over time. At the end of any assessment, some questions will remain unanswered that stakeholders may wish to be addressed in the next phase of the assessment. Stakeholders also may have new questions, either because of the insights gained from the assessment process or because of changes in other factors unrelated to the assessment process. Some of the unanswered questions may be no longer relevant given evolving stakeholder needs. For this reason, stakeholder needs and concerns should be elicited for every phase of the assessment process.

Identification of stakeholder questions is only one step in formulating a research agenda. There are often many alternative research projects that could be undertaken to try to fill the remaining knowledge gaps. Unfortunately, limited resources are available for conducting research, so investments in research needs must be prioritized. Research resources should be invested in those activities that will most likely yield the greatest amount of useful information, that is information that will provide the most useful insights to stakeholders in the timeliest fashion. This requires value of information calculations that yield insights into the incremental value to stakeholders of information anticipated from an investment in a particular research activity. The results of these calculations depend on changing stakeholder needs and values, and the timeliness and relevance of information. Value of information exercises may be costly to undertake but are essential to any assessment process.

Integrated assessment is one approach for conducting value of information exercises. This includes stakeholder involvement as an integrating mechanism rather than sole reliance on a model. It is also a valuable approach that has wider applicability than relying solely on an integrated assessment model (which is not always available). Integrated assessment has been defined as "an interdisciplinary process of combining, interpreting, and communicating knowledge from diverse scientific disciplines in such a way that the whole set of cause-effect interactions of a problem can be evaluated from a synoptic perspective with two characteristics: it should have added value compared to single disciplinary oriented assessment; and it should provide useful information to decision-makers" (66). As noted by Bernard and Ebi (35), integrated assessment is a synthesis of knowledge across disciplines with the purpose of informing decisions rather than advancing knowledge for its intrinsic value. The outcome can be used to prioritize decision-relevant uncertainties and research needs. The multidisciplinary nature of this research challenges the more traditional, single discipline focused research.

Assessment and scientific research are ongoing activities. To ensure that they are complementary, an ongoing feedback process between assessment activities and research activities is essential. When successfully implemented such a process permits scientific research—whether in the health, physical, biological or social sciences—to identify new risks or opportunities and provide information and data required for an assessment. At the same time, assessments identify and prioritize research needs that must be filled in order to better answer questions being asked by the stakeholder community.

Increasing public awareness: importance of communicating assessment results

Communication with stakeholders should be an ongoing process that keeps stakeholders engaged throughout an assessment process. Also, once an assessment has been completed, the results must be communicated to stakeholders in a timely and meaningful fashion to inform decision-making (the communication of results at the end of an assessment also can be useful in engaging more stakeholders in future assessments). This requires the development of a communication strategy that is part of the assessment process.

A communication strategy must ensure access to information, presentation of information in a usable form and guidance on how to use the information (67). Risk communication is a complex, multidisciplinary, multidimensional and evolving process. It is most successful and efficient when focused on filling knowledge gaps and misconceptions that are most critical to the decisions people face (68). Often, information has to be tailored to the specific needs of risk managers in specific geographical areas and demographic groups in order to be effective (69). This requires close interaction between information providers (e.g. researchers, assessors) and those who need the information to make decisions.

Good reporting of assessment results can enhance the public's ability to evaluate science/policy issues and the individual's ability to make rational personal choices. Poor reporting can mislead and disempower a public that increasingly is affected by science and technology and by decisions determined by technical expertise (70).

Assessments are a valuable source of information for risk managers, decision-makers and the public: the results can be used to promote education, training and public awareness of the potential health impacts of climate change. Maintaining the credibility of the assessment is central to maintaining the confidence of decision-makers. The degree of oversight and other procedural features of assessment, such as public participation or access to data and reports and the weight accorded to formal assessment in policy-making, may vary from culture to culture (71). Yet, most successful and effective assessments will contain communication strategies that possess the following key characteristics (72):

- assessment process is open and inclusive. Representatives of all key stakeholder groups invited to participate, regardless of their views;
- non-threatening atmosphere is created at the consultation table: all parties are encouraged to be candid, respectful and supportive;
- information about the assessment is shared early, continuously and candidly. Any perceived refusal to share information will be interpreted negatively by stakeholders and destroy trust in the assessment process;
- principal focus of the assessment process is on issues of concern to stakeholders. In this sense, they should control the consultative process. Assessors should not prejudge what those issues are;

- stakeholders are consulted even about the design of the consultative process;
- cooperative approach to the collection and interpretation of project-related data from the outset of the assessment process;
- expected "deliverable" is clearly identified, and public's role in the final decision is clarified;
- project-design options provided wherever possible;
- ongoing communications with grass roots constituencies maintained through the stakeholders' representatives in the public involvement process.

Information about the assessment, including assessment results, are disseminated through a variety of mechanisms, including:

- workshops and seminars;
- electronic systems that provide decision-makers with data and information, including metadata which summarizes the data;
- tools for the decision-making process;
- different formats and contents of information (videos, printed materials, CD-ROMs, etc.) designed for different target audiences.

Conclusions

Policy-focused assessment is a valuable process for providing timely and useful information to decision-makers, resource managers and other stakeholders in the public health community. To be successful on an ongoing basis, an assessment process must have the following characteristics:

- remain relevant by continually focusing on questions and effects of concern to stakeholders;
- assessment team and approach should be multidisciplinary;
- entail constant interactions between research and assessment communities to ensure key research gaps are filled;
- carefully characterize uncertainties and explain their implications for stakeholder decisions;
- provide relevant information in a useful format that has meaning to stakeholders. Risk management options should be explored, including cost, effectiveness, and potential barriers to implementation;
- an interactive process that builds on previous assessments to provide information required to support decisions and policy development;
- models and tools to support decision-making should be developed, wherever possible.

It has been argued that the existence of scientific uncertainties precludes policymakers from taking action today in anticipation of climate change. This is not true. In fact policymakers, resource managers and other stakeholders make decisions every day, despite the existence of uncertainties. The outcomes of these decisions may be affected by climate change or the decisions may foreclose future opportunities to adapt to climate change. Hence, the decision-makers would benefit from information about climate change and its possible effects. The entire process of policy-focused assessment is premised on the need to inform risk managers who must make decisions every day despite the existence of uncertainties. It is also noteworthy that policy-focused assessments already have influenced policy and resource management decisions of interest to the public health community.

Care must be taken to respect the boundary between assessment and policy formation. Policy-focused assessment's goal is to inform decision-makers, not to make specific policy recommendations or decisions. Policy decisions depend on more than the science, and involve societal attitudes towards risk, social values and other factors affecting decision-making. But the information provided by policy-focused assessments is invaluable. An informed decision is always better than an uninformed decision.

References

1. Scheraga, J.D. & Furlow, J. From assessment to policy: lessons learned from the U.S. National Assessment. *Human and Ecological Risk Assessment* 7(5) (2001).
2. Scheraga, J.D. & Smith, A.E. Environmental policy assessment in the 1990s. *Forum for Social Economics* 20(1) (1990).
3. McMichael, A.J. *Human frontiers, environments and disease: past patterns, uncertain futures.* Cambridge, UK, Cambridge University Press, 2001.
4. Beauchamp, T.L. & Childress, J.F. *Principles of biomedical ethics.* Fifth Edition. Oxford, UK, Oxford University Press, 2001.
5. Intergovernmental Panel on Climate Change (IPCC), *Climate change 2001: synthesis report.* Watson, R.T. et al. eds. Cambridge, UK, Cambridge University Press, 2001a.
6. National Academy of Sciences (NAS), *Policy implications of greenhouse warming: mitigation, adaptation, and the science base.* Washington, DC, USA, National Academy Press, 1992.
7. Office of Technology Assessment (OTA), *Preparing for an uncertain climate.* Washington, DC, USA, United States Government Printing Office, 1993.
8. Tamburlini, G. & Ebi, K.L. Searching for evidence, dealing with uncertainties, and promoting participatory risk-management. In: *Children's health and environment: a review of evidence.* Tamburlini, G. et al. eds. A joint report from the European Environment Agency and WHO Regional Office for Europe, Copenhagen, Denmark, EEA, 2002.
9. United Nations Conference on Environment and Development, *Rio Declaration on environment and development.* Rio de Janeiro, 1992.
10. Scheraga, J.D. & Sussman, F.G. Discounting and environmental management. In: *The international yearbook of environmental and resource economics.* Tietenberg, T. & Folmer, H. eds. Massachusetts, USA, Edward Elgar, 1998.
11. Kolb, J.A. & Scheraga, J.D. A suggested approach for discounting the benefits and costs of environmental regulations. *Journal of Policy Analysis and Management* 9(3): 381–390 (1990).
12. Scheraga, J.D. Perspectives on government discounting policies. *Journal of Environmental Economics and Management* 18: 65–71 (1990).
13. Herrod-Julius, S. & Scheraga, J.D. The TEAM model for evaluating alternative adaptation strategies, In: *Research and practice in multiple criteria decision making,* Haimes, Y.Y. & Steuer, R.E. eds. New York, USA, Springer-Verlag, pp. 319–330, 2000.
14. Glass, G.E. et al. Using remotely sensed data to identify areas at risk for hantavirus pulmonary syndrome. *Emerging Infectious Diseases* 6(3): 238–247 (2000).
15. Intergovernmental Panel on Climate Change (IPCC), *Climate change 1995: impacts, adaptations and mitigation of climate change: scientific-technical analyses.* Watson, R.T. et al. eds. Cambridge, UK, Cambridge University Press, 1996.
16. Smith, J.B. & Lenhart, S.S. Climate change adaptation policy options. *Climate Research* 6: 193–201 (1996).
17. Intergovernmental Panel on Climate Change (IPCC), *Climate change 2001: impacts, adaptation and vulnerability.* Contribution of Working Group II, McCarthy, J.J. et al. eds. Cambridge, UK, Cambridge University Press, 2001b.

18. World Health Organization (WHO), *Decision-making in environmental health: from evidence to action*. Corvalán, C. et al. eds. London, UK, published on behalf of WHO by E & FN Spon, 2000b.

19. National Academy of Sciences (NAS), *Conference on human health and global climate change: summary of the proceedings*. National Science and Technology Council and the Institute of Medicine, Washington, DC, USA, National Academy Press, 1996.

20. International Joint Commission (IJC), *Great Lakes water quality agreement of 1978*, Agreement, with annexes and terms of reference, between the United States and Canada signed at Ottawa, November 22, 1978, as amended by Protocol signed November 18, 1987, Office Consolidation, IJC United States and Canada, Reprint, February 1994.

21. Sousounis, P.J. & Bisanz, J.M. *Preparing for a changing climate: the potential consequences of climate variability and change, Great Lakes*. Michigan, USA, University of Michigan, Ann Arbor, 2000.

22. Mortsch, L.D. et al. eds. *Adapting to climate change and variability in the Great Lakes-St. Lawrence Basin*. Proceedings of a binational symposium, Environment Canada, Waterloo, University of Waterloo Graphics, 1998.

23. McGeehin, M.A. & Mirabelli, M. The potential impacts of climate variability and change on temperature-related morbidity and mortality in the United States. *Environmental Health Perspectives* 109(2): 185–189 (2001).

24. Kalkstein, L.S. & Greene, J.S. An evaluation of climate/mortality relationships in large US cities and the possible impacts of climate change. *Environmental Health Perspectives* 105: 84–91 (1997).

25. Chestnut, L.G. et al. Analysis of differences in hot-weather-related mortality across 44 United States metropolitan areas. *Environmental Science Policy* 1: 59–70 (1998).

26. Blum, L.N. et al. Heat-related illness during extreme weather emergencies. *Journal of the American Medical Association* 279(19): 1514 (1998).

27. Semenza, J.C. et al. Risk factors for heat-related mortality during the July 1995 heat wave in Chicago. *New England Journal of Medicine* 335: 84–90 (1996).

28. Kilbourne, E.M. et al. Risk factors for heat-related stroke: a case-control study. *Journal of the American Medical Association* 247: 3332–3336 (1982).

29. Kalkstein, L.S. & Smoyer, K.E. The impact of climate change on human health: some international implications. *Experiencia* 49: 469–479 (1993).

30. Kalkstein, L.S. A new approach to evaluate the impact of climate upon human mortality. *Environmental Health Perspectives* 96: 145–150 (1991).

31. Kalkstein, L.S. et al. A new spatial synoptic classification: application to air-mass analysis. *International Journal of Climatology* 16: 983–1004 (1996a).

32. Kalkstein, L.S. et al. The Philadelphia hot weather-health watch warning system: development and application. *Bulletin of the American Meteorological Society* 77: 1519–1528 (1996b).

33. Bernard, S.M. Media-disseminated heat wave warnings: suggestions for improving message content. unpublished manuscript, 2002.

34. Shriner, D.S. & Street, R.B. North America. In: *The regional impacts of climate change: an assessment of vulnerability*. Report of the Intergovernmental Panel on Climate Change, Watson, R.T. et al. eds. Cambridge, UK, Cambridge University Press, 1997.

35. Bernard, S.M. & Ebi, K.L. Comments on the process and product of the health impacts assessment component of the National Assessment of the potential consequences of climate variability and change for the United States. *Environmental Health Perspectives* 109(2): 177–184 (2001).

36. World Health Organization (WHO). *Climate change and human health*. McMichael, A.J. et al. eds. Geneva, Switzerland, 1996.

37. Gubler, D.J. & Kuno, G. eds. *Dengue and dengue haemorrhagic fever*. London, UK, CAB International, 1997.

38. U.S. Environmental Protection Agency (USEPA). *The potential effects of global climate change on the United States*. Washington, DC, USA, 1989.

39. Salm, R.V. et al. Mitigating the impact of coral bleaching through marine protected area design. In: *Coral bleaching: causes, consequences and response.* Schuttenberg, H. ed. Papers presented at the 9th International Coral Reef Symposium session on coral bleaching: assessing and linking ecological and socioeconomic impacts, future trends and mitigation planning. Coastal Management Report #2230. ISBN # 1-885454-40-6. Coastal Resources Center, Narragansett, RI. 2001.

40. West, J.M. Environmental determinants of resistance to coral bleaching: implications for management of marine protected areas. In: *Coral bleaching and marine protected areas.* Salm, R. & Coles, S.L. eds. Proceedings of the workshop on mitigating coral bleaching impact through MPA design. Bishop Museum, Honolulu, 29–31 May, 2001. Asia Pacific Coastal Marine Program Report No. 0102: pp. 40–52. The Nature Conservancy, Honolulu: 2001. http://www.conserveonline.org.

41. Westmacott, S. et al. *Management of bleached and severely damaged coral reefs.* Switzerland and Cambridge, UK, IUCN Gland, vii + 36 pp., 2000. http://www.iucn.org/places/usa/resources.html.

42. Spalding, M.D. et al. *World atlas of coral reefs.* Berkeley, California, USA, University of California Press, pp. 47–51, 2001.

43. Muscatine, L. The role of symbiotic algae in carbon and energy flux in reef corals. *Coral Reefs* 25: 1–29 (1990).

44. Hoegh-Guldberg, O. Climate change, coral bleaching and the future of the world's coral reefs. *Marine and Freshwater Research* 50(8): 839–866 (1999).

45. Done, T.J. Phase shifts in coral reef communities and their ecological significance. *Hydrobiologia* 247(1–3): 121–132 (1992).

46. Hughes, T.P. Catastrophes, phase shifts and large-scale degradation of a Caribbean coral reef. *Science* 265(5178): 1547–1551 (1994).

47. Goreau, T.J. Mass expulsion of zooxanthellae from Jamaican reef communities after Hurricane Flora. *Science* 145: 383–386 (1964).

48. Egana, A.C. & DiSalvo, L.H. Mass expulsion of zooxanthellae by Easter Island corals. *Pacific Science* 36: 61–63 (1982).

49. Jones, R.J. Zooxanthellae loss as a bioassay for assessing stress in corals. *Marine Ecology Progress Series* 149: 163–171 (1997).

50. Jones, R.J. & Steven, A.L. Effects of cyanide on corals in relation to cyanide fishing on reefs. *Marine and Freshwater Research* 48: 517–522 (1997).

51. Meehan, W.J. & Ostrander, G.K. Coral bleaching: a potential biomarker of environmental stress. *Journal of Toxicology and Environmental Health* 50(6): 529–552 (1997).

52. Kushmaro, A. et al. Bleaching of the coral *Oculina patagonica* by Vibrio AK-1. *Marine Ecology Progress Series* 147(1–3): 159–165 (1997).

53. Benin, E. et al. Effect of the environment on the bacterial bleaching of corals. *Water, Air and Soil Pollution* 123(1–4): 337–352 (2000).

54. Lesser, M.P. et al. Bleaching in coral reef anthozoans: effects of irradiance, ultraviolet radiation and temperature on the activities of protective enzymes against active oxygen. *Coral Reefs* 8: 225–232 (1990).

55. Gleason, D.F. & Wellington, G.M. Ultraviolet radiation and coral bleaching. *Nature* 365: 836–838 (1993).

56. Glynn, P.W. Coral reef bleaching: ecological perspectives. *Coral Reefs* 12: 1–17 (1993).

57. Brown, B.E. Coral bleaching: causes and consequences. *Coral Reefs* 16 (supplement): S129–S138 (1997).

58. Wilkinson, C. ed. *Status of coral reefs of the world: 1998.* Queensland, Australia, Australian Institute of Marine Science, pp. 184, 1998.

59. Wilkinson, C. ed. *Status of coral reefs of the world: 2000.* Queensland, Australia, Australian Institute of Marine Science, pp. 363, 2000.

60. Kelly, P.M. & Adger, W.N. Theory and practice in assessing vulnerability to climate change and facilitating adaptation. *Climatic Change* 47: 325–352 (2000).

61. Focks, D.A. et al. Transmission thresholds for dengue in terms of Aedes aegypti pupae per person with discussion of their utility in source reduction efforts. *American Journal of Tropical Medicine & Hygiene* 62: 11–18 (2000).

62. Focks, D.A. et al. The use of spatial analysis in the control and risk assessment of vector-borne diseases. *American Entomologist* 45: 173–183 (1998).

63. Focks, D.A. & Chadee, D.D. Pupal survey: an epidemiologically significant surveillance method for Aedes aegypti: an example using data from Trinidad. *The American Journal of Tropical Medicine & Hygiene* 56: 159–167 (1997).

64. Intergovernmental Panel on Climate Change (IPCC). *Emissions Scenarios*. A special report of Working Group III, Cambridge, UK, Cambridge University Press, 2000.

17. Intergovernmental Panel on Climate Change (IPCC), *Climate change 2001: impacts, adaptation and vulnerability*. Contribution of Working Group II, McCarthy, J.J. et al. eds. Cambridge, UK, Cambridge University Press, 2001b.

65. World Health Organization (WHO). *Workshop report: climate variability and change and their health effects in Pacific Island countries*. Report on workshop held in Apia, Samoa, July 25–28, 2000, Geneva, Switzerland, World Health Organization, December 2000a.

66. Rotmans, J. & Dowlatabadi, H. Integrated assessment modelling. In: *Human choice and climate change (Volume 3): The tools for policy analysis*, Rayner, S. & Malone, E. eds. Columbus, USA, Battelle Press, pp. 291–377, 1998.

67. Moreno, A.R. & Arrazola, M. *Environmental education and telecommunications*. Proceedings of the 6th International *auDes* Conference bridging minds & markets. Bridging environmental education & employment in Europe. Venice, Italy, 5–7 April, 2001.

68. Read, D. et al. What do people know about global climate change? Survey studies of educated laypeople. *Risk Analysis* 14(6): 971–982 (1994).

69. Staneva, M.P. *Public perception of global change: intra-national spatial comparisons*. The 2001 open meeting of the human dimensions of global environmental change. Research Community. Rio de Janeiro, Brazil. October 6–8, 2001.

70. Myers, N.J. & Raffensperger, C. When science counts: a guide to reporting on environmental issues. *The Networker:* the media and environment 3(2) (1998). www.sehn.org/Volume_3-2_2.htm.

71. Miller, C. et al. *Shaping knowledge, defining uncertainty: the dynamic role of assessments*. Working Group 2—assessment as a communications process, 1997. www.grads.iges.org/geaproject1997/gea4.html.

72. Kruk, G. Introduction: An overview of risk communication. In: *EMF risk perception and communication*. Repacholi, M.H. & Muc, A.M. eds. Proceedings, international seminar on EMF risk perception and communication. Ottawa, Ontario, Canada, 31 August–1 September 1998. Geneva, Switzerland, World Health Organization, pp. 109–135, 1999.

CHAPTER 13

Conclusions and recommendations for action

C.F. Corvalán,[1] H.N.B. Gopalan,[2] P. Llansó[3]

Introduction

"It is in the interest of all the world that climatic changes are understood and that the risks of irreversible damage to natural systems, and the threats to the very survival of man, be evaluated and allayed with the greatest urgency" (1).

His Excellency Maumoon Abdul, Gayoom, President of the Republic of the Maldives

These words were delivered in 1987 at the United Nations General Assembly on the Issues of Environment and Development. Although much has been learned about the expected impacts of climate change on ecosystems and people, and some important actions are emerging, much is still to be done to diminish further the gap between knowledge and action.

Global climate change of the magnitude and rate seen in the past hundred years is a relatively recent, unfamiliar threat to the conditions of the natural environment and human health. It is one of a set of large-scale environmental changes now underway, each reflecting the increasing impacts of human activities on the global environment (2). These changes, including: stratospheric ozone depletion; biodiversity loss; worldwide land degradation; freshwater depletion; and the global dissemination of persistent organic pollutants have great consequences for the sustainability of ecological systems, food production, human economic activities and human population health (3).

In turn, these global environmental changes are the result of a complex set of drivers. These include: population change (population growth, movement and rapid urbanization); unsustainable economic development (manifested in current production and consumption patterns); energy, agricultural and transport policies; and the current state of science and technology (4). Economic and technological developments have contributed to a remarkable improvement in the global health status since the industrial revolution. The unwanted side effect of this development has been a range of harmful changes to the environment, initially at local level but now extending to the global scale. Many of these large-scale environmental changes threaten ecosystems and human health. Indeed, scientists are concerned that current levels and types of human economic activities may be impairing the planet's life-support systems at a global level (5).

Various global environmental threats have been followed by concerted actions in the form of international conventions, global assessments and global agendas

---

[1] World Health Organization, Geneva, Switzerland.
[2] United Nations Environment Programme, Nairobi, Kenya.
[3] World Meteorological Organization, Geneva, Switzerland.

BOX 13.1 Selected international conventions

Biodiversity

Convention on Biological Diversity (1992) and its Cartagena Protocol on Biosafety (2000) (http://www.biodiv.org).

Climate change

The United Nations Framework Convention on Climate Change (UNFCCC, 1992) and its Kyoto Protocol (1997) (http://www.unfccc.int).

Desertification

Convention to Combat Desertification (1992) (http://www.unccd.int).

Hazardous chemicals

Rotterdam Convention on the Prior Informed Consent (PIC) Procedure for Certain Hazardous Chemicals and Pesticides in International Trade (1998) (http://www.pic.int).

Hazardous wastes

Basel Convention on Transboundary Movements of Hazardous Wastes and their Disposal (1989) and its Protocol on Liability and Compensation (1999) (http://www.basel.int).

Ozone

Vienna Convention for the Protection of the Ozone Layer (1985) and its Montreal Protocol on Substances that Deplete the Ozone Layer (1987) (http://www.unep.org/ozone).

Persistent organic pollutants

The Stockholm Convention on Persistent Organic Pollutants (2001) (http://www.pops.int).

Wetlands

The Convention on Wetlands (commonly known as the Ramsar Convention 1971) (http://www.ramsar.org).

---

for action, with the support of many nations (Box 1). Some have been more successful than others; some need more time to reach consensus. All have a sense of urgency and need commitment by governments and people, at all levels.

The United Nations Conference on the Human Environment in 1972 reflected the major concerns of the times: chemical contamination, depletion of natural resources, environmental and social impacts of rapid urbanization and the threat of nuclear weapons. Its Declaration also acknowledged emerging concerns about rapid environmental change and potentially global threats as the result of human induced activities: "A point has been reached in history when we must shape our actions throughout the world with a more prudent care for their environmental consequences. Through ignorance or indifference we can do massive and irreversible harm to the earthly environment on which our life and well-being depend" (6).

Two decades later (1992) at the United Nations Conference on Environment and Development (the Earth Summit), traditional environmental problems were addressed along with a new set of problems: emerging global environmental threats. Thus, the Earth Summit not only gave rise to a plan of action for sustainable development towards the twenty-first century (Agenda 21, which included both well known and newer global threats) but also opened for signature the United Nations Framework Convention on Climate Change. Important agreements also were reached regarding the Convention on Biological Diversity (the conservation of biological diversity; the sustainable use of its components; and the fair and equitable sharing of the benefits from the use of genetic resources), and the Convention to Combat Desertification.

At its core, sustainability is about maintaining functional ecological and other biophysical life-support systems on Earth. If these systems decline, eventually human population health indices also will begin to turn down. Technology can buy time, it is possible to buffer against immediate impacts and indeed extract more "goods and services" from the natural world, but nature's bottom-line accounting cannot be evaded. That is, the continuing health of human populations depends on not exceeding (or staying within) the environment's "carrying capacity". The global climate system is a prime determinant of ecological sustainability and thus of Earth's capacity to sustain healthy human life. Viewed in this way, the sustainability of human population health becomes a central criterion in the transition to sustainable development (7).

The Rio Declaration on Environment and Development at the Earth Summit (8) adopted the precautionary principle to protect the environment. This principle states that where there are threats of serious or irreversible damage, lack of full scientific certainty shall not be used as a reason for postponing cost-effective measures to prevent environmental degradation. The precautionary principle has been incorporated within the United Nations Framework Convention on Climate Change. This principle addresses the dilemma that while uncertainties still surround climate change, to wait for full scientific certainty before taking action would almost certainly be to delay actions to avert serious (perhaps irreversible) impacts until they are too late. This precautionary approach is well understood in public health (9). As stated by WHO's Director-General, Dr. Brundtland, in this context, "having unintentionally initiated a global experiment, we cannot wait decades for sufficient empirical evidence to act. That would be too great a gamble with our children's future" (10).

The First World Climate Conference, sponsored by The World Meteorological Organization (WMO) in 1979, recognized climate change as a problem of increasing significance. The WMO also established the World Climate Programme in that year. In 1988 the United Nations Environment Programme (UNEP) and WMO established the Intergovernmental Panel on Climate Change (IPCC). The Panel was given a mandate to assess the state of existing knowledge about the climate system and climate change; the environmental, economic, and social impacts of climate change; and the possible response strategies. Its First Assessment Report was released in 1990, the latest (Third Assessment Report, or TAR) in 2001.

The ultimate objective of the United Nations Framework Convention on Climate Change (UNFCCC) is the stabilization of greenhouse gas concentration in the atmosphere, achieved within a time frame sufficient to allow ecosystems to adapt naturally to climate change. It accepts that some change is inevitable and therefore adaptive measures are required. Impacts will be felt on ecosystems and directly or indirectly on human health. Such impacts include agriculture and food security; sea level rise and coastal areas; biological diversity; water resources; infrastructure; industry; and human settlements. The Kyoto Protocol to the UNFCCC strengthens the international response to climate change (Box 2).

The evidence relating to the existence of human-induced global climate change has become compelling: the IPCC's Third Assessment Report concluded that human-induced warming has begun. Global average temperature is rising faster than at any time in the last 10 000 years, and there is new and stronger evidence to indicate that most of the warming observed over the last 50 years is due to human activities (*11*). This has been associated with climate-attributable changes in simple physical and biological systems including: the retreat of glaciers; thinning of sea ice; thawing of permafrost; earlier nesting and egg-laying by birds; pole-wards extension of insect and plant species; and earlier flowering of plants.

This general scientific elucidation of climate change has sensitized the general public and policy-makers to the potential that climate change will cause various adverse health impacts. While the full range of projected climate change would have both beneficial and adverse effects, adverse effects will dominate if the larger climate changes and rates of change occur. That adds an extra dimension of concern and weight to the international discourse on this topic.

Figure 13.1 illuminates the basic components of the complex process: global driving forces, environmental changes, human exposures and the better-known, potential health effects. Upstream actions (mitigation mechanisms to reduce emissions of GHGs) are the most appropriate and effective but climate change is a process that will take many decades to control and reverse. Co-ordinated adaptation mechanisms will be required from all sectors, including health. This is a new challenge for the health sector, and research needs are many and large: better understanding of human exposures in this context; contamination pathways or transmission dynamics; expected and lesser-known health impacts; and adaptation options and their evaluation. Table 13.1 gives examples of priority research areas for the examples of health effects outlined in Figure 13.1.

As a response to the requirements stated in Agenda 21 and the UNFCCC, a number of organizations carrying out significant climate related activities jointly developed the Inter-Agency Committee on the Climate Agenda (IACCA). WHO, WMO and UNEP jointly contribute towards the health aspects of the Climate

FIGURE 13.1 Climate change and health: pathway from driving forces, through exposures to potential health impacts. Lines under research needs represent input required by the health sector. *Source: adapted from reference (12).*

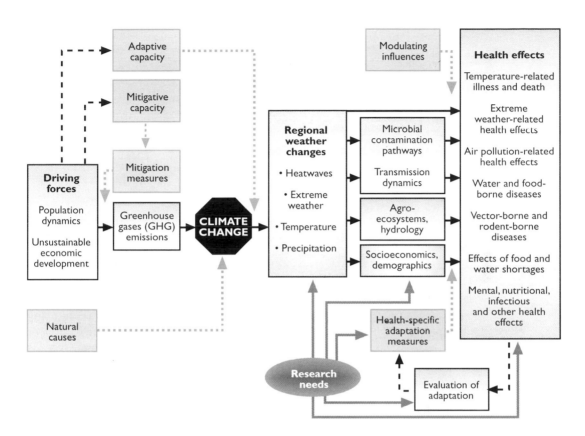

Agenda within the general field of "climate impact assessment and response strategies to reduce vulnerability". The second IACCA session, held in Geneva in 1998, proposed the establishment of an Inter-Agency Network on Climate and Human Health, with a secretariat coordinated by WHO. Joint activities of the Interagency Network between WHO, WMO and UNEP began in 1999. The collaborative work focuses on three areas: capacity building, information exchange and research promotion (Box 3).

Conclusions and recommendations

Several major conclusions emerge from this book. They are listed below in the order of the chapters from which they emerge, not in order of priority. All are selected for their global relevance:

Climate related exposures

Climate is an important determinant for human health. Both weather and climatic variables can be seen as human exposures that directly or indirectly impact on human health. Moreover, these are not expected to remain constant, and overall likely to increase their impacts on human health.

TABLE 13.1 Priority health research areas for different risk factors resulting from climate change.

Risk Factors	Health effects	Priority research areas
Extreme heat or cold; Stagnant air masses	Temperature related illness and death	Improved prediction, warning and response
Strong precipitation variability	Extreme weather related health effects	Assessment of past impacts and effectiveness of warnings
Local air pollution; Stagnant air masses	Air pollution related health effects	Combined effects of climate factors and air pollution; Weather related allergens
Precipitation; Water temperature	Water-borne and food-borne diseases	Climate and marine-related diseases; Climate, land-use impacts on water quality and health
Temperature, humidity, precipitation	Vector-borne and rodent-borne diseases	Climate related disease transmission dynamics; Improved surveillance
Temperature, water scarcity, land use	Nutritional deficiencies	Health and agricultural sector adaptation strategies
Extreme events, population displacement	Mental health	Assessment of past interventions related to emergencies and population displacement

Source: adapted from reference 12.

BOX 13.3 Inter-agency network on climate and human health

WHO, WMO and UNEP collaborate in:

Capacity building activities: assisting Member States in: a) undertaking national assessments of climate-induced human health impacts; b) determining and meeting capacity-building and research needs in order to identify and address priority areas; c) identifying and implementing adaptation strategies and preventive and mitigating measures, designed effectively to reduce adverse health impacts.

Information exchange: a) provision of information to Member States, national and international training and research institutions, and the public at large, on the state-of-the-art in the global research effort on climate and health interactions, their consequences for population health and for public health response; b) fulfilling "clearing house" functions to ensure free access to information including databases needed for research on climate variability and climate change on human health in developing countries.

Research promotion: a) to serve as the UN-based lead group of institutions and experts for the guidance of research programmes on the human health implications of climate and of global environmental change, including the impacts of climate variability, climate change and stratospheric ozone depletion.

The IPCC's Third Assessment Report concluded that most of the warming observed over the past 50 years is attributable to human activities. This implies that reversing this trend also will need to be the result of human actions. However, it is known that human influences will continue to change atmospheric composition throughout the twenty-first century. Global average temperature is projected to rise by 1.4 to 5.8°C from 1990 to 2100. Global climate change will not likely be spatially uniform, and is expected to include changes in temperature and the hydrologic cycle.

Studying the natural complexities of weather and climate variability in relation to health outcomes offers unique challenges. Weather and climate can be summarized over various spatial and temporal scales. The appropriate scale of analysis will depend on the study hypothesis. Each study needs to define the exposure of interest and the lag period between exposure and effect. Analogue studies are one means of estimating risks of climate sensitive diseases for future climate change although the predictive value of these studies may be limited. Future events may differ from historical events and the extent of vulnerability of a population changes over time. For these and other reasons, scenario-based modelling is used to project what might happen under different climate conditions.

Current knowledge is limited in many areas. Research is needed, particularly in the following:

- developing innovative approaches to analysing weather and climate in the context of human health;
- setting up long-term data sets to answer key questions, such as whether infectious diseases are likely to change their range;
- improved understanding of how to incorporate outputs from multiple Global Climate Models into health studies better to address the range of uncertainties associated with projected future health impacts.

One important consideration for researchers is the ongoing change in the climate baseline. Many archives of meteorological data contain baseline parameters, or climatological normals, based on the average data for three decades, which are re-calculated every 10 years. While such archives may have desirable densities in terms of time and space, analysis methods need to take into account the changing climate baseline such archives provide to avoid erroneous, potentially conservative, conclusions. The WMO addresses this by maintaining the Climatological Standard Normals (CLINO) (13). Re-computed each 30 years, the CLINO provide a stable baseline throughout that period. As all CLINO datasets reflect the same input period, the differences in values between observing locations reflect the differences in climate rather than differences caused by non-common time periods.

Reaching consensus on the science

The science of climate change increasingly has achieved consensus among scientists, in particular through the assessments of the IPCC. Although much research still is needed on climate change and health links, there is increasing evidence that human health will be affected by climate change in many and diverse ways.

Higher temperatures and changes in precipitation and climate variability are likely to alter the geographical range and seasonality of some climate sensitive vectors—potentially extending the range and season of some vector-borne diseases, contracting them for others. Heavy rainfall and increases in water surface temperature are associated with contamination of marine and fresh water with water-borne diseases. Countries lying in the tropics are highly vulnerable to climate sensitive infectious diseases and are likely to experience greater impacts than those in the colder regions. The greatest increases in thermal stress are forecasted in the mid to high latitude cities especially in populations with non-adapted architecture and no air conditioning. There has been increasing frequency of natural disasters in certain regions since the 1990s; these could continue to increase with a higher frequency and severity.

Knowledge is still limited in many areas, for example:

- the contribution of short-term climate variability to disease incidence needs further research;
- early warning systems for prediction of disease outbreaks, heatwaves and other extreme events need to be developed further and validated;
- there is a risk that frequent extreme events may lead over time to weakened adaptive capacity. To evaluate and correct this deficiency methods are needed to carry out national and regional level assessments for adaptive capacity.

Remaining challenges for scientists

In many respects, climate change is a different kind of problem to those that health scientists are accustomed to studying. To respond, new approaches and thoughtful recasting of existing scientific methods are needed. For example, to appreciate the possible impacts of climate change it is necessary (although not sufficient) to understand the present-day effects of weather and climate variability on health. From this understanding, and given the rate and extent of global climate change, there may be opportunities to study directly the early effects on human health. These impacts will depend largely on how successfully human societies can adapt to this threat. Processes of adaptation therefore comprise a key research area in their own right.

Further, a particular challenge is posed by uncertainties. Conventional quantitative risk assessment, based on well-documented risks in today's world and applied to existing population exposure profiles, can yield reasonable precision: that is, statistical and situational uncertainties often can be greatly reduced. However, forecasting future risks to health from the complex processes of large-scale environmental change entails an unusual range of uncertainties (*14*).

Uncertainties in climate change science come from many sources. Some depend on data or arise from uncertainties about the structure of models or their key parameter values, some result from differing projections of social, demographic and economic futures, others reflect divergence of values and attitudes. The precautionary principle assumes importance because of these uncertainties in forecasting the consequences (health and otherwise) of climate change. Where scientific knowledge is uncertain and the situation complex, and where there is a finite (though perhaps small) risk of serious (possibly irreversible) damage to population health, then preventive action should be taken. That is, in such potentially serious situations, scientific uncertainty does not justify policy inaction. Some key areas to address in current and future research include:

- identifying areas where first effects of climate change on human health will be apparent;
- improving estimates of climate change impacts by a combination of anticipated trends in adaptive capacity and climate scenarios;
- identifying the most helpful ways of expressing uncertainties associated with studies of climate change and health.

Extreme climate events

It can be difficult to understand climate change's relevance to health. The impact of a small magnitude change in mean temperature can seem insignificant as a respectively small magnitude change in sea level. The importance of such changes is in the way that they can indicate shifts in the distribution of health-relevant climate elements, especially their extreme values. For example, an increase in the mean temperature implies that new record high temperatures are likely but does not imply a corresponding change in the range between hottest and coldest temperatures (11) that comes with an increase in the variability. Other characteristics become apparent when both the mean and the variability change.

Analysis of the potential extremes associated with climate change is both complex and important: extreme weather conditions influence the health outcomes that the public health community will have to handle. The IPCC's Third Assessment Report projected changes in extreme climate events that include higher maximum temperatures; more hot days and heatwaves; more intense precipitation events; increased risk of drought; increase in wind and precipitation regimes of tropical cyclones (over some areas); intensified droughts and floods with El Niño events; and increased variability in Asian summer monsoon precipitation.

In assessing future health impacts, analogue studies of extreme events and human health can provide important clues about the interactions between climate, ecosystems and human societies that may be triggered by future climate trends. Localized effects of simple climate extremes are readily quantifiable in many situations. This is not the case for complex climate extremes although these would provide important qualitative insights into these relationships, and the factors affecting population vulnerability, and thus be of greatest value for public health.

There is good evidence of associations between important communicable diseases and climate on several temporal and geographical scales. The increasing trend of natural disasters is due partly to more complete reporting, partly to an increase in population vulnerability. Poverty, population growth and migration are major contributory factors affecting this vulnerability. Especially in poor countries, the impacts of major vector borne diseases and disasters can limit or even reverse improvements in social development. Research gaps to be addressed include:

- further modelling of relationships between extreme events and health impacts, especially in poor countries;
- improved understanding of factors affecting vulnerability to climate extremes;
- assessment of the effectiveness of adaptation measures in different settings.

Infectious diseases

Indirectly transmitted infectious diseases (e.g. via insect vectors or water) are highly susceptible to a combination of ecological and climatic factors because of the numerous components in their transmission cycles, and the interaction of each of these components with the external environment. Insights specifically related to climate changes' influence on infectious diseases can be derived from past disease epidemics and seasonal fluctuation; long term disease trend analysis; and predictive models capable of estimating how future scenarios of different climatic conditions will affect the transmissibility of particular infectious diseases. Over the past 30 years, observed intensification of El Niño in the Bay of Bengal paralleling increasing proportions of cholera cases may be one of the first pieces of evidence that warming trends are affecting human infectious diseases (15, 16).

Pathways through which weather conditions affect infectious diseases are variable and far greater than simply warmer temperatures. Precipitation extremes, humidity, and sea level change all can influence particular diseases. Certain moderating factors are important in examining how climate affects disease. These include socio-demographic influences such as human migration and transportation, drug resistance, and nutrition; as well as environmental influences such as deforestation or habitat fragmentation, agricultural development and water projects, biodiversity loss, and urbanization.

There remain many research and data needs. Disease incidence data is needed to provide a baseline for epidemiological studies. The lack of precise knowledge of current disease incidence rates makes it difficult to comment about whether incidence is changing as a result of climatic conditions. Research teams should be international and interdisciplinary, including epidemiologists, climatologists and ecologists to assimilate the diversity of information from these respective fields. The MARA project (17) serves as an excellent example for a useful validated predictive malaria model. Incorporating socioeconomic understanding of the adaptive capacity for societies affected by health outcomes of climate change will help predict population vulnerability, thereby optimizing preventive policies.

The burden of disease

It is not easy to assess and evaluate the risks posed to population health by global climate change. For a start, the empirical evidence implicating recent trends in regional and global climate in altered health outcomes is sparse. More generally, incomplete knowledge of how changes in climatic conditions affect a range of health outcomes makes it impossible to predict with confidence the range, timing and magnitude of likely future health impacts of global environmental changes. In spite of these limitations, estimation of the potential burden of disease due to climate change would help decision-makers to assess the potential magnitude of the problem (in human health terms) and assist the decision-making process concerned with adaptation. Considering only better-studied climate and health causal relationships, 150 000 deaths and 5.5 million DALYS can be attributed to climate change in 2000 (18).

Under conservative assumptions, the best models currently available indicate potential increases in the risk of diarrhoea, malaria, dengue, malnutrition, deaths and injuries from flooding, and heat-related mortality in tropical countries. These would be large enough to constitute a significant public-health impact within the next 30 years. Although there are predictions of some benefits of climate change in terms of reduced cardiovascular and respiratory mortality associated with low winter temperatures, mainly at higher latitudes, probably these will be small compared to negative effects elsewhere. The risks are heavily concentrated in populations of poorer tropical regions, mainly because poor socioeconomic conditions and infrastructure offer less protection against increased risks. Unless measures are taken to reduce overall risks in these populations, there will be unequal impacts on the health of the poor and marginalized groups.

Although the impacts potentially are large, estimates remain imprecise due to the uncertainties around climate predictions, climate-health relationships and possible future adaptation. Models that are driven mainly by changes in precipitation (inland flooding and malnutrition) are particularly uncertain. Key research gaps remain. There is a need, for example, for all impact models to be based on larger baseline datasets from a wider range of climate and socioeconomic conditions, and to be validated routinely against past and present disease distributions. More quantitative measures of adaptation and vulnerability also are needed in order to give better estimates of the degree to which potential risks of climate change are likely to be offset, or exacerbated, by changing socioeconomic conditions and biological adaptation. Improved probabilistic representation of all sources of uncertainty around predictions, including climate scenarios, disease-climate relationships and future adaptation is needed. There is also a need for improved methods of estimating the health impacts of changes in the frequency of extreme events and indirect causal pathways (e.g. population displacement due to flooding) which are difficult to model quantitatively.

Stratospheric ozone depletion, climate change and health

Stratospheric ozone depletion essentially is a different process from climate change. However, the phenomenon of the climate's greenhouse-warming shares many of the chemical and physical processes involved in the depletion of stratospheric ozone (19). Some greenhouse gases contribute to stratospheric ozone destruction. Agreements among nations (Montreal Protocol) have achieved reductions in ozone-depleting gases and it is expected that stratospheric ozone depletion will begin to decline in a decade or so. With all other sources assumed constant, the recovery of stratospheric ozone is expected to be underway by the second quarter of this century, and should be substantially complete by the third quarter. Unfortunately for climate change, that result is not all positive. Substitutes for the CFCs are known to contribute to greenhouse-warming, confirmed by global observations. Another confounding effect is that as ozone depletion acts to cool the climate system, ozone recovery is actually expected to contribute to climate warming.

"Failure to comply with the Montreal Protocol and its amendments would significantly delay or even prevent the ozone layer's future recovery jeopardizing, among others, public health." (20). A review of the evidence has been presented, for three main categories of health impact: skin cancers (adverse impact); eye disorders (adverse impact); immune-related processes and disorders (mixed impact—some adverse, some potentially beneficial). The epidemiological evidence for causation is best for skin cancers, less consistent for eye disorders and incomplete for immune-related disorders. Further research is required to understand this important risk factor. With knowledge of the changing climate (in addition to information and education campaigns in countries) patterns of individual and community sun-seeking and sun-avoiding behaviour will change, variably, around the world with consequent impacts on the received personal doses of ultraviolet radiation.

National assessments

Several countries, developed and developing, have undertaken national assessments of the potential health impacts of climate change that provide important information about future impacts on vulnerable areas and populations. However there is a need to improve and standardize the health impact assessment tools and methods. Quantitative epidemiological methods can allow some estimation of the range of magnitude of potential impacts, but the uncertainties surrounding those estimates are, so far, poorly described. Health impact assessments should provide information for, or feed into the process of, the UNFCCC national communications and other climate change assessments, strategies or action plans.

Much remains to be done: few comprehensive national assessments have been undertaken, even fewer in the most vulnerable populations and countries. Assessments in these populations would require some basic research on climate and health relationships, particularly in key areas such as diarrhoeal disease, vector-borne disease and malnutrition. More accurate climate information at the local level, particularly on climate variability and extremes, also is a prerequisite for impact assessment. Methods and tools for such assessments are still being developed.

Monitoring climate change impacts on human health

Climate change is likely to cause incremental changes in the frequency or distribution of diseases that also are affected by several other factors. In their report on malaria early warning systems Thomson and Connor stressed that the commonest causative factor in malaria epidemics is "abnormal meteorological conditions, which temporarily change the ecological equilibrium between host(s), vector(s) and parasite(s)" (21). Monitoring to assess climate-change impacts on health therefore requires data gathering coupled with analytical methods for quantifying the climate-attributable part of such diseases.

Monitoring and surveillance systems in many parts of the world currently are unable to provide data on climate-sensitive diseases (e.g. diarrhoea or vector-borne diseases) that are sufficiently standardized and reliable to allow comparisons over long time periods or between locations. Some aspects of disease systems (e.g. seasonality of biting patterns of disease vectors) are likely to be affected by climate and little else. Standardized long-term monitoring of such indicators could provide direct evidence of climate-change impacts on health. For the majority of impacts, however, simple long-term measurement of climate and disease may be uninformative as there will be many alternative explanations for any observed trends. In such cases, evidence on short-term meteorology-disease relationships applied to measured change in climate can be used as an indirect means of estimating climate change-related impacts on health. In addition, long-term surveillance should encompass variables that may confound observed associations between climatic changes and disease incidence.

Current research gaps include the need for more standardized surveillance of climate-sensitive health states, especially in developing countries. This is an argument for strengthening systems designed to meet current needs rather than creating new systems specifically to detect climate change impacts on human health. Also there is a need to facilitate both access to health data (especially from developing countries) and its linkage to information on climate and determinants of vulnerability.

Adapting to climate change

Since changes in the world's climate are occurring already and will continue in future decades, if not centuries, there is now a clear need for adaptation policies to complement mitigation policies. Efficient implementation of adaptation strategies can significantly reduce adverse health impacts of climate change.

Outside the health sector, adaptation options for managed systems such as agriculture and water supply generally are increasing because of technological advances. However, many developing countries have limited access to these technologies, appropriate information, finance or adequate institutional capacity. The effectiveness of adaptation strategies will depend upon cultural, educational, managerial, institutional, legal and regulatory practices that are both domestic and international in scope. In order to adapt for health impacts there is a critical general need for a sound and broadly-based public health infrastructure (including environmental management, public education, food safety regimes, vaccination programmes, nutritional support, emergency services and health status monitoring). This must be supplemented by health-directed policies in other sectors, including transport, urban planning, industry, agriculture, fisheries, energy, water management and so on.

Human populations vary in their vulnerability (or susceptibility) to certain health outcomes. The vulnerability of a population depends on factors such as population density, level of economic development, food availability, local environmental conditions, pre-existing health status and the availability of public health-care. It also depends on various structural and politically determined characteristics. Adaptive capacity in health systems varies among countries and socioeconomic groups.

The poorest groups in the poorest countries have the least ability to cope with climate change. Poor populations will be at greatest health risk from climate change because of their lack of access to material and information resources and because of their typically lower average levels of health and resilience (nutritional and otherwise). Long-term improvement in the health of impoverished populations will require income redistribution, increased employment opportunities, better housing and stronger public health infrastructure. Services with a direct impact on health, such as primary care, disease control, sanitation and disaster preparedness and relief, also must be improved. Development plays an important role in determining the adaptive capacity of communities and nations; enhancing adaptive capacity is necessary to reduce vulnerability. The reduction of socioeconomic vulnerability remains a top priority.

Implementation of adaptation measures usually will have near-term as well as future benefits due to reduction in impacts associated with current climate variability. In addition, adaptation measures can be integrated with other health objectives and programmes. For example, basic adaptation to climate change can be facilitated by improved environmental and health monitoring and surveillance systems. Basic indices of population health status (e.g. life expectancy) are available for most countries. However, disease (morbidity) surveillance varies widely depending on locality and the specific disease. To monitor disease incidence or prevalence (which may often provide a sensitive index of impact), low-cost data from primary care facilities could be collected in sentinel populations.

Although many actions need to be implemented with some urgency, decision-makers will benefit from research that answers specific questions. There is a key need for research on barriers and opportunities for enhancing adaptive capacity in order to protect human health, as well as potential interactions with ongoing development projects and programmes. Research also is needed on the processes of "adaptation decision-making", including identifying the roles and responsibilities of individuals, communities, nations, institutions and the private sector in adaptation. In addition, research on the costs and effectiveness of autonomous and planned adaptation measures is needed to assist in evaluating adaptation options.

Responses: from science to policy

It is clear that the magnitude and character of the problem of global climate change is such that a community-wide understanding and response is required, albeit guided by policy-makers provided with comprehensive advice from the international scientific community. Policy-focused assessment is a valuable process for providing timely and useful information to policy-makers, resource managers and other stakeholders in the public health community. Such assessments have already influenced—and are continuing to influence—policy and resource management decisions of interest to the public health community. A successful policy-focused assessment of the potential health impacts of climate change should have several key characteristics. These include the following:

- multidisciplinary assessment team;
- each assessment to answer in a timely fashion questions asked by stakeholders in the public health community;
- evaluation of risk management adaptation options;
- identification and prioritization of key research gaps;
- characterization and explanation of uncertainties and their implications for decision-making;
- development of tools in support of decision-making processes.

In addition, care must be taken to respect the boundary between assessment and policy formulation. Policy decisions are based on scientific assessments but also should include other considerations that reflect societal values (e.g. equity considerations) and other factors that affect decision-making (e.g. political feasibility).

The existence of scientific uncertainties about climate change and its potential health impacts does not preclude policy-makers from taking actions in anticipation of the health impacts of climate change. There are numerous examples in public health where improvements in health were achieved through effective actions before full scientific evidence was produced. Perhaps one of the most cited examples is that of John Snow's investigation and actions to stop a cholera epidemic in London, in 1854, a time when there were widely divided opinions on the causes of infectious diseases. Applying the yet unnamed precautionary principle, Snow managed to convince the relevant authorities to close down the water pump he identified as the source of the epidemic. The words of another famous epidemiologist note that action need not follow comprehensive scientific proof: "All scientific work is incomplete—whether it be observational or experimental. All scientific work is liable to be upset or modified by advancing knowledge. This does not confer upon us a freedom to ignore the knowledge that we already have, or to postpone the action that appears to demand at a given time" (9).

Numerous research questions about the potential impacts of climate change on human health have been identified. These need to be assessed and prioritized in order to be able to answer clearly the most important research gaps in each particular case and location. There is also an urgent need to perform assessments of adaptation strategies to reduce the risks to public health from climate change. For each adaptation option, assessments must evaluate the costs, benefits, effectiveness (in practice), barriers to implementation and risks of maladaptation. Finally better decision-support tools must be identified or developed to help public health officials make decisions under uncertainty, given available assessment results.

Concluding remarks

International agreements to deal with global environmental issues such as climate change should invoke the principles of sustainable development proposed in Agenda 21 and the UNFCCC. These include the precautionary principle described earlier, the principle of costs and responsibility, implying that the cost of pollution or environmental damage should be borne by those responsible, and that of equity. Considerations of equity or fairness can apply within and between countries and over time (between generations). Equity implies having equal or similar opportunities, allowing all to maintain an acceptable level of living conditions or quality of life. The balance of benefits and costs of climate change, for example, is likely to differ between affluent communities in wealthy countries and marginal populations in poor countries, and between current generations (some may benefit from early stages of warming) and future generations (costs will outweigh benefits, if forecast warming trends continue). Box 4 shows how these key principles are described in Agenda 21 and UNFCCC.

BOX 13.4 Key principles in Agenda 21 and UNFCCC

Precautionary approach

Agenda 21:

In order to protect the environment, the precautionary approach shall be widely applied by States according to their capabilities. Where there are threats of serious or irreversible damage, lack of full scientific certainty shall not be used as a reason for postponing cost-effective measures to prevent environmental degradation.

UNFCCC:

The Parties should take precautionary measures to anticipate, prevent or minimize the causes of climate change and mitigate its adverse effects. Where there are threats of serious or irreversible damage, lack of full scientific certainty shall not be used as a reason for postponing such measures, taking into account that policies and measures to deal with climate change should be cost-effective so as to ensure global benefits at the lowest possible cost.

Costs and responsibility

Agenda 21:

National authorities should endeavour to promote the internationalization of environmental costs and the use of economic instruments, taking into account the approach that the polluter should, in principle, bear the cost of pollution, with due regard to the public interest and without distorting international trade and investment.

UNFCCC:

The Parties should protect the climate system for the benefit of present and future generations of humankind on the basis of equity and in accordance with their common but differentiated responsibilities and respective capabilities. Accordingly, the developed countries should take the lead in combating climate change and the adverse effects thereof.

Continued

Adherence to these principles would make a substantial contribution towards the prevention of any future global environmental threat and the reduction of existing ones. As climate change processes already are underway, efforts also must focus on assessing current and future vulnerabilities and identifying necessary interventions or adaptation options. Adaptation has the potential to reduce adverse effects of climate change but is not expected to prevent all damages (22). Therefore, early planning for health is essential to reduce, hopefully avoid, near future and longer-term health impacts of global climate change. The optimal solution, however, is in the hands of governments, society and every individual—a commitment for a change in values to enable a full transition to sustainable development.

References

1. His Excellency Maumoon Abdul, Gayoom, President of the Republic of the Maldives. *Address to the United Nations General Assembly on the issues of environment and development.* New York, USA, 19 October 1987.
2. Watson, R. et al. *Protecting our planet securing our future: linkages among global environmental issues and human needs.* UNEP, NASA, World Bank, 1998.
3. McMichael, A.J. *Human frontiers, environments and disease: past patterns, uncertain futures.* Cambridge, UK, Cambridge University Press, 2001.
4. Corvalán, C.F. et al. Health, environment and sustainable development: identifying links and indicators to promote action. *Epidemiology* 10: 656–660 (1999).
5. McMichael, A.J. Population, environment, disease, and survival: past patterns, uncertain futures. *Lancet* 359: 1145–1148 (2002).
6. United Nations (UN). *Report of the UN conference on the human environment.* New York, USA, United Nations, 1973.
7. McMichael, A.J. et al. The sustainability transition: a new challenge. (Editorial). *Bulletin of the World Health Organization* 78: 1067 (2000).
8. United Nations (UN). *Agenda 21: the United Nations programme of action from Rio.* New York, USA, United Nations, 1993.
9. Hill, A.B. The environment and disease: association or causation? *Proceedings of the Royal Society of Medicine* 58: 295–300 (1965).
10. Brundtland, G.H. Director-General, World Health Organization. World Ecology Awards Ceremony, St Louis, Missouri, USA, 27 June 2001.
11. Intergovernmental Panel on Climate Change (IPCC). *Climate change 2001, the scientific basis.* Contribution of Working Group I to the Third Assessment Report of the Intergovernmental Panel on Climate Change. Cambridge, UK, Cambridge University Press, 2001.

12. Patz, J.A. et al. The potential health impacts of climate variability and change for the United States: executive summary of the report of the health sector of the U.S. National Assessment. *Environmental Health Perspectives* 108(4): 367–376 (2000).

13. World Meteorological Oorganization (WMO), Climatological Normals (CLINO) for the Period 1961–1990, World Meteorological Organization, 1996. (WMO No. 847; 92-63-00847-7).

14. O'Riordan, T. & McMichael, A.J. In: *Climate, environmental change and health: concepts and research methods.* Martens, W.J.M. & McMichael, A.J. eds. Cambridge, UK, Cambridge University Press, 2002.

15. Rodo, X. et al. ENSO and cholera: a nonstationary link related to climate change? *Proceedings of the National Academy of Sciences* 99(20): 12901–12906 (2002).

16. Patz, J. A human disease indicator for the effects of recent global climate change. *Proceedings of the National Academy of Sciences* 99: 12506–12508 (2002).

17. Cox, J. et al. *Mapping malaria risk in the highlands of Africa.* MARA/HIMAL technical report, 1999.

18. World Health Organization (WHO), *The World Health Report 2002.* Geneva, Switzerland, World Health Organization, 2002.

19. World Meteorological Organization (WMO)/United Nations Environment Programme(UNEP). *Scientific Assessment of Ozone Depletion,* 2002.

20. Obasi, G.O.P. Secretary-General, World Meteorological Organization. Press release: *The battle to repair the ozone layer is far from over,* WMO, 16 September 2002.

21. World Health Organization (WHO), *Malaria Early Warning Systems,* WHO, 2001. (WHO/CDS/RBM2001.32.).

22. Intergovernmental Panel on Climate Change (IPCC). *Climate Change 2001: impacts, adaptation and vulnerability.* Contribution of Working Group II to the Third Assessment Report. Cambridge, UK, Cambridge University Press, 2001.

Glossary[1]

absolute humidity: The mass of water vapour in a given volume of air.

acclimatization: Physiological and/or behavioural adaptation to climate.

acid rain (deposition): Precipitation that has a **pH** lower than about 5.0, the value produced when naturally occurring **carbon dioxide**, sulfate and **nitrogen oxide** dissolve into water droplets in clouds. Increases in acidity may occur naturally (e.g. following emissions of aerosols during volcanic eruptions) or as a result of human activities (e.g. emission of sulfur dioxide during **fossil fuel combustion**).

acute effect: Short-lived effect (in contrast to **chronic effect**).

adaptability: *See adaptive capacity.*

adaptation: Adjustment in natural or human systems to a new or changing environment. Adaptation to **climate change** refers to adjustment in response to actual or expected climatic stimuli or their effects, which moderates harm or exploits beneficial opportunities. Various types of adaptation can be distinguished, including anticipatory and reactive adaptation, public and private adaptation, and autonomous and planned adaptation.

adaptation assessment: The practice of identifying options to adapt to **climate change** and evaluating them in terms of criteria such as availability, benefits, costs, effectiveness, efficiency, and feasibility.

adaptation costs: Costs of planning, preparing for, facilitating, and implementing **adaptation** measures, including transition costs.

adaptive capacity: The ability of a system to adjust to **climate change** (including **climate variability** and **extreme events**) to moderate potential damages, take advantage of opportunities or cope with the consequences.

aeroallergen: Any of various airborne substances, such as pollen or spores, which can cause an allergic response.

aerosol: A collection of airborne solid or liquid particles with a typical size of 0.01–10 µm which are present in the atmosphere. Aerosols are an important source of negative **radiative forcing** and **acid rain**.

age related macular degeneration (AMD): An acquired degenerative disease which affects the central retina of patients most commonly over the age of 60.

air mass: Synoptic meteorological characterization of the entire body of air and its qualities. Air masses can be determined empirically using a combination of meteorological variables which include temperature, **relative humidity**, wind speed, wind direction, and barometric pressure.

albedo (whiteness): The fraction of solar radiation reflected by a surface or object, often expressed as a percentage. Snow covered surfaces have a high

---

[1] The glossary was prepared by Katrin Kuhn, London School of Hygiene and Tropical Medicine, London, England.

albedo; the albedo of soils ranges from high to low; vegetation covered surfaces and oceans have a low albedo. The Earth's albedo varies mainly through varying cloudiness, snow, ice, leaf area, and land cover changes.

algal blooms: Abnormally increased **biomass** of algae in a lake, river, or ocean.

amoebiasis: An infection of the large intestine by the protozoan parasite *Entamoeba histolytica*. The disease is frequently asymptomatic and varies from **dysentery** with fever, chills, and bloody or mucoid diarrhoea to mild abdominal discomfort with diarrhoea containing blood or mucus alternating with periods of constipation or remission.

amplification: The sharp increase in the size of a pathogen population, usually occurring in an amplifying host. *See also reservoir host.*

Annex I countries: Group of countries included in Annex I (as amended in 1998) to the **UNFCCC**, including all the developed countries in the Organisation for Economic Cooperation and Development, and economies in transition. By default, the other countries are referred to as non-Annex I countries.

anomaly: An event which is a deviation from normal behaviour that has a finite but very low probability of occurring.

a posteriori: *See a priori.*

a priori: Type of knowledge which is obtained independently of experience. A proposition is known *a priori* if one does not refer to experience to declare it true or false. Conversely, *a posteriori*, means knowledge gained through the senses and experience.

anthropogenic: Caused or produced by human activity. For example sulfate aerosols which are present in the troposphere due to the industrial emission of sulfur dioxide are called **anthropogenic**.

anthropogenic emissions: Emissions of **greenhouse gases** and **aerosols** associated with human activities. These include **fossil fuel burning** for energy, deforestation and land use changes that result in net increase in emissions.

anthroponosis: A disease of humans which can be transmitted to other animals.

arbovirus: Viruses transmitted by arthropods (arbo = <u>ar</u>thropod <u>bor</u>ne). Examples include the **dengue** virus, St. Louis encephalitis, Western equine encephalitis and **yellow fever**.

arid region/zone: Ecosystem which receives less than 250 mm precipitation per year.

atmosphere: The gaseous envelope surrounding the Earth. The dry atmosphere consists almost entirely of nitrogen and oxygen, together with a number of trace gases such as argon, helium and radiatively active **greenhouse gases** such as **carbon dioxide** and **ozone**. In addition, the atmosphere contains water vapour, clouds, and **aerosols**.

AOGCMs: Atmosphere-Ocean general circulation models. *See climate models.*

atopic eczema: Excess inflammation (dermatitis) of the skin and linings of the nose and lungs. It is very common in all parts of the world, mainly affecting infants and young adults.

attribution: *See detection and attribution.*

AIMS: Australian Institute of Marine Science.

autoimmune diseases: Any disorder in which the immune system mistakenly attacks the cells, tissues, and organs of a person's own body. There are many different autoimmune diseases, and they can each affect the body in different ways. For example, the autoimmune reaction is directed against the brain in multiple sclerosis and the gut in Crohn's disease.

basal metabolic rate: The minimal caloric requirements needed to sustain life in a resting person (i.e. a measure of the energy used by the body to maintain those processes necessary for life).

basic reproduction rate (R_0): A quantitative measure of the ability of a **vector-borne disease** to spread through a population. It is defined as the number of new cases of a disease which will arise from one current case when introduced into a non-immune host population during a single transmission cycle. A disease with a basic reproduction rate less than 1 will not spread in the community (i.e. become endemic). This rate will apply during the initial stages of spreading as the rate of disease spread will slow once the population has acquired some immunity.

billion: One thousand million, or 10^9.

biodiversity: The numbers and relative abundances of different genes (genetic diversity), species, and **ecosystems** (communities) in a particular area.

biofuel: A fuel produced from dry organic matter or combustible oils produced by plants.

biological model: a mathematical approach to determine the relationship between environmental variables and an outcome of interest (e.g. the distribution of disease vectors) using biological associations between the environment and aspects of population dynamics (e.g. how insect development rates change with temperature). Unlike **statistical models**, this approach requires detailed understanding of disease population dynamics.

biomass: The total mass of living organisms in a given area or volume: recently dead plant material is often included as dead biomass.

biome: A grouping of similar plant and animal communities that reflects the ecological and external character of the fauna in question. Biomes correspond approximately climatic regions, e.g. tropical rain forest biome, desert biome, and tundra biome.

biosphere: The part of the Earth's system comprising all **ecosystems** and living organisms in the atmosphere, on land (terrestrial biosphere), or in the oceans (marine biosphere), including derived dead organic matter such as litter, soil organic matter, and oceanic detritus.

biotoxin: Toxin produced by a living organism.

capacity building: In the context of **climate change**, capacity building is a process of developing the technical skills and institutional capability in developing countries and economies in transition to enable them to participate in all aspects of **adaptation** to, **mitigation** of, and research on climate change.

carbon dioxide (CO_2): A naturally occurring gas as well as a by-product of burning **fossil fuels** and land-use changes and other industrial processes. It is the principal **greenhouse gas** which affects the Earths **radiative balance** and the reference gas against which other greenhouse gases are measured.

carbon sink: Repository for **carbon dioxide** removed from the atmosphere. Oceans appear to be major sinks for storage of atmospheric CO_2.

carrying capacity: The number of individuals in a population that the resources of a habitat can support at a given point in time.

cataract: A clouding of the natural lens (the part of the eye responsible for focusing light and producing clear, sharp images) which is a natural result of the ageing process. Cataracts are the leading cause of visual loss in adults of 55 years and older.

CDC: Centers for Disease Control and Prevention.

chlorofluorocarbons (CFCs): Halogenated chemicals which are used for refrigeration, air conditioning, packaging, insulation, solvents, or aerosol propellants. They are all covered under the 1987 **Montreal Protocol**. Since they are not destroyed in the lower **atmosphere**, CFCs drift into the upper atmosphere where, given suitable conditions, they break down **ozone**. These gases are being replaced by other compounds, including hydrochlorofluorocarbons, covered under the **Kyoto Protocol**.

chloroquine: Medication used to treat and prevent **malaria**.

cholera: An intestinal infection, caused by the bacterium *Vibrio cholerae*, which results in frequent watery stools, cramping abdominal pain, and eventual collapse from dehydration. It is thought that **zooplankton** in cold waters may carry large number of cholera vibrios on their bodies. Zooplankton feed by grazing on **phytoplankton** which bloom with sunshine and warm conditions. Thus, a phytoplankton (algal) bloom may lead to an increase in the population of zooplankton which carry the vibrios.

chronic effect: Long-lasting effect (in contrast to **acute effect**).

CI: Conservation International (a non-governmental organisation).

ciguatera fish poisoning: Food-borne disease caused by ingestion of neurotoxins in certain fish. The toxins may become concentrated in higher predators, such as reef fish, which may remain toxic for more than two years after becoming contaminated. The symptoms of acute poisoning include gastrointestinal distress, followed by neurological and cardiovascular problems which are rarely fatal. Ciguatera is considered a major health and economic problem on many tropical islands where fish forms a large part of the diet.

CITES: Convention on International Trade in Endangered Species of Wild Fauna and Flora.

climate: Usually defined as the "average weather" or more rigorously as the statistical description in terms of the mean and variability of relevant quantities over a period of time ranging from months to thousands or millions of years. The classical period is 30 years as defined by the **WMO**. These relevant quantities are most often surface variables such as temperature, precipitation and wind.

climate change: Refers to a statistically significant variation in either the mean state of the climate or in is variability, persisting for an extended period (typically decades or longer). Climate change may be due to natural internal processes or external **forcings**, or to persistent **anthropogenic** changes in the composition of the **atmosphere**. The **UNFCC** defines climate change as "a change of climate which is attributed directly or indirectly to human activity that alters the composition of the global atmosphere and which is in addition to natural climate variability observed over comparable time periods". *See also climate variability.*

climate models: A numerical representation of the **climate system** based on the physical, chemical, and biological properties of its components, their interactions and feedback processes, and accounting for all or some of its known properties. The climate system can be represented by models of varying complexity, differing in such aspects as number of spatial dimensions and the extent to which physical, chemical, or biological processes are represented. Coupled **atmosphere–ocean general circulation models** (AOGCMs) provide a comprehensive representation of the climate system. There is an evolution towards more complex models with active chemistry and biology. Climate models are applied as a research tool to study and simulate the climate but also for operational purposes including climate **predictions**.

climate system: The highly complex system consisting of five major compo-nents: the **atmosphere**, the **hydrosphere**, the **cryosphere**, the land surface, and the **biosphere** and the interactions between them. The climate system evolves in time under the influence of its own internal dynamics and because of external (e.g. volcanic eruptions) and human **forcings** (e.g. changing com-position of the atmosphere).

climate variability: Variations in the mean state and other statistics (e.g. stan-dard deviations, the occurrence of **extreme events** etc) of the climate on all temporal and spatial scales beyond that of individual weather events. Vari-ability may be due to natural internal processes within the **climate system** or to variations in natural or **anthropogenic** external forcing.

confidence interval: An estimated range of values which is likely to include an unknown population parameter, the estimated range being calculated from a given set of sample data. If independent samples are taken repeatedly from the same population, and a confidence interval calculated for each sample, then a certain percentage (confidence level) of the intervals will include the unknown population parameter. Confidence intervals are usually calculated so that this percentage is 95%.

confounding factors: Any factor associated with the **exposure** under study and considered risk factors for the disease in their own right (i.e. not just inter-mediate variables on the pathway between exposure and disease).

copepods: Small crustaceans which are the most numerous multi-celled animals in the aquatic community. Their habitats range from the highest mountains to the deepest ocean trenches, and from the cold polar ice-water interface to the hot active hydrothermal vents. Copepods may be free-living, symbiotic, or internal or external parasites on almost every phylum of animals in water. They also have the potential to control malaria by consuming mosquito larvae and are thought to be intermediate hosts for many human and animal parasites.

coping ability/capacity: The variation in climatic stimuli that a system can absorb without producing significant impacts.

coral bleaching: The paling in colour of corals resulting from a loss of symbi-otic algae. Bleach occurs in response to physiological shock in response to abrupt changes in temperature, salinity, and turbidity. Mass-bleaching events have been associated with small changes in sea temperature.

Coriolis force: A bending force causing any movement on the Northern hemi-sphere to be diverted to the right when the globe is rotating (in the Southern hemisphere it is bent to the left).

cost-effective: A criterion that specifies whether a technology of measure deliv-ers a service at an equal or lower cost than current practice, or the least-cost alternative for the achievement of a given target.

CRED: The Centre for Research on the Epidemiology of Disasters.

cryosphere: The component of the climate system consisting of all snow, ice, and **permafrost** on and beneath the surface of the earth and ocean.

cryptosporidium: Genus of parasites of the intestinal tracts of fishes, reptiles, birds, and mammals. A particular species isolated in humans has been identi-fied as *Cryptosporidium parvum*. Cryptosporidiosis, or Cryptosporidium infec-tion, is today recognized as an important opportunistic infection, especially in immunocompromised hosts.

CMM: Cutaneous malignant melanoma.

Cytomegalovirus (CMV): A virus related to the herpes virus. It is so common that almost 100 percent of adults in developing countries and 50 percent to

85 percent of adults in developed world are infected. Usually the virus causes no serious problems, however in immunocompromised hosts and newborns of infected mothers cytomegalovirus can be fatal.

demography: The study of populations, especially with reference to size and density, fertility, mortality, growth, age distribution, migration, and the interaction of all these factors with social and economic conditions.

dengue/dengue haemorrhagic fever (DHF): An acute febrile syndrome caused by dengue arbovirus type 1–4, commonly transmitted by the mosquitoes *Aedes aegypti* and *Ae. albopictus* which breed in small water bodies in containers, car tyres etc. Dengue is often called breakbone fever because it is characterized by severe pain in joints and back. Subsequent infections of dengue virus may lead to dengue haemorrhagic fever which can be fatal.

desert: An **ecosystem** which receives less than 100 mm precipitation per year.

detection and attribution: Detection of climate change is the process of demonstrating that climate has changed in some defined statistical sense, without providing a reason for that change. Attribution of causes and effects of climate change is the process of establishing the most likely causes for the detected change or effect with some defined level of confidence.

diphtheria: An acute toxin-mediated bacterial disease which usually affects the tonsils, throat, nose and/or skin. It is passed from person to person by droplet transmission, usually by breathing in diphtheria bacteria after an infected person has coughed, sneezed or even laughed. Diphtheria can lead to breathing problems, heart failure, paralysis and sometimes death.

direct transmission: Transmission of an infectious disease from human to human (or animal to animal), without the involvement of **intermediate** hosts or **reservoirs** (e.g. **TB**, sexually transmitted diseases).

Disability Adjusted Life Year (DALY): An indicator of life expectancy combining mortality and morbidity into one summary measure of population health to account for the number of years lived in less than optimal health. It is a health measure developed for calculating the global burden of disease which is also used by WHO, the World Bank and other organizations to compare the outcomes of different interventions.

diurnal temperature range: The difference between minimum and maximum temperature over a period of 24 hours.

dose-response: Association between dose and the incidence of a defined biological effect in an exposed population. Dose-response relationships are used to determine the probability of a specific outcome or disease, or risk of a disease, by extrapolating from high doses to low doses and from laboratory animals to humans, and using **mathematical models** that define risk as a function of exposure dose.

drought: The phenomenon that exists when precipitation has been significantly below normal recorded levels, causing serious hydrological imbalances.

dysentery: An infection of the gut caused by shigella bacteria. Symptoms include acute bloody diarrhoea, vomiting, stomach pains and fever.

early warning systems (EWS): A system consisting of **Mathematical models** and **surveillance** measures designed for the early detection, prevention and control of an **epidemic** of infectious disease or other abnormal event (e.g. famine or heat waves).

economies in transition: A type of national economy in the process of changing from a planned economic system to a market economy.

ecological study: Study in which the analysis of a relationship is based on aggregate or grouped data (such as **rates**, proportions and means)—i.e. no data are collected at the individual level.

ecological system/ecosystem: A system of living organisms together in their physical environment, with specific interactions and exchange of matter, energy and information. The boundaries of an ecosystem depend on the focus of interest and can range from very small **spatial scales** to the entire Earth.

effect modifier: A factor that modifies (alters), by variation in intensity or magnitude, the effect of a risk factor under study; a generic term which includes interaction, synergism, and antagonism.

El Niño/Southern Oscillation (ENSO): El Niño, in its original sense, is a warm water current that periodically flows along the coast of Ecuador and Peru. This event is associated with a fluctuation of the intertropical surface pressure patterns and circulation in the Indian and Pacific Oceans, called the Southern Oscillation. This coupled atmosphere–ocean phenomenon is collective known as the El Niño Southern Oscillation or ENSO. During an El Niño event, the prevailing trade winds weaken and the equatorial counter current strengthens, causing warm surface waters in the Indonesian area to flow eastward to overlie the cold waters of the Peru current. This event has great impact on the wind, **sea surface temperature**, and precipitation patterns in the tropical Pacific. It has climatic effects throughout the Pacific region and in many other parts of the world. The opposite of an El Niño event is called La Niña.

EM-DAT: The Emergency Events Database—EM-DAT, created and maintained by the Université Catholique de Louvain in Belgium. The main objective of the database is to serve the purposes of humanitarian action at national and international levels. For example, it allows one to decide whether floods in a given country are more significant in terms of humans impact than are earthquakes, or whether a country is more vulnerable than another.

emerging infectious disease: A disease which is new in the population or rapidly increasing in **incidence** or expanding in geographical range.

emission: In the **climate change** context, the release of **greenhouse gases** and/or **aerosols** into the atmosphere over a specified area and period of time.

emission scenario: A possible pattern of net **greenhouse gas** and **aerosol** emissions for the next hundred years or more. Emission scenarios provide input for **climate models** and contribute to the evaluation of future radiative forcing on the atmosphere. Emission scenarios are not predictions of the future but illustrate the effect of a wide range of economic, demographic, and policy assumptions. *See also SRES.*

endemic: Term applied to describe sustained, relatively stable transmission pattern of an infection within a specified population in a certain locality or region.

entomological inoculation rate (EIR): The number of infectious mosquito bites a person is exposed to in a certain time period, typically a year.

enzootic disease: An **endemic** and **zoonotic** disease—i.e. affecting mainly animals but with an occasional spill-over to humans such as **Rift Valley Fever, West Nile virus**.

epidemic: Occurrence in a community or region of cases of an illness, specific health-related behaviour, or other health-related events clearly in excess of normal expectancy. The community or region and the period in which the cases occur are specified precisely. The number of cases indicating the presence of an epidemic varies according to the agent, size, and type of population exposed; previous experience or lack of exposure to the disease; and time and place of occurrence. *See also outbreak.*

epidemiology: Study of the distribution and determinants of health-related states or events in specified populations. Epidemiology is the basic quantitative science of public health.

epizootic: An out break (**epidemic**) of disease in an animal population, often with the implication that it may also affect humans. *See also enzootic.*

erythema: Reddening of the skin.

evaporation: The process by which a liquid becomes a gas.

evapotranspiration: The sum total of water lost from land through physical **evaporation** and plant transpiration.

exposure: Amount of a factor to which a group or individual was exposed; sometimes contrasted with dose (the amount that enters or interacts with the organism). Exposures may be either beneficial or harmful.

extreme event: A weather event that is rare within its statistical reference distribution at a particular place. Definitions of "rare" vary, but an extreme event would normally be as rare or rarer than the 10th or 90th percentile. By definition, the characteristics may vary from place to place. An extreme **climate** event is an average of a number of weather events over a certain period of time, an average which is itself extreme.

extrinsic incubation period: In blood-feeding arthropod vectors, the time between acquisition of the infectious blood meal and the time when the vector becomes capable of transmitting the agent. In the case of **malaria**, this is the life stages of the parasite spent within the female mosquito vector.

falciparum malaria: *See* Plasmodium falciparum *and malaria.*

FAO: Food and Agricultural Organisation of the United Nations.

feedback: an interaction mechanism between processes in the **climate system** is called a climate feedback when the result of an initial process triggers changes in a second process that in turn influences the initial one. A positive feedback intensifies the original process, and a negative feedback reduces it.

first principle: The first law of thermodynamics says that the total *quantity* of energy in the universe remains constant. This is the principle of the conservation of energy. The first principle establishes the equivalence of the different forms of energy (radiant, chemical, physical, electrical, and thermal), the possibility of transformation from one form to another, and the laws that govern these transformations. This first principle considers heat and energy as two magnitudes of the same physical nature.

flood: Temporary partial or complete inundation of normally dry areas caused by rapid runoff or overflow from lakes, rivers, or tidal waters.

forcings: *See climate system.*

forecast: *See prediction.*

fossil fuels: Carbon-based burning materials from fossil carbon deposits, including coal, oil and natural gas.

fossil CO_2 (carbon dioxide) emissions: The action of giving off **carbon dioxide** resulting from the combustion of fuels from fossil carbon deposits such as oil, natural gas and coal.

Fourier analysis (spectral analysis): A mathematical analysis that describes variations within a time series of data in terms of cycles, of different frequencies and amplitudes. It is often used for describing seasonal and longer-term cyclical variations in disease incidence.

free radical: Any highly reactive chemical molecule that has at least one unpaired electron.

fuzzy logic/suitability: An approach to mathematics and computing based on "degrees of truth" rather than the usual "true or false" (1 or 0) Boolean logic on which the modern computer is based. Fuzzy logic includes 0 and 1 as extreme cases of truth (or "the state of matters" or "fact") but also includes the various states of truth in between.

GEF: Global Environment Facility.

general circulation: The large scale motions of the atmosphere and the ocean as a consequence of differential heating on a rotating earth, aiming to restore the energy balance of the system through transport of heat and momentum.

genetic engineering: The techniques used to manipulate genes in an organism in order to study their functions and their interactions in an environment different from the original one.

Geographical Information System (GIS): System of hardware, software and procedures designed for integrated storing, management, manipulation, analysing, modelling and display of spatially referenced data for solving planning and management problems.

glacier: A mass of land ice flowing downhill and constrained by the local topography. A glacier is maintained by accumulation of snow at high altitudes, balanced by melting at low altitudes or discharge into the sea.

Global Positioning System (GPS): A hand-held radio navigation system that allows land, sea, and airborne users to determine their exact location, velocity, and time 24 hours a day, in all weather conditions, anywhere in the world.

global warming: Observed and **projected** temperature increases.

gonotrophic cycle: For blood-feeding arthropods, the interval between blood meal and egg-laying.

GLWQA: Great Lakes Water Quality Agreement.

greenhouse effect: Greenhouse gases absorb **infrared radiation**, emitted by the Earth's surface, the **atmosphere** itself due to the same gases and by clouds. Atmospheric radiation is emitted to all sides, including downward to the Earth's surface. Thus greenhouse gases trap heat within the surface–troposphere system. This is called the "natural greenhouse effect". Atmospheric radiation is strongly coupled to the temperature of the level at which it is emitted. An increase in the concentration of greenhouse gases leads to an increased infrared opacity of the atmosphere and therefore to an effective radiation into space from a higher altitude at a lower temperature. This causes a radiative forcing, an imbalance that can only be compensated for by an increase of the temperature of the surface-troposphere system. This is the "enhanced greenhouse effect".

greenhouse gases (GHGs): Those gases in the **atmosphere** which absorb and emit radiation at specific wavelengths within the spectrum of **infrared radiation** emitted by the Earth's surface, the atmosphere and clouds. **Water vapour**, **carbon dioxide**, **nitrous oxide**, **methane** and **ozone** are the primary greenhouse gases in the atmosphere. Moreover, there are a number of entirely human-made gases in the atmosphere, such as the **halocarbons** and others dealt with under the **Montreal** and **Kyoto Protocols**.

Gross Domestic Product (GDP): The sum of gross value added, at purchaser's prices, by all resident and non-resident producers in the economy in a country or region for a given period of time (normally 1 year), plus any taxes and minus any subsidies not included in the value of the products. GDP is an often used measure of welfare.

Gulf stream (thermohaline current): A well-defined western boundary current of the North Atlantic, which carries warm, saline tropical water north and north-eastward along the eastern coast of the United States, joining the Labrador Current at the Grand Banks, about 40°N and 50°W, to become the North Atlantic Current; generally swift and deep, it transports a very large volume of water.

haemorrhagic: Causing or characterised by haemorrhage or bleeding.

halocarbons: A group of human-made chemicals that contain carbon and members of the halogen family (fluorine, chlorine or bromide). Halocarbons include **chlorofluorocarbons**, substances that deplete stratospheric ozone.

hantavirus pulmonary syndrome: A recently identified **zoonotic** disease, caused by a virus (hantavirus), carried by rodents. Infection in humans occur via inhalation or ingestion of materials contaminated with rodent excreta, although a tick vector may be involved. Early symptoms include fever, fatigue and muscle aches while late symptoms consist of coughing and shortness of breath (hence the name).

hard ticks: Ticks of the family Ixodidae, characterised by the presence of a scutum (dorsal plate) and visible mouthparts from the dorsal side. *See also soft ticks.*

Health For All: A global health policy aimed at meeting the major challenges in health during the next decades, that has been developed by the World Health Organization in consultation with all its national and international partners. This policy for the 21st century evolves from the Health-For-All policy which has been a common aspirational goal since its inception in 1979.

health impact assessment (HIA): A combination of procedures, methods and tools by which a policy, project or hazard may be judged as to its potential effects on the health of a population, and the distribution of those effects within the population.

heat budget: Heat budgets are a way of studying atmospheric processes to indicate the sources and sinks of energy. The atmospheric heat budget shows where the atmospheric heat energy comes from and where it goes.

heat island effect: Local human-induced climate conditions (high temperatures) in urban areas caused by heat adsorption in concrete, brick and pavement surfaces, reduction of convective cooling due to presence of tall buildings, and reduced evaporative cooling.

helminths: Specific type of parasitic worm.

historical analogue studies: Studies that use a past event to elucidate factors pertaining to current or future events.

Humboldt current: A cold ocean current of the South Pacific, flowing north along the western coast of South America. Also called Peru Current.

hydrological cycle: Movement and circulation of water in the atmosphere, on land surfaces, and through the soils and subsurface of rocks. About 97% of the world's water is in the oceans and about 75% of its fresh water takes the form of glaciers and polar ice. Water vapour in the atmosphere condenses and appears as dew or precipitation. Liquid water, ice and snow evaporate.

hydroxyl ions/radicals: One of the most toxic and reactive type of free radicals.

hypothermia: Condition which occurs when the human body temperature drops below 35.5 °C or 96 °F due to exposure to cold. Symptoms include slow or irregular speech, shallow or very slow breathing, fatigue, confusion, slow pulse, weakness or drowsiness, shivering, cold, pale skin.

ice sheet: A glacier of more than $50\,000\,km^2$ in area forming a continuous cover over a land surface or resting on a continental shelf. There are only two large ice sheets in the modern world, on Greenland and Antarctica.

impacts: Consequences of **climate change** on natural systems and human health. Depending on the consideration of **adaptation**, we can distinguish between potential impacts and residual impacts:

- Potential impacts are all impacts that may occur given a projected change in **climate**, with no consideration of adaptation.
- Residual impacts are the impacts of climate change that can occur after adaptation.

immunosuppression: Reduction in the effectiveness of a person's immune system. Local immunosuppression occurs at the site of **exposure** or disturbance. Systemic immunosuppression involves a reduction of the body's immune response at a site distant from the exposure.

incidence: The number of cases of illness commencing, or of persons falling ill, during a given time period within a specified population. *See also prevalence.*

indirect transmission: transmission of an infectious disease with the involvement of **intermediate** hosts, **vectors** or **reservoirs** (e.g. **malaria**, **hantavirus**).

Industrial Revolution: A period of rapid industrial grown beginning in England during the second half of the 18th century and spreading to Europe and other countries. The invention of the steam engine was an important trigger of this development. The Industrial Revolution marks the beginning of a strong increase in the use of **fossil fuels**.

inertia: Delay, slowness, or resistance in the response of the **climate**, biological or human systems to factors that alter their rate of change, including continuation of change in the system after the cause of that change has been removed.

infection rate: Proportion of all individuals in a population infected with a specific disease agent.

infrared radiation: Radiation emitted by the Earth's surface, the atmosphere, and clouds. It is also known as terrestrial or long-wave radiation. Infrared radiation has a distinctive range of wavelengths (spectrum) longer than the wavelength of the red colour in the visible part of the spectrum.

insecticide: A pesticide used for controlling or eliminating insects.

integrated assessment: A method of analysis that combines results and models from the physical, biological, economic, and social sciences, and the interactions between these components, in a consistent framework, to evaluate the status and the consequences of environmental change and the policy responses to it.

intermediate host: Host of a disease agent other than the one in which sexually mature forms of the pathogen occur. *See also reservoir host.*

Intergovernmental Panel on Climate Change (IPCC): A group of experts established in 1988 by the **World Meteorological Organization (WMO)** and the **United Nations Environment Programme (UNEP).** Its role is to assess the scientific, technical and socio-economic information relevant for the understanding of the risk of human-induced **climate change**, based mainly on peer reviewed and published scientific/technical literature. The IPCC has three Working Groups and a Task Force.

interquartile range: The distance between the 75th percentile and the 25th percentile. The interquartile range is essentially the range of the middle 50% of the data.

in vitro: A process that takes place under artificial conditions or outside of the living organism.

in vivo: A process that takes place inside the living organism.

IHD: Ischaemic heart disease, i.e. heart problems caused by narrowed heart arteries (also called coronary artery disease and coronary heart disease). Often causes chest pain known as angina pectoris and can ultimately lead to heart attack.

Kyoto Protocol: An agreement which was adopted at the third session of the **UNFCCC** conference in Japan in 1997. It contains legally binding commitments, in addition to those included in the **UNFCCC**.

Landsat Thematic Mapper: A U.S. **remote sensing satellite** used to acquire images of the Earth's land surface and surrounding coastal regions.

La Niña: *See El Niño Southern Oscillation.*

leishmaniasis: Infection with *Leishmania* parasites, resulting in a group of diseases classified as cutaneous, mucocutaneous or visceral. Transmission is by sandflies of the genus *Phlebotomus* or *Lutzomyia*. In most regions, the transmission cycle involves **reservoir hosts** (primarily wild or domestic canines and rodents) but parasites can also be transmitted from person to person by the bite of sandflies.

leptospirosis: Bacterial infection of humans by the genus *Leptospira*. Symptoms include high fever, jaundice, severe muscular pains and vomiting. Transmission is associated with contact with infected animals or water contaminated with rat urine. Also known as Weil's disease.

Lyme disease: A zoonotic bacterial infection caused by the spirochaete *Borrelia burgdoferi* and transmitted by **hard ticks** of the genus *Ixodes*. The main animal **reservoir hosts** for Lyme disease are wild deer as well as domesticated pets.

lupus erythematosus: An auto-immune illness that affects the skin and internal organs.

lymphatic (Bancroftian) filariasis: Parasitic disease common in tropical and subtropical countries. Microscopic parasitic worms (*Wuchereria bancrofti* or *Brugia malayi*) are transmitted to humans by several mosquito species (including *Anopheles*, *Aedes* and *Culex*). The worms cause inflammation and eventual blocking of lymph vessels, resulting in a swelling of the surrounding tissue, often referred to as elephantiasis.

lymphocytes: White blood cells circulating in blood and lymph and involved in antigen-specific immune reactions. Lymphocytes are subdivided into B-lymphocytes, which produce circulating antibodies, and T-lymphocytes, which are primarily responsible for cell-mediated immunity. T-lymphocytes are divided into cytotoxic lymphocytes which bind to and kill foreign cells, helper T-lymphocytes which assist antibody production, and suppressor T-lymphocytes which inhibit this immune response.

maladaptation: Any changes in natural or human systems that inadvertently increase vulnerability to climatic stimuli; an adaptation that does not succeed in reducing vulnerability but increases it instead.

malaria: Endemic or epidemic parasitic disease caused by four species of the protozoan genus *Plasmodium* which are transmitted to humans by the bite of female *Anopheles* mosquitoes. Disease is characterised by high fever attacks and systemic disorders and is responsible for approximately 2 million deaths every year, 90% of which occur in Sub-Saharan Africa. Malaria is the most serious and common vector-borne disease in the world.

MARA: Mapping Malaria Risk in Africa. A collaboration initiated to provide an atlas of African malaria, containing relevant information for rational and targeted implementation of malaria control.

MPAs: Marine Protected Areas.

mathematical models: Representation of a system, process or relationship in mathematical form in which equations are used to describe the behaviour of the system or process under study. See also **biological** and **statistical models** and **predictive modelling**.

melatonin: A hormone produced by the pineal gland in the brain which can help regulate the symptoms of jet lag or insomnia.

meningococcal meningitis: Cerebrospinal meningitis or fever, characterised by infection of the fluid (cerebrospinal fluid, or CSF) and tissues (meninges) that surround the brain and spinal cord. Meningococcal meningitis is caused by infection with the meningococcus bacteria.

mesosphere: Region of the Earth's **atmosphere** above the **stratosphere.**

meta analysis: Process of using statistical methods to combine the results of different independent studies.

methane (CH$_4$): A hydrocarbon that is a **greenhouse gas** produced through anaerobic (without oxygen) decomposition of waste in landfills, animal digestion, decomposition of animal wastes, coal production, and incomplete fossil-fuel combustion. It is one of the six gases to be mitigated under the **Kyoto Protocol**.

micro climate: (i) In climatology: localised climate, incorporating physical processed in the atmospheric boundary layer. The boundary layer is the lowest 100–200 m of the **atmosphere** and the part of the **troposphere** that is directly influenced by Earth's surface. For example, atmospheric humidity is influenced by vegetation, ambient air temperatures by building and roads etc. (ii) in ecology: climatic conditions in the environmental space occupied by a species, a community of species or an **ecosystem**. For example, on mountain slopes, temperatures experienced by plants differ depending on the direction of the slope. Similarly, in forests, air temperature varies according to canopy cover and height. In many cases, such differentials are crucial for species survival and longevity.

minimum erythemal dose (MED): Minimal dose of ultraviolet radiation sufficient to cause **erythema**.

mitigation: Human intervention to reduce emissions or enhance the sinks of **greenhouse gases**.

monitoring: Performance and analysis of routine measurements aimed at detecting changes in the environment or health status of populations. Not to be confused with **surveillance** although surveillance techniques may be used in monitoring.

Montreal Protocol: The international agreement signed in 1987 to limit the production and emission of substances that deplete stratospheric **ozone**. The Parties to the Protocol further agreed to the London and Copenhagen Adjustments and Amendments in 1990 and 1992, respectively, aimed at accelerating the phasing out of ozone-depleting substances by 1 January 1996 (although concessionary delays have been applied to developing countries).

morbidity: Rate of occurrence of disease or other health disorder within a population, taking account of the age-specific morbidity rates. Health outcomes include: chronic disease incidence/prevalence, hospitalisation rates, primary care consultations and **Disability-Adjusted-Life-Years (DALYs).**

mortality: Rate of occurrence of death within a population within a specified time period.

Newton's second law of motion: A physics law which explains how an object will change velocity if it is pushed or pulled upon. Firstly, this law states that if you do place a force on an object, it will accelerate, i.e., change its velocity, and it will change its velocity in the direction of the force. Secondly, this acceleration is directly proportional to the force. For example, if you are pushing on an object, causing it to accelerate, and then you push, say, three times harder, the acceleration will be three times greater. Thirdly, this acceleration is inversely proportional to the mass of the object. For example, if you are pushing equally on two objects, and one of the objects has five times more mass than the other, it will accelerate at one fifth the acceleration of the other.

nitrous oxide (N_2O): A powerful **greenhouse gas** emitted through soil cultivation practices, especially the use of commercial and organic fertilizers, fossil fuel combustion, nitric acid production, and **biomass** burning. One of the six greenhouse gases to be curbed under the **Kyoto Protocol.**

non-Hodgkin's Lymphomas (NHL): A type of cancer of the lymphatic system. There are two main types of lymphoma: (1) Hodgkin's disease and (2) non-Hodgkin's lymphoma. The two are only distinguishable by microscopic examination. There are approximately 20 different types of non Hodgkin's lymphoma each with a different characteristic and cell invasion behaviour.

Normalized Difference Vegetation Index (NDVI): A remotely sensed index which is used to classify the greenness (i.e. vegetation coverage) of an area. It is related to the proportion of photosynthetically absorbed radiation, and calculated from atmospherically corrected reflectances from the visible and near infrared channels detected by **remote sensing satellites.**

North Atlantic Oscillation (NAO): Opposing variations of barometric pressure near Iceland and the Azores. On average a westerly current between the Icelandic low pressure area and the Azores high pressure area carries cyclones with their associated frontal systems towards Europe. However, the pressure difference between Iceland and the Azores fluctuates and can be reversed at any time. It is the dominant mode of winter **climate variability** in the North Atlantic region.

onchocerciasis: Also known as river blindness. A parasitic disease in the tropical regions of Africa and America, caused by infestation by a filarial worm (usually *Onchocerca volvulus*) and transmitted by the bite of various species of blackfly. Infection causes subcutaneous nodules and, if worms migrate to the eye, very often blindness.

oropouche: A virus of the Orthobunyavirus genus causing disease in humans in the Caribbean and Central and South America. The pathogen is transmitted by the tiny biting midge *Culicoides paraensis* and causes a self-limiting, acute, dengue-like febrile illness called ORO fever.

outbreak: An **epidemic** limited to localised increase in the incidence of a disease, e.g., in a village, town or closed institution.

ozone: Form of the element oxygen with three atoms instead of the two that characterise normal oxygen molecules. Ozone is an important **greenhouse gas**. The **stratosphere** contains 90% of all the ozone present in the **atmosphere** which absorbs harmful ultraviolet radiation. In high concentrations, ozone can be harmful to a wide range of living organisms. Depletion of stratospheric ozone, due to chemical reactions that may be enhanced by **climate change**, results in an increased ground-level flux of **ultraviolet-B-radiation**.

ozone layer: *See stratospheric ozone layer.*

Pacific Decadal Oscillation (PDO): A long-lived **El Niño**-like pattern of Pacific **climate variability.** Two main characteristics distinguish PDO from **ENSO:** (1) PDO "events" persist for 20-to-30 years, while typical ENSO events persist for 6 to 18 months and (2) the climatic fingerprints of the PDO are most visible in the North Pacific/North American sector, while secondary signatures exist in the tropics—the opposite is true for ENSO.

PAHO: Pan American Health Organization.

paleoclimatology: Study of past climates based on data from fossils and ice cores.

pandemic: Epidemic occurring over a very wide area, crossing international boundaries and usually affecting a large number of people.

permafrost: Perennially frozen ground that occurs wherever the temperature remains below 0°C for several years.

pH: Measure of the acidity or alkalinity of a solution, ranging from 0 (acidic) to 7 (neutral) to 14 (alkaline).

photochemical oxidants: *See secondary air pollutants.*

photoconjuctivitis: Acute inflammation of the conjunctiva caused by prolonged exposure to intense solar radiation.

photokeratitis: Acute reversible inflammation of the cornea caused by prolonged exposure to intense solar radiation, usually in highly reflective environments. Temporary visual loss associated with ultraviolet radiation reflected from the surface of snow is known as "snow blindness".

photoperiod: The period during every 24 hours when an organism is exposed to daylight.

photosynthesis: Process by which the energy of sunlight is used by green plants to build up complex substances from carbon dioxide and water.

phytoplankton: The plant form of **plankton** (e.g. diatoms). Phytoplankton are the dominant plants in the sea, and are the base of the entire marine food web. These single-celled organisms are the principal agents for photosynthetic carbon fixation in the ocean. *See also cholera and zooplankton.*

PICCAP: Pacific Islands Climate Change Assistance Programme.

(bubonic) plague: Infectious disease caused by the bacterium *Yersinia pestis* and transmitted from rodent to rodent by infected fleas. Rat-borne epidemics continue to occur in some developing countries, particularly in rural areas. Symptoms include fever, headache, and general illness, followed by the development of painful, swollen regional lymph nodes. Once a human is infected, a progressive and potentially fatal illness generally results unless specific antibiotic therapy is given.

plankton: Aquatic organisms that drift or swim weakly. *See also phytoplankton and zooplankton.*

Plasmodium falciparum: One of the four species of *Plasmodium* that cause human malaria and the one associated with the highest morbidity and mortality. *See also malaria.*

Plasmodium vivax: One of the four species causing human malaria, associated with less severe but prolonged symptoms. *See also malaria.*

polio (poliomyelitis): An inflammation of the grey matter of the spinal cord, caused by a virus which results in an acute infection. It is believed that the virus is transmitted by contact with the faeces of an already infected person. The majority of infected individuals experience only mild symptoms or non-paralytic polio. The virus penetrates the nervous system in a small number of cases, causing varying degrees of muscle weakness and paralysis—i.e. true polio.

population health: A measure of the health status of populations, proposed during the 1990s to selectively replace the use of the terms *human health* which is more restrictive, and *public health* which also encompasses preventive and curative measures and infrastructures.

positive radiative forcing: *See radiative forcing.*

ppb: Parts per billion. One ppb is 1 part in one billion by volume. *See also parts per million.*

ppm: Parts per million; unit of concentration often used when measuring levels of pollutants in air, water, body fluids, etc. One ppm is 1 part in one million by volume.

precautionary principle: The adoption of prudence when outcomes are uncertain but potentially serious.

prediction: In the context of climate, a prediction or forecast is the result of an attempt to produce a most likely description or estimate of the actual evolution of the climate in the future (e.g. at seasonal, interannual or long term time scales). *See also projection.*

prevalence: The number of events, e.g. instances of a given disease or other condition, in a given population at a designated time. *See also incidence.*

primary air pollutants: Air pollutants produced as a result of the combustion of **fossil** and **biomass** fuels. They include: carbon monoxide, nitrogen oxides and sulfur dioxide.

primary health care: Essential health care made accessible at a cost the relevant country and community can afford, incorporating methods that are practical, scientifically sound and socially acceptable. This may include community education, promotion of adequate food supplies, basic sanitation and water, family planning and the prevention and control of locally **endemic** diseases.

principal component analysis: A mathematical transformation of a sample of points in N-dimensional space, so that they are measured on axes (principal components) that maximize the amount of variation in the dataset. These principal components are usually weighted combinations of the original measured values (e.g. a combination of temperature and rainfall information).

projections: A potential future evolution of a quantity or set of quantities, often computed with the aid of a model. Projections are distinguished from "predictions" in order to emphasize that projections involve assumptions concerning, for example, future socio-economic and technological developments that may or may not be realised and are therefore subject to substantial uncertainty.

PROMED: Global Outbreak and Response Network Programme for Monitoring Emerging diseases, run by the Federation of American Scientists and sponsored by WHO. It provides a framework, via the Internet, for electronic data exchange on outbreaks of emerging diseases.

proxy: The context of climate; a local record that is interpreted using physical and biophysical principles to represent some combination of climate-related variations back in time. Climate-related data derived in this way are referred to as proxy data. Examples are tree ring records, characteristics of corals and various data derived from ice cores.

pterygium: Wing-shaped growth of the conjunctiva epithelium.

qualitative analysis: An attempt to describe the non-numerical relationship between an outcome of interest and possible **exposures**. *See also quantitative analysis.*

Quality Adjusted Life Year (QALY): The arithmetic product of life expectancy and a measure of the quality of the remaining life years. QALY places a weight on time in different health states. A year of perfect health is worth 1; however, a year of less than perfect health life expectancy is worth less than 1. Death is considered to be equivalent to 0, however, some health states may be considered worse than death and have negative scores. QALYs provide a common currency to assess the extent of the benefits gained from a variety of interventions in terms of health-related quality of life and survival for the patient.

quantitative analysis: An attempt to model (i.e. quantify) the numerical relationship between an outcome of interest and possible **exposures**. *See also qualitative analysis.*

radiative forcing: A simple measure of the importance of a potential climate change mechanism. Radiative forcing is the amount of perturbation of the energy balance of the Earth-atmosphere system (W/m^2) following, for example, a change in carbon dioxide concentrations or a change in the output of the sun. The climate system responds to radiative forcing so as to re-establish the energy balance. Positive radiative forcing tends to warm the Earth's surface and negative radiative forcing tends to cool it. Radiative forcing is normally quoted as a global or annual mean.

rangeland: Unimproved grasslands, shrublands, savannas, hot and cold deserts, tundra.

red tide: Algal bloom which causes the seawater to become discoloured by the sheer concentration of algae seeking the sunlight. This discolouration is a result of the various pigments the plants use to trap sunlight; depending on the species of algae present, the water may reflect pink, violet, orange, yellow, blue, green, brown, or red. Since red is the most common pigment, the phenomenon has come to be called red tide.

relative humidity: The ratio of the mass of water vapour in a given volume of air to the value of saturated air at the same temperature.

remote sensing satellites: Polar-orbiting satellites which observe the Earth's surface, producing images of various temporal and spatial resolutions. *See also Normalized Difference Vegetation Index and Sea Surface Temperatures.*

reservoir/reservoir host: Any animal, plant, soil or inanimate matter in which a pathogen normally lives and multiplies, and on which it depends primarily for survival; e.g. foxes are a reservoir for rabies. Reservoir hosts may be asymptomatic.

Rift Valley Fever: A viral zoonosis which mainly affects live stock in many areas of the world but which occasionally causes severe epidemics in humans, leading to high morbidity and mortality. The death of RVF-infected livestock often leads to substantial economic losses. The virus is transmitted to animals and humans by a range of mosquitoes including *Aedes* and *Culex*.

river blindness: *See onchocerciasis.*

run-off: Water from precipitation or irrigation that does not evaporate or seep into soil but flows into rivers, streams or lakes, and that may carry sediment.

salinization: The accumulation of salts in the soil.

salmonellosis: Bacterial food-poisoning caused by *Salmonella* species, most frequently reported in North America and Europe. Most people become infected by ingesting foods contaminated with significant amounts of *Salmonella* and the poisoning typically occurs in outbreaks in the general population or hospitals, restaurants etc. Improperly handled or undercooked poultry and eggs are the foods which most frequently cause *Salmonella* food poisoning. Chickens are a major carrier of *Salmonella* bacteria, which accounts for its prominence in poultry products.

saturation deficit: The degree of saturation in the 1000–500 hPa layer. If the amount of moisture in a layer is held constant while the thickness decreases, the air will become more saturated until precipitation begins. The thickness at which precipitation is expected to begin for a given amount of moisture in the atmosphere is known as the saturation thickness. The difference between the saturation thickness and the actual thickness defines the saturation deficit.

scenarios: A plausible and often simplified description of how the future may develop, based on a coherent and internally consistent set of assumptions about key driving forces and relationships. Scenarios are neither predictions nor forecasts and may sometimes be based on a narrative storyline. *See also SRES scenarios and emission scenarios.*

schistosomiasis: A parasitic disease, also known as bilharziasis, caused by five species of flatworms, or blood flukes, known as schistosomes throughout the tropics. The eggs of the schistosomes in the excreta of an infected person hatch on contact with water and release larvae, the miracidia which penetrate a fresh water **intermediate** snail host and produce new parasites (cercariae). The cercariae are excreted by the snail into the water and penetrate human skin. Disease due to schistosomiasis is indicated either by the presence of blood in the urine (urinary schistosomiasis) leading eventually to bladder cancer or kidney problems or, in the case of intestinal schistosomiasis, by initial diarrhoea, which can lead to serious complications of the liver and spleen.

sea surface temperature (SST): The water temperature at 1 meter below the sea surface. However, there are a variety of techniques for measuring this parameter that can potentially yield different results because different things are actually being measured. **Remote sensing satellites** have been increasingly utilized to measure SST and have provided an enormous leap in our ability to view the spatial and temporal variation in SST. The satellite measurement is made by sensing the ocean radiation in two or more wavelengths in the infrared part of the electromagnetic spectrum which can be then be empirically related to SST.

seasonality/seasonal variation: Seasonal fluctuations in disease **incidence** or **prevalence** or other phenomena (e.g. abundance of **vectors**).

secondary air pollutants: Air pollutants formed by chemical and photochemical reactions of primary air pollutants and atmospheric chemicals.

sensitivity: Degree to which a system is affected by climate-related changes, either adversely or beneficially. The effect may be direct (e.g. a change in crop yield in response to temperature change) or indirect (e.g. damages caused by increases in the frequency of coastal flooding).

sentinel site: Specific health facility, usually a general/family practice, which undertakes to maintain surveillance and report certain specific predetermined events such as cases of certain infectious diseases.

serotype: Identifiable factor or factors in blood serum detected by serological tests.

seroprevalence: Prevalence of a specified serotype in a specified population.

SIDS: Small Island Developing States.

snow blindness: *See photokeratitis.*

snowshoe hare virus (SHV): A **zoonotic** viral infection which persists in cycles of transmission among wild mammals and mosquitoes. A wide range of wild mammal species can be infected with SSH virus and the snowshoe hare is thought to be important in some areas of Canada. Disease in people, when it occurs, takes the form of infection and inflammation of the brain (meningitis and encephalitis).

soft ticks: Ticks of the family Argasidae, characterised by the absence of a scutum (dorsal plate) and lack of visible mouthparts from the dorsal view. *See also hard ticks.*

solar activity: Variations in the energy output of the sun, measurable as numbers of sun spots as well as radiative output, magnetic activity, and emissions of high energy particles.

solar radiation: Radiation emitted by the sun, also referred to as short-wave radiation.

spatial and temporal scale/resolution: Climate may vary on a large range of spatial and temporal scales. Spatial scales may range from local or high resolution (less than $100\,000\,km^2$), to continental or low resolution (10 to 100 million km^2). Temporal scales may range from seasonal to geological (up to hundreds of millions of years).

spirochaete: Bacterium with a spiral shape. *See also Lyme disease.*

SRES: Emissions scenarios used as a basis for the climate projections in the IPCC TAR in 2001. There are four scenario families, comprising A1, A2, B1 and B2:

- A1: a future world of very rapid economic growth, low population growth, and the rapid introduction of new and more efficient technologies. Major underlying themes are convergence among regions, capacity building, and increased cultural and social interactions, with a substantial reduction in regional differences in per capita income.
- A2: a very heterogeneous world. The underlying theme is self-reliance and preservation of local identities. Fertility patterns across regions converge very slowly, which results in high population growth. Economic development is primarily regionally oriented and per capita economic growth and technological change are more fragmented and slower than in other storylines.
- B1: a convergent world with low population growth but rapid changes in economic structures toward a service and information economy. The emphasis is on global solutions to economic, social, and environmental sustainability, including improved equity, but without additional climate initiatives.
- B2: a world in which the emphasis is on local solutions to economic, social, and environmental sustainability. It is a world with moderate population

growth, intermediate levels of economic development, and less rapid and more diverse technological change.

stakeholder: Person or entity that has an interest or 'stake' in the outcome of a particular action or policy.

statistical model: a mathematical approach to determine the relationship between environmental factors and an outcome of interest (e.g. the distribution of disease vectors) using statistical procedures such as regression or discriminant analysis. In comparison to **biological models**, this approach requires extensive records of vector distribution through time and/or space.

stratosphere: The region of the atmosphere above the troposphere extending from about 10 km to about 50 km.

stratospheric ozone depletion: The reduction of the quantity of **ozone** contained in the **stratosphere** due principally to the release of halogenated chemicals, such as **Chlorofuorocarbons (CFCs)**.

stratospheric ozone layer: The stratosphere contains a layer in which the concentration of ozone is greatest, the "ozone layer". The layer extends from about 12 to 40 km. A very strong depletion of the ozone layer takes place over the Antarctic region, caused by human-made chlorine and bromine compounds.

stressor: Single condition or agent that contributes to stress of an organism, population or **ecosystem**.

surveillance: Continuous analysis, interpretation and feedback of systematically collected data for the detection of trends in the occurrence or spread of a disease, based on practical and standardized methods of notification or registration. Sources of data may be related directly to disease or factors influencing disease.

susceptibility: Probability that an individual or population will be affected by an external factor.

sustainability: A characteristic of human activity that is undertaken in such a manner that it does not adversely affect environmental conditions and which means that that activity can be repeated in the future.

synoptic: Any of the methods used to analyse relationships between total atmospheric conditions and the surface environment. Usually expressed in two forms: "**air mass** identification" which assesses the meteorological quality of the entire **atmosphere** and "weather type evaluation" which identifies various weather systems and their impact.

T helper cells: *See lymphocytes.*

TNC: The Nature Conservancy.

thermal expansion: In connection with sea level, the increase in volume (and decrease in density) that results from warming water. A warming of the ocean leads to an expansion of the ocean volume and hence an increase in sea level.

thermohaline circulation: Large-scale density driven circulation in the ocean, caused by differences in temperature and salinity.

threshold: Abrupt change in the slope or curvature of a **dose-response** graph.

tick-borne encephalitis (TBE): A viral infection of the central nervous system transmitted by the bite of *Ixodes ricinus* ticks. The disease occurs in Scandinavia, western, eastern and central Europe, and countries that made up the former Soviet Union. Initial symptoms include fever, headache, nausea and vomiting and in some cases the disease progresses into a neurological infection resulting in paralysis and coma.

tide gauge: A device at a coastal location which continuously measures the level of the sea with respect to the adjacent land.

time-series analysis: Statistical methods used to describe events that are measured in an ordered sequence at equally-spaced time intervals, and often to analyse their variations as functions of other variables (e.g. analyses of daily records of daily **mortality** rates, as a function of concurrent variation in temperature).

trachoma: A form of bacterial conjunctivitis caused by *Chlamydia trachomatis*.

tropopause: The boundary between the **troposphere** and the **stratosphere**.

troposphere: The lowest part of the atmosphere from the surface to about 10 km in altitude in mid-latitudes where clouds and "weather" phenomena occur. In the troposphere, temperatures generally decrease with height.

trypanosomiasis: Parasitic disease caused by protozoans of the genus *Trypanosoma*. In the Americas, American trypanosomiasis (Chagas disease) is caused by *T. cruzi* and transmitted by reduviid (kissing) bugs of the genus *Triatoma* and *Rhodnius*. In Africa, human African trypanosomiasis (sleeping sickness) is caused by *T. brucei rhodesiense* and *T. b. gambiense* and transmitted by **tsetse flies**.

tsetse fly: Any of several bloodsucking African flies of the genus *Glossina* which transmit African sleeping sickness (trypanosomiasis) to humans. The tsetse fly also carries the parasites that cause nagana in cattle and other diseases of wild and domestic animals.

tuberculosis (TB): A bacterial infection caused by *Mycobacterium tuberculosis*. TB is highly contagious, spreading through the air in infected droplets from coughing, sneezing, talking or spitting of infectious people.

typhoid: Infectious fever usually spread by food, milk, or water supplies which have been contaminated with *Salmonella typhi*, either directly by sewage, indirectly by flies, or as a result of poor personal hygiene.

UKCIP: The UK Climate Impacts Programme.

USEPA: United States Environmental Protection Agency.

ultraviolet radiation (UVR): Solar radiation within a certain wavelength, depending on the type of radiation (A, B or C). Ozone absorbs strongly in the UV-C (<280 nm) and solar radiation in these wavelengths does not reach the earth's surface. As the wavelength is increased through the UV-B range (280 nm to 315 nm) and into the UV-A (315 nm to 400 nm) ozone absorption becomes weaker, until it is undetectable at about 340 nm.

UN Framework Convention on Climate Change (UNFCCC): Convention signed at United Nations Conference on Environment and Development in 1992. Governments that become Parties to the Convention agree to stabilize **greenhouse gas** concentrations in the **atmosphere** at a level that would prevent dangerous **anthropogenic** interference with the **climate system**.

UNEP: United Nations Environment Programme.

uncertainty: An expression of the degree to which a value is unknown. This can result from lack of information or disagreement about what is known. Uncertainty can be represented by **quantitative** measures (e.g. a range of values calculated by mathematical models) or **qualitative** statements (e.g. reflecting the judgement of a team of experts).

unstable malaria: Haphazard transmission of malaria, occurring only during "favourable" episodes. *See also malaria.*

urban heat island: *See heat island effect.*

vector: An organism that acts as an essential **intermediate** host or definite host for a human pathogen and that plays an active role in its transmission; for example *Anopheles* mosquitoes are vectors of **malaria**. This definition excludes mechanical carriers of infective materials (such as houseflies and cockroaches), strictly passive intermediate hosts (e.g. the snail hosts of **schistosomiasis**) and reservoir species (e.g. foxes carrying rabies).

vector-borne diseases: Range of infectious diseases which are transmitted between hosts by **vectors** such as mosquitoes or ticks (e.g. **malaria, dengue fever, Lyme disease**).

vivax malaria: See *Plasmodium vivax* and malaria.

vulnerability: The degree to which a system is susceptible to, or unable to cope with, adverse effects of **climate change**, including **climate variability** and **extremes**. Vulnerability is a function of the character, magnitude and rate of climate variation to which a system is exposed, its **sensitivity** and its **adaptive capacity**.

watershed: The region draining into a river, river system or body of water.

water stress: A country is water stressed if the available freshwater supply relative to water withdrawals acts as an important constraint on development. Withdrawals exceeding 20% of renewable water supply has been used as an indicator of water stress.

water vapour: Also called humidity; the largest single greenhouse gas. Water vapour also forms an important link between the land and ocean; it is a carrying mechanism that transports energy around the globe and, therefore, a major driver of weather patterns—a fact demonstrated spectacularly by typhoons and hurricanes powered by tropical evaporation.

WHO: World Health Organization.

WMO: World Meteorological Organization.

West Nile Virus (WNV): A zoonotic virus transmitted by mosquitoes (normally *Culex*) and maintained in a wildlife cycle involving birds. Occasional spill-over to the human population results after virus **amplification** and can cause large epidemics. Symptoms may be mild and include fever, headache and malaise while symptoms of severe infection include high fever, neck stiffness, coma and paralysis.

WWF: World Wildlife Fund.

yellow fever: A mosquito-borne viral disease occurring in sub-Saharan Africa and tropical South Americ. Several different species of the *Aedes* and *Haemagogus* (S. America only) mosquitoes transmit the yellow fever virus. Infection causes a wide spectrum of disease, from mild symptoms to severe illness and death.

zoonosis: An infectious disease of vertebrate animals, such as rabies, which can be transmitted to humans.

zooplankton/zooxanthellae: The animal forms of **plankton**. They consume **phytoplankton** and other zooplankton.

Index

extreme events, 275
limitations, 39
animal diseases, 126
animal reservoirs, 106
Annex I countries, 285
anomaly, 285
Anopheles, 53, 191, 296
Anopheles albimanus, 118
Anopheles freeborni, 118
Anopheles gambiae, 112, 118
Antarctic region
ice cover, 7, 30
ozone hole, 160
anthropogenic, 285
anthropogenic emissions, 285
anthroponoses, 105, 106–7, 285
antibiotic resistance, 122
Antigua and Barbuda, health impact
assessment, 184, 186, 190, 194,
195
antimalarial drugs, 122, 230
antioxidants, 169
AOGCMs *see* Atmosphere-Ocean
general circulation models
aquarium trade, 254
arbovirus infections, 53, 111–12, 285
see also dengue/dengue haemorrhagic
fever
Arctic Climate Impact Assessment, 56
Arctic region
effects of climate change, 1–2, 6
ozone hole, 160
arid region/zone, 285
Asia, IPCC-TAR, 52
asthma, 212
atmosphere, 19–20, 162, 285
Atmosphere-Ocean general circulation
models (AOGCMs), 33–5, 36, 40,
285, 287
atopic eczema, 174, 285
attributable burden of disease,
estimating, 134–5
attribution, 289
monitoring climate change impacts
and, 205–6
Australia
health impact assessment, 186
IPCC-TAR, 52–3
Australian Institute of Marine Science
(AIMS), 255
autoimmune diseases, 169, 173, 285
autonomy, respect for, 238

Bangladesh, 93, 94, 114, 230
basal metabolic rate, 286

basic reproduction rate (R$_0$), 116–17,
286
BCG vaccine, 171–2
bed nets, 222
beneficence, 238
benefit-cost criterion, 240–1
benefits
adaptation, 231–3
ancillary, 70
measurement, 241
bilharziasis *see* schistosomiasis
billion, 286
biodiversity, 286
international convention, 268, 269
loss, 13
biofuel, 286
biomass, 286
biome, 286
biosphere, 286
biotoxins, 49–50, 286
black flies, 50
bridge species, 107
burden of disease, 134
adaptive capacity and pre-existing,
231
attributable, estimating, 134–5
avoidable, estimating, 135
conclusions and recommendations,
276–7
evidence available for estimating,
135–7
global estimates, 134
impact estimates for 2000, 136, 152
monitoring, 207
in WHO regions (2001), 232

Cameroon, health impact assessment,
184, 186, 190
campylobacter infections, 107
Canada, policy-focused assessments,
183, 248
cancer, 174
skin *see* skin cancer
capacity building, 286
carbon dioxide (CO$_2$), 22, 286
atmospheric concentrations, 22, 23,
24
fossil emissions, 291
projected emissions, 31–2
carbon sink, 286
cardiovascular disease (CVD)
adaptive capacity and, 231
temperature-related mortality, 141–3
carrying capacity, 286
cataracts, 163, 167–9, 286

Centre for Research on the Epidemiology of Disasters (CRED), 92
cereal grain yields, 11, 13, 48, 65
see also crop yields
CFCs *see* chlorofluorocarbons
Chagas' disease, 50
chemical pollution, 13, 268
children
adaptive responses to heat, 249
sunlight exposure, 175
China, famines, 9
Chlamydia trachomatis, 304
chlorofluorocarbons (CFCs), 24, 286–7
Montreal Protocol, 161, 166, 167
stratospheric ozone depletion, 159–60
chloroquine, 122, 287
cholera (*Vibrio cholerae*), 280, 287
climate extremes and, 87
climate sensitivity, 111
ecological influences, 126
effects of climate change, 114, 275
IPCC-TAR, 49, 50, 51, 54
national health impact assessment, 190
predicting epidemics, 121
seasonal fluctuations, 107
transmission, 106–7
chronic effect, 287
ciguatera fish poisoning, 49–50, 191, 287
climate, 24–6
cycles, 2
data requirements and sources, 208–9
definition, 18, 287
effects on disease transmission, 45–7
exposure assessment, 36–9, 40
extreme events *see* extreme climate events
normals, 25, 273
variables, 25
climate change, 26–9
20th century, 29–30
anthropogenic, 31–3, 39, 270
conclusions and recommendations, 271–3
definitions, 287
evidence of, 29–30, 204–5
health impacts *see* health impacts of climate change
impact assessments *see* impact assessments
international conventions, 268

past, 2, 26–7, 28
projections, 31–3, 39, 272
associated with specific scenarios, 138, 139
rates, 36
recent scientific assessments, 5–8, 270
regional variations, 6–7, 29, 33
research *see* research
scientific consensus, 273–4
vs climate variability, 18
vs global warming, 223
see also global warming
climate models, 33–6, 287
for deriving scenarios, 256
in national assessments, 194
climate scenarios
baseline, 137–8
in impact assessment, 183, 184
national, 194
see also emission scenarios
climate sensitivity, 21, 32
evidence of, for monitoring, 207
infectious diseases, 104, 107–11
species variations, 124–6
climate system, 19–24, 288
climate variability, 24–6, 79
20th century, 29–30
adaptation difficulties, 247
coping ability and, 223
definitions, 18, 24, 288
modes, 25–6
past, 26
research needs, 274
Climatological Standard Normals (CLINO), 273
CLIMEX model, 115
cloud effects, 20, 22
cockroaches, 54
cold-related mortality, 86–7, 89–90
cardiovascular, 141–3
factors affecting vulnerability, 89
impact of climate change, 90, 143
see also winter mortality
communicable diseases *see* infectious diseases
communication
assessment results, 261–2
as policy response, 243
Comparative Risk Assessment (CRA), 134–5, 140, 183
see also quantifying health impacts
confidence interval, 288
conflict resolution, 245
confounding factors, 211, 288

conjunctiva, solar radiation effects, 169
consensus, scientific, 273–4
Conservation International (CI), 255
contact hypersensitivity (CH), 170
control, disease, 122, 211
Convention on Biological Diversity, 268, 269
copepods, 49, 87, 288
coping ability/capacity, 223–4
  characterizing uncertainties, 256
  definition, 223, 288
  see also adaptive capacity
coral bleaching, 254–5, 288
coral reef fisheries, 254–5
Coriolis force, 288
cornea, solar radiation effects, 169
COSMIC, 194
cost-effective, 288
costs
  adaptation, 231–3, 251–2, 284
  ancillary, 70
  in benefit-cost criterion, 240–1
  measurement, 241
  and responsibility, principles, 281
crop yields, 13, 145, 146
  see also cereal grain yields
cryosphere, 21, 288
cryptosporidiosis (*Cryptosporidium* infection), 123, 288
  IPCC-TAR, 49, 50, 54, 56
  rainfall sensitivity, 85, 111
Culex, 107, 305
Culex nigripalpus, 112
Culex pipiens, 112, 123–4
Culex tarsalis, 116
Culicoides paraensis, 297
cyclones, 48, 52–3, 94
  coping ability, 224
cyclospora infections, 107
cytokine profiles, epidermal, 170
cytomegalovirus (CMV), 170, 288–9

DALYs *see* disability-adjusted life years
data requirements/sources, for monitoring, 208–11
decision-support tools, 241, 259
deer mice, 85, 242
deforestation, 22, 122–3
delayed type hypersensitivity (DTH), 170
demography, 289
  adaptive capacity and, 231
dengue/dengue haemorrhagic fever (DHF), 12, 72, 106, 289
  climate extremes and, 83–4

coping ability, 224
  early warning system, 121
  eradication programme, 122
  IPCC-TAR, 49, 50, 53, 55
  national health impact assessments, 189, 191–2, 193
  simulation modelling and risk reduction, 117, 258–9
  time series analysis, 39
desert, 289
desertification, 268
detection, 289
developed countries
  coping ability, 224
  national assessments of health impacts, 186, 187–91
  National Communications to UNFCCC, 197
developing countries
  adaptive capacity, 226–7, 228, 230–1
  coping ability, 224
  impact of weather extremes, 91
  national assessments of health impacts, 185, 186
  National Communications to UNFCCC, 197, 198–9
diabetes mellitus, insulin-dependent (IDDM), 173
diarrhoeal diseases, 103
  climate extremes and, 85–6, 93, 94
  climate sensitivity, 110–11
  environmental influences, 123, 124
  evidence for effects of climate, 113, 144
  exposure assessment, 37
  IPCC-TAR, 50, 52, 54
  monitoring, 215–16
  national health impact assessments, 191
  quantifying impact of climate change, 142–5, 152, 153
  relationship with temperature, 63–4
  seasonal fluctuations, 107
  see also water-borne diseases
diphtheria, 289
disability-adjusted life years (DALYs), 134, 289
  estimated impact of climate change, 152
  in WHO regions (2001), 232
discounting, 241
disease burden *see* burden of disease
disease control programmes, 122, 211
diurnal temperature range, 25, 289
dose-response, 289

droughts, 48
  definition, 94, 289
  diarrhoeal illness and, 85, 94
  El Niño-related, 91
  health impacts, 94
  historical, 9–10, 30
  malaria outbreaks, 82, 86, 94
  mechanisms of health effects, 81
  water-borne disease and, 50
drug development, 230
drug donations, 230
drug resistance, 122
Dutch Country Studies Programme,
  190
dysentery, 289

early warning systems (EWS), 222, 289
  ENSO-related disease outbreaks,
    120–1
  future needs, 274
  hot weather, 250–1
  vs monitoring long-term trends,
    211–13
Earth Summit (1992), 269
  see also Agenda 21
ecological footprint, 4–5
ecological influences
  higher ultraviolet radiation
    exposures, 163
  infectious disease/climate change
    links, 124–6
ecological perspective, 3–5
ecological studies, 61–2, 290
ecological system/ecosystem, 290
economic development/growth
  environmental impact, 4–5, 267
  health and, 227
economic disruptions/losses, 51, 92
economic resources, adaptive capacity
    and, 226–7
economies in transition, 289
  adaptive capacity, 225, 230
eczema, atopic, 174, 285
education and training, 228–9
effect modification, 206–7
effect modifier, 290
efficiency aspects, public health policy,
  238
El Niño/Southern Oscillation (ENSO),
  2, 25–6, 79, 290
  20th century events, 27, 30
  coral reef damage, 255
  diarrhoeal disease and, 85, 113,
    215
  early warning systems, 120–1

evidence for health impacts, 113,
  114
exposure assessment, 37, 38, 39
hantavirus pulmonary syndrome
    and, 242
in history, 9
infectious diseases and, 81–6
IPCC-TAR, 51, 54–5
malaria and, 39, 41, 113
mechanisms of health impacts, 79–80
natural disasters related to, 91
elderly
  adaptive responses to heat, 249
  temperature-related mortality, 87,
    89–90
EM-DAT database, 91–2, 148, 290
emerging infectious diseases, 290
emission, 290
emission scenarios, 290
  baseline, 137–8
  considered for 2030, 138
  IPCC-TAR (SRES), 30–1, 302
  predicted climate changes using, 138,
    139
  see also climate scenarios
encephalitis, 111–12
  see also tick-borne encephalitis; other
    specific types
endemic, 290
energy, principle of conservation,
  291
ENSO see El Niño/Southern Oscillation
enteric diseases see diarrhoeal diseases
entomological inoculation rate (EIR),
  290
environmental changes, global, 2, 7,
  267
  ecological perspective, 3–5
  health impacts, 13
  international responses, 267–9,
    281–2
  interrelationships between, 7, 8
environmental influences
  infectious disease/climate change
    links, 122–6
  monitoring, 210–11
enzootic disease, 290
epidemic potential (EP), 191–2
epidemics, 290–1
  effects of climate, 103, 107
  forecasting, 96
epidemiological methods, 192
epidemiology, 291
episode analysis, 37
epizootic, 291

equity
adaptive capacity and, 230–1
in international agreements, 281, 282
in public health policy, 238
eradication, disease, 122
erythema, 291
Escherichia coli, 111
ethics, public health, 238–9
Europe, 7
historical famines, 9
IPCC-TAR, 53–4
national assessments of health impacts, 185, 186
predicted health impacts of ozone depletion, 166, 167
European Ministerial Conference on Environment and Health, 182
evaporation, 291
evapotranspiration, 291
evolution, in response to climate change, 125–6
experimental studies, 62
exposure(s), 291
assessment, 36–9, 40, 137–9
conclusions and recommendations, 271–3
difficulties in studying, 61–2
directness of associations with disease, 62
scenarios, 137–8
see also emission scenarios
extreme climate events, 15, 79–102, 291
adaptive capacity, 225, 230
categories, 79
conclusions and recommendations, 274–5
coping ability, 224
forecasting, 96
IPCC-TAR, 47–8, 275
mechanisms of health impacts, 79–80, 81
monitoring, 212, 217
projections, 33, 34
quantifying future health impacts, 147–9
see also specific types
extrinsic incubation period, 291
eye disorders, 163, 167–9

famines, 9–10, 94
El Niño-related, 91
malaria mortality and, 82
see also malnutrition

feedback, 291
Fiji
diarrhoeal disease, 85, 215–16
health impact assessment, 186, 191–2, 193
filariasis, 191
lymphatic (Bancroftian), 50, 124, 295
see also onchocerciasis
fires, 48, 95
first principle, 291
fish, toxic poisoning, 49–50
flies, 54
floods, 291
adaptive capacity, 229–30
coastal, 147–8, 149
diarrhoeal illness and, 85, 93
El Niño-related, 91
health impacts, 93–4
inland, 147, 148, 149
IPCC-TAR, 48, 50, 52–3, 54, 55
mechanisms of health impacts, 80
quantifying health impacts, 147–9, 152, 153
US federal insurance, 252
food-borne diseases (food poisoning), 12, 54
climate sensitivity, 110–11
monitoring, 212
relationship with temperature, 63–4
food production/supply
adaptive capacity, 230
coral reef fisheries, 254–5
drought-related impairment, 94
future estimates, 145, 146
higher ultraviolet radiation exposures and, 163
impairments, 9–10, 11, 13
IPCC-TAR, 48
forcing, radiative, 21–2, 300
by greenhouse gases, 22–4
forecast (weather), 26, 299
fossil fuels, 22, 291
Fourier analysis, 291
free radical, 292
freshwater depletion, 13
future effects, modelling, 66–8
fuzzy logic/suitability, 292

gases
greenhouse *see* greenhouse gases
ozone-destroying, 159–60, 161
gastroenteritis, acute, 63–4
general circulation, 24, 292
generalized additive models (GAM), 38

generalized estimating equations (GEE), 38
genetic engineering, 292
geographical information system (GIS), 208, 292
giardiasis, 85, 111, 123
  IPCC-TAR, 49, 50, 56
glacial-interglacial transitions, 27
glaciers, 29, 30, 292
Global Positioning System (GPS), 292
global warming, 292
  20th century, 2, 3, 27–9
  early indicators of health effects, 114–15
  interactions with ozone depletion, 11–12, 161
  physical and biological effects, 5, 114
  projections, 3, 5–6, 31–2, 39, 272
  recent scientific assessments, 5–8, 270
  vs global climate change, 223
  see also climate change; temperature
gonotrophic cycle, 292
Great Lakes Water Quality Agreement (GLWQA), 248
greenhouse effect, 1, 20–1, 292
  vs stratospheric ozone depletion, 161–2
  see also climate change; global warming
greenhouse gases (GHGs), 2, 19–24, 292
  exposure scenarios *see* emission scenarios
  mitigation *see* mitigation
  projected emissions, 31–2
Greenland, Viking settlements, 9
Gross Domestic Product (GDR), 292–3
Gulf Stream, northern Atlantic, 7, 293

HadCM2 global climate model (GCM), 138
Haemagogus, 305
haemorrhagic, 293
halocarbons, 293
halogenated chemicals, 159–60
hantavirus infection/hantavirus pulmonary syndrome (HPS), 106, 293
  El Niño events and, 55–6, 84–5, 120–1
  in south-western USA, 120–1, 242
health
  adaptive capacity and, 231
  economic resources and, 226–7
  factors affecting, 238
  global environmental change and, 7, 8, 13
  markers, for monitoring, 208
  outcomes, for quantifying health impacts, 139–40
  see also population health
Health Canada, guidelines to impact assessment, 200
Health for All policy, 293
health impact assessments (HIA), 181, 293
  adaptation and, 221–2
  global, 181
  key concepts, 182–3
  methods, 183–5, 192
  national *see* national assessments of health impacts
  need for guidelines, 200
  policy-focused *see* policy-focused assessment
  see also quantifying health impacts
health impacts of climate change, 7, 10–12, 270–1
  in ancient history, 8–10
  detecting evidence of early, 12, 64–6
  difficulties in researching, 45
  direct, 11, 47–8
  establishing baseline relationships, 63–4
  indirect, 11, 48–51
  main pathways, 10
  modelling future, 66–8
  monitoring *see* monitoring
  national assessments, 181–203
  priority threats, 207–8
  quantifying *see* quantifying health impacts
  by region, IPCC-TAR, 51–6
  research *see* research
  scientific consensus, 273–4
  types, 44
heat budget, 293
heat island effect, 22, 293
heat-related mortality, 87–9
  assessing prevention strategies, 249, 250–1
  attribution, 205
  cardiovascular, 141–3
  causes, 87
  IPCC-TAR, 47–8, 52, 53, 55
  monitoring systems, 211–13
  risk factors, 89
heat stress, 47–8, 52, 53, 273
heat stroke, 52, 87

heatwaves, 86–9
  adaptation difficulties, 247
  adaptation strategies, 249, 250–1
  definitions, 37, 88
  exposure assessment, 37
  impact of climate change, 90
  IPCC-TAR, 47–8, 52–3
  lag periods, 37
  mortality *see* heat-related mortality
  see also hot weather
helminths, 293
hepatitis B vaccine, 171
herpes simplex infections, orolabial, 171
"highest-valued," 238
Highland Malaria Project (HIMAL), 119–20
Hippocrates, 8
historical analogue studies *see* analogue studies, historical
historical records
  characterizing uncertainty, 255–6
  infectious disease/climate links, 111–14
history, human
  climate change and human health, 8–10
  climatic cycles and, 2
HIV infection, 171
homelessness, 231
hot weather
  adaptation difficulties, 247
  adaptation strategies, 249, 250–1
  mortality related to *see* heat-related mortality
  watch/warning systems, 250–1
  see also heatwaves
human capital, 228, 229
Humboldt current, 293
humidity
  absolute, 284
  relative, 300
  sensitivity, vector-borne diseases, 110
Hurricane Andrew, 224
Hurricane Mitch, 95, 229
hurricanes, 66
hydrological cycle, 293
hydroxyl ions/radicals, 293
hypothermia, 293

ice
  glacial, 29
  sea, 29
ice caps, 30
ice sheet, 294

ice storms, 55
"if . . . then . . ." questions, 256
illiteracy, 228
immune-related disorders, 169–74
immunosuppression, 294
  ultraviolet radiation-induced, 163, 169–74
impact assessments, 183–5
  health *see* health impact assessments
  recent, 5–8
  see also under Intergovernmental Panel on Climate Change
impact fraction (IF), 134, 135
impacts, 294
  potential, 294
  residual, 294
in vitro, 295
in vivo, 295
incidence, 294
India, famines, 9–10
Industrial Revolution, 294
industrialization, 4–5
inertia, 294
infection rate, 294
infectious diseases, 103–32, 273
  classification, 104, 105–7
  climate extremes and, 79–80, 93
  conclusions and recommendations, 275–6
  control programmes, 122, 211
  directly transmitted, 105–6
  documenting effects of climate change, 104, 111–21
    evidence from long-term trends, 114–15
    historical evidence, 111–14
  El Niño and, 81–6
  emerging, 290
  food-borne *see* food-borne diseases
  indirectly transmitted, 106–7
  integrated monitoring, 210
  modifying influences, 104–5, 121–6
    environmental, 122–6
    sociodemographic, 121–2
  national health impact assessments, 187, 189, 190
  predictive modelling, 11, 115–21
    ENSO early warning systems, 120–1
    landscape based approach, 117–18
    process-based (mathematical) approach, 116–17
    statistical based models, 115–16
  research needs, 127
  resurgence, 14, 225

seasonality, 107
ultraviolet-induced
    immunosuppression and, 170–1
vector-borne see vector-borne
    diseases
water-borne see water-borne diseases
see also specific diseases
influenza, 89, 103, 107, 170
information
    adaptive capacity and, 228–9
    communication, 243, 261–2
    dissemination methods, 262
infrared radiation, 20, 21, 162, 294
infrastructure
    adaptive capacity and, 229–30
    public health see public health
        infrastructure
insecticides, 52, 228, 253, 294
institutional inertia, 230
institutions, social, 230
insurance, US federal flood, 252
integrated assessment, 67–8, 194–6,
    294
    adaptation options, 233
    horizontal, 67
    for prioritizing research needs, 260
    regional, 195–6
    between sectors, 194–5
    vertical, 67
Inter-Agency Committee on the
    Climate Agenda (IACCA), 270–1
Inter-Agency Network on Climate and
    Human Health, 271, 272
Intergovernmental Panel on Climate
    Change (IPCC), 43–60, 181, 294–5
    adaptation and adaptive capacity,
        220–1, 222, 226, 230
    emission scenarios, 30–1, 137–8, 302
    establishment, 43, 269
    guidelines on impact assessment, 183
    policy-making and, 239, 242–3
    post-TAR assessments, 56
    Second Assessment Report, 43,
        242–3
    Third Assessment Report (TAR), 5–6,
        43, 47–56, 272
        20th century climate change, 29
        anthropogenic climate change, 2,
            31–3, 39, 270
        direct effects on health, 47–8
        extreme climate events, 47–8, 275
        health impacts by region, 51–6
        indirect effects on health, 48–51
        projected temperature increase, 3,
            5–6, 31–2, 272

uncertainty issues, 45, 74, 192,
    193
vector-borne diseases see under
    vector-borne diseases
Working Group II (WGII), 44, 47
working groups and task force, 44
intermediate host, 294
international conventions/agreements,
    267–9, 281–2
International Joint Commission (IJC),
    United States-Canada, 248
interquartile range, 295
invasive species, 13
IPCC see Intergovernmental Panel on
    Climate Change
Irian Jaya, 86
irrigation, crop, 123–4
ischaemic heart disease (IHD), 37, 295
Ixodes persulcatus, 53
Ixodes ricinus, 53, 303
Ixodes scapularis, 118

Japanese encephalitis, 50
justice, 238

Kiribati, health impact assessment, 186,
    191–2, 193
Korea, Democratic People's Republic of,
    230
Kyoto Protocol, 270, 295

La Niña, 26, 84, 290
lag periods, 37
land-use changes
    data sources, 211
    effects on climate, 22
    infectious diseases and, 122–4
Landsat Thematic Mapper, 295
landscape based models, infectious
    diseases, 117–18
Latin America, IPCC-TAR, 54–5
latitudinal gradients
    climate, 25
    human disease, 165–6, 173, 174
    human skin pigmentation, 164
    vaccine efficacy, 171
leishmaniasis, 11, 50, 53, 123, 189, 295
lens opacities (cataracts), 163, 167–9,
    286
leptospirosis, 54, 85, 109–10, 189, 295
life-support systems, global, 1, 4–5
Listeria monocytogenes, 170, 171
literacy, 228
literature review, 192–3
"little ice age", 27

Saint Louis encephalitis virus (SLEV), 111–12, 116
salinization, 301
Salmonella typhi (typhoid), 49, 50, 54, 304
salmonellosis (*Salmonella* infections), 64, 301
sandflies, 50, 118, 295
sanitation, 225, 229
satellites
    meterological data collection, 208
    for predictive modelling, 117–18, 120–1
    remote sensing, 117–18, 211, 300, 301
saturation deficit, 301
scenario-based models, 39, 66–8
scenarios, 30–1, 256, 301
    analysis, 256–7
    see also climate scenarios; emission scenarios
schistosomiasis, 11, 50, 301
schizophrenia, 174
scientific assessments, recent, 5–8
scientists, 229
    challenges for, 61–78, 274
    main tasks, 63–75
        dealing with uncertainty, 71–5
        developing scenario-based models, 66–8
        establishing baseline relationships, 63–4
        estimating ancillary benefits/costs, 70
        evaluating adaptation options, 68–70
        evidence for early health effects, 64–6
        informing policy, 70–1
sea ice, 29
sea level rise, 6, 29–30, 33
    adaptation strategies, 252
    effects on vector-borne diseases, 109, 110
    health impacts, 51, 56, 147–8
    projections, 138
sea surface temperature (SST), 301
seasonal affective disorder, 174
seasonal forecasts, 96
seasonal variations/seasonality, 25, 301
    infectious diseases, 107
sensitivity, 301
    analyses, 67, 256
    climate *see* climate sensitivity

sentinel site, 302
seroprevalence, 302
serotype, 302
shellfish poisoning, 49–50, 111
skills, 228–9
skin cancer, 163–7
    ozone depletion and, 166–7
    ultraviolet radiation (UVR) and, 165–6
skin disorders, 163–7, 174
skin pigmentation, human, 164
sleeping sickness, 50, 304
Small Island Developing States (SIDS)
    IPCC-TAR, 56
    national assessments of health impacts, 191–2
    National Communications to UNFCCC, 197, 198–9
smog, photochemical, 12, 55, 161–2
smoking, 169
snails, water, 50, 301
Snow, John, 280
snow blindness (photokeratitis), 163, 169, 298
snow cover, 29
snowshoe hare virus (SHV), 302
social disruptions, 51
social scientists, 229
sociodemographic influences, infectious diseases, 121–2
socioeconomic status
    data sources, 210–11
    vulnerability and, 14, 142
solar activity, 2, 21, 302
solar radiation (sunlight), 20, 21, 162, 302
    health impacts of higher exposures, 163–74
    lens opacities and, 167–9
    public health message re exposure, 174–5
    skin cancer and, 165–6
    see also ultraviolet radiation
solar retinopathy, 169
soot, 24
Southern Oscillation, 26
    see also El Niño/Southern Oscillation
spatial and temporal scale/resolution, 302
Special Report on Emission Scenarios (SRES), 30–1, 302
spectral analysis, 291
spirochaete, 302
SRES (Special Report on Emission Scenarios), 30–1, 302

Sri Lanka, health impact assessment, 186, 190
stakeholders, 246, 303
  communication with, 261–2
  conflict resolution, 245
  conflicting desires, 239
  engagement, in health impact assessment, 246–7
  needs assessment, 194, 260
statistical based models, infectious diseases, 115–16
storms, 48
stratosphere, 20, 159, 162, 303
  ozone depletion see ozone depletion, stratospheric
  ozone layer, 159, 303
stressor, 303
suicide, 94
sulphate particles, 24
summer see heatwaves; hot weather
Sun, 19
sun-bathing, 164–5
sunlight see solar radiation
surveillance, 14, 303
  future needs, 127, 278
  systems, developing and implementing, 243–4
susceptibility, 303
  see also vulnerability
sustainability (sustainable development), 239, 269, 281, 303
synoptic, 303
synoptic climatological analysis, 37
synoptic scale, 24
systemic lupus erythematosus (SLE), 173–4, 295

T lymphocytes, 295
  helper type 1 (T_H1), 169–70, 173
  helper type 2 (T_H2), 169–70, 173–4
TAR see Intergovernmental Panel on Climate Change (IPCC), Third Assessment Report
technological advances
  adaptive, 227–8, 253
  environmental impact, 267
temperature
  -related mortality, 87–90, 141–3, 153
    see also cold-related mortality; heat-related mortality
  ambient
    diarrhoeal illness and, 85
    food poisoning and, 64
    mosquito-borne diseases and, 82
  in atmospheric layers, 20

diurnal range, 25, 289
extremes, 30, 86–90
  monitoring, 212, 217
  projections, 33
  see also heatwaves
global average, 2, 3, 5, 6
  increasing see global warming
  past changes, 26–7, 28
  projected changes, 31–2, 39, 272
sea surface (SST), 301
sensitivity
  vector-borne diseases, 108, 109
  water-borne diseases, 111
spatial gradients, 25
thermal expansion, 29–30, 303
thermohaline circulation, 303
threshold, 303
tick-borne encephalitis (TBE), 303
  changes in distribution, 53–4, 65, 114–15
  effects of climate change, 114–15
  integrated monitoring, 210
ticks, 110, 125
  hard, 46, 53–4, 118, 293
  soft, 302
tide gauge, 303
time-series analyses, 38–9, 86–7, 205, 304
trachoma, 304
trade, international, 121–2
training and education, 228–9
transmission (disease)
  classification, 105–7
  direct, 105–6, 289
  effects of climate, 45–7
  human movements and, 121–2
  indirect, 106–7, 294
  potential, 116–17
transport policy, 70
transportation infrastructure, 229–30
travel, human, 121–2
triatome bugs, 50
Trichinella spiralis, 170
trophosphere, 19–20, 162, 304
tropopause, 20, 162, 304
trypanosomiasis, 124, 304
  African, 50, 304
  American, 50, 304
tsetse flies, 46, 50, 304
tuberculosis (TB), 174, 304
typhoid (Salmonella typhi), 49, 50, 54, 304

ultraviolet A radiation (UVA), 162, 165, 167, 304